MULTI-TERMINAL DIRECT-CURRENT GRIDS

MULTI-TERMINAL DIRECT-CURRENT GRIDS

Modeling, Analysis, and Control

Nilanjan Ray Chaudhuri

Balarko Chaudhuri

Rajat Majumder

Amirnaser Yazdani

IEEE
IEEE PRESS

WILEY

For general information on our other products and services or for technical support, please contact our Customer Care Department within the United States at (800) 762-2974, outside the United States at (317) 572-3993 or fax (317) 572-4002.

Wiley also publishes its books in a variety of electronic formats. Some content that appears in print may not be available in electronic formats. For more information about Wiley products, visit our web site at www.wiley.com.

Library of Congress Cataloging-in-Publication Data:

Chaudhuri, Nilanjan Ray, 1981-
Multi-terminal direct-current grids : modeling, analysis, and control / Nilanjan Ray Chaudhuri, Balarko Chaudhuri, Rajat Majumder, Amirnaser Yazdani.
pages cm
Includes bibliographical references and index.
ISBN 978-1-118-72910-6 (hardback)
1. Electric power distribution–Direct current. I. Chaudhuri, Balarko, 1977- II. Majumder, Rajat, 1977- III. Yazdani, Amirnaser, 1972- IV. Title.
TK3111.C44 2014
621.31′2–dc23

2014017303

Printed in the United States of America

10 9 8 7 6 5 4 3 2 1

To our parents and families

CONTENTS

FOREWORD

In recent years, the electric power delivery system has been changing dramatically. Concern for the environment, the economy, and power system security has placed stringent demands on power transmission. We must acknowledge the inevitable depletion of fossil fuels and the growing importance of sustainable and environmental awareness. Electricity derived from renewable energy sources offers advantages in terms of economical efficiency, especially when compared with oil, gas and coal.

Existing rights-of-way for high voltage transmission lines are generally underutilized and some transmission circuits are nearing their end of life. New rights-of-way for overhead transmission lines are becoming difficult if not impossible to acquire. Voltages Sourced Converter (VSC) transmission is the way forward with its capability to operate with many terminals, allow for greater power density in its circuits, and replace aging AC transmission lines or existing AC lines that may be converted to DC. DC overhead transmission lines can be compacted to share right-of-way with roads and railway lines. VSC transmission is more favorable for use with underground and undersea cable transmission where it is expedient to do so.

In the future, issues of the combination of diversity of loads, wind and solar energy across wide geographical areas will become crucial. DC grids are critical to solving these major concerns.

This book initiates the technical background that applies to multi-terminal DC transmission as well as the evolution through to the VSC converter configurations and the anticipated and inevitable DC grids. It covers basic theory setting the stage for on-going developments. At this present time, this book is likely to be a useful resource and of great benefit to the student of DC transmission, both in graduate studies and in the workplace.

DENNIS WOODFORD

Electranix Corporation
Winnipeg, Canada

PREFACE

Electric power transmission primarily relies on the alternating current (AC) technology with the direct current (DC) alternative being used only in very specific (e.g., very long distance overhead lines and subsea or underground cables for moderate distances) applications. The majority of the presently operational and planned high voltage direct current (HVDC) links are point-to-point with two points of connection with the AC networks. However, it is envisaged that DC electric networks embedding more than two terminal converter stations, which are commonly referred to as *multi-terminal* direct current (MTDC) grids, would offer the possibility of meshed subsea interconnections between countries and continents to share the diversity in renewable energy portfolios for better supply reliability. It is not clear what the exact topology of such DC grids would be and where they are going to be installed. However, the vision is that there would be an MTDC grid around the North Sea to tap into the rich offshore wind resources in the region and also strengthen the interconnection among the UK, Scandinavia, and continental Europe. In the long-run, the plan is to build a pan-European subsea grid to interconnect the major generation and load centers and eventually, interconnect this grid with the solar power generation sites in sub-Saharan Africa. Although some of these visions are quite ambitious and futuristic, the need for a subsea DC grid is now almost established.

There is a strong case for a subsea MTDC grid to share the diverse renewable energy resources across different geographical regions (e.g., UK, Scandinavia, and continental Europe; parts of North America) and also interconnect offshore energy (e.g., WFs) sources for better supply reliability and operational flexibility. However, at present, there is no such MTDC grid in operation except the first phase of the Zhoushan project commissioned in China in 2013. Despite the need, the uptake is delayed due to technological barriers like unavailability of fast protection system, DC circuit breakers and efficient voltage source converters (VSCs) with DC-side fault-clearing capability. Alongside protection and DC fault current interruption, another major cause of concern for the system operators is the unknown interaction between the DC grid and the host AC systems. An MTDC grid based on VSC technology is unprecedented with no operational experience. Hence, there is a lack of understanding of the unknown interaction between the MTDC grid and the host AC systems. For instance, what kind of services (e.g., frequency support and power oscillation damping) can an MTDC grid provide for the host AC systems. What are the implications of such service provisions on the overall stability of the AC–DC grid system. In order to analyze these system level issues, a systematic modeling framework for stability studies of combined AC-MTDC grid system is essential. These have received little attention until recent times, when a number of research papers were published on DC grid modeling, control, power quality, and so on, by and large in the context of offshore WFs. However, a single and comprehensive source of reference which includes the modeling, analysis, and control of MTDC grids has not yet been published.

This monograph is the first of its kind on the topic "multi-terminal direct current (MTDC) grids." Although several research papers have been published recently on this important topic area, a comprehensive text is yet to be compiled as a single source of reference. This monographs aims to fill this gap by presenting the following:

1. A generic modeling framework for combined AC-MTDC grids which can be used by system planners and operators for stability analysis.

2. Validation of the averaged model of an AC-MTDC grid in MATLAB®/SIMULINK® against switched model in EMTDC/PSCAD.

3. Modal analysis to study the interactions between the MTDC grid and surrounding AC system as observed in time-domain simulation.

4. Design of control to enable the MTDC grid to provide AC system support (e.g., reduced loss-of-infeed and frequency support).

5. State-of-the-art research challenges in protection of MTDC grids.

The main body of this monograph begins with an overview of the voltage-sourced converter (VSC) systems that are the basic building blocks of MTDC grids. A generic modeling framework for MTDC grid is developed, enabling easy integration of the MTDC grid model with a multi-machine AC system model for stability studies. One particular concern is how an MTDC grid would react to loss/outage of one or more converter stations and the resulting power imbalance. Sharing the burden of such a loss/power mismatch has to be appropriate in order to minimize the impact on the converters and the neighboring AC systems. The concept of droop control for autonomous power sharing in MTDC grids is covered. With increasing penetration of asynchronous generation (wind farms) and transmission (HVDC), reduction in system inertia and its effect on frequency control is a growing concern. Hence, provision for frequency support among the surrounding AC systems across the MTDC grid is critical. Modification of the autonomous power sharing (droop) control to exercise such frequency support across the MTDC grids is described. Towards the end of the monograph, a chapter is dedicated to a review of the protection issues of MTDC grids including the present state of research in high power DC circuit breakers and fault blocking VSC systems. Also, the challenges in MTDC grid protection are highlighted along with an overview of some of the protection strategies that could be used in future.

This monograph is targeted at the academicians with research and/or teaching interest in DC electric power transmission, graduate students focusing on DC power transmission, manufacturers of DC power transmission systems, developers of offshore transmission networks and utilities/system operators who have or are planning to have VSC DC links connected to their system or contemplating being part of a DC grid in future.

N. Ray Chaudhuri

Niskayuna, NY, USA
April, 2014

B. Chaudhuri

London, UK
April, 2014

R. Majumder

Raleigh, NC, USA
April, 2014

A. Yazdani

Toronto, ON, Canada
April, 2014

ACRONYMS

A2MC	Alternate-arm multi-level converter
AC	Alternating current
AGC	Automatic generation control
BW	Bandwidth
CSC	Current source converter
DAEs	Differential algebraic equations
DC	Direct current
DFIG	Doubly fed induction generator
EWIP	East-West Interconnector Project
FC	Full converter
FDWT	Fast Dyadic Wavelet transform
HIL	Hardware in loop
HVDC	High voltage direct current
IGBT	Insulated gate bipolar transistor
IGCT	Integrated gate commutated thyristor

JWG	Joint working group
LCC	Line commutated converter
MMC	Modular multi-level Converter
MTDC	Multi terminal direct current
NETS	National electricity transmission system
NPC	Neutral point clamped
NR	Newton Raphson
ODEs	Ordinary differential equations
OPF	Optimal power flow
PCC	Point of common coupling
PI	Proportional integral
PLL	Phase-locked loop
PWM	Pulse width modulation
SACOI	Sardinia Corsica Italy
SG	Synchronous generator
SHE	Selective harmonic elimination
SVCs	Static VAr compensators
SVM	Space vector modulation
TMR	Triple modular redundancy
VCO	Voltage controlled oscillator
VSC	Voltage source converter
WF	Wind farm
WGs	Working groups
WTGs	Wind turbine generators

SYMBOLS

C_c	Shunt capacitance at each end of pi section of DC cable
C_{dc}	Converter DC-side filter capacitor
E_{fd}	Field voltage of synchronous generator
E_d'	Transient emf due to flux-linkage in q-axis damper coil of synchronous generator
E_q'	Transient emf due to field flux-linkage of synchronous generator
H	Available headroom in the converter rating
$H(s)$	Compensator used in the PLL
I_{dc}	Current flowing into DC side of converter
I_{gX}	X-axis component of current phasor $\overline{I}_g$
I_{gY}	Y-axis component of current phasor $\overline{I}_g$
I_l	Total DC current flowing into the converter and corresponding DC link capacitor from DC grid side
I_d	d-axis component of stator current of synchronous generator
I_q	q-axis component of stator current of synchronous generator
$K_I(s)$	Compensator for inner current control loop of converter
$K_{Vac}(s)$	Converter AC voltage compensator
$K_{Vdc}(s)$	Converter DC link voltage compensator
L	Aggregated inductance of converter transformer and phase reactor
L_c	Series inductance of each pi-section of DC cable

P, Q, V	Real power, reactive power, and voltage magnitude in generator or load bus
P_g	Real power injection into the PCC of the converter station
Q_g	Reactive power injection into the PCC of the converter station
$\Re$	Converter real power rating
R	Aggregated resistance of converter transformer and phase reactor
R_c	Series resistance of each pi-section of DC cable
R_{sc}	Effective series resistance of DC link capacitor of converter
R_f	DC-side fault resistance
R_g	DC grid grounding resistance
R_s	Armature resistance of synchronous generator
R_{go}	Droop coefficient of governor
$T_{do}^{'}, T_{do}^{''}$	d-axis open-circuit transient and sub-transient time constants of synchronous generator
$T_{qo}^{'}, T_{qo}^{''}$	q-axis open-circuit transient and sub-transient time constants of synchronous generator
V_{dc}	DC bus voltage of converter
V_{dc_com}	Common DC voltage feedback signal
V_g	Magnitude of AC system voltage phasor at the PCC of converter station
V_{gX}	X-axis component of AC system voltage phasor at the PCC of converter station
V_{gY}	Y-axis component of AC system voltage phasor at the PCC of converter station
V_{tX}	X-axis component of voltage phasor at the AC terminal of converter
V_{tY}	Y-axis component of voltage phasor at the AC terminal of converter
V_i	Terminal voltage magnitude of ith synchronous generator
V_{tr}	Measured voltage state variable after sensor lag block of synchronous generator
$X_d, X_d^{'}, X_d^{''}$	Synchronous, transient, and sub-transient reactances along d-axis of synchronous generator
$X_q, X_q^{'}, X_q^{''}$	Synchronous, transient, and sub-transient reactances along q-axis of synchronous generator
X_{ls}	Armature leakage reactance of synchronous generator
$\overline{Y}_{bus}$	Network bus admittance matrix
$\overline{Y}_{ii} = G_{ii} + jB_{ii}$	Self admittance of ith bus
$\overline{Y}_{ij} = G_{ij} + jB_{ij}$	Mutual admittance between ith and jth bus
$\overline{V}_g$	Phasor of AC system voltage at the PCC of converter station
$\overline{V}_i$	Phasor of AC system voltage at ith bus
$\overline{V}_t$	Voltage phasor at the AC terminal of converter
$[CCI]$	Incidence matrix indicating connectivity of branches of DC grid

$[CCU]$	Incidence matrix indicating connectivity between nodes of DC grid
$[I_c]$	Vector of currents flowing through shunt capacitance of DC cable
$[I_{dc}]$	Vector of currents flowing into DC side of converter
$[I_f]$	Vector of DC fault currents
$[I_{line}]$	Total node current injection vector of DC grid
$[I_{qr}]$	Vector of $R-L$ branch currents flowing from any general node q to r of DC cable
$[I_s]$	Vector of currents flowing toward metallic return through DC link capacitor
$[V_{dc}]$	Vector of DC bus voltage at converter
$[V_q]$	Vector of node voltages of DC grid
$[V_{qr}]$	Vector of series branch voltage drops between any general node q to r of DC grid
$f_{(k)}$	Frequency at the PCC of the kth converter station
f_{pi}	pth power balance equation at ith bus for AC-MTDC load flow
f_s	Nominal AC system frequency
i_{gabc}	Three-phase instantaneous current flowing from converter to AC grid
i_{tabc}	Three-phase instantaneous current at the converter AC terminal
ℓ	Number of AC areas connected to MTDC grid
m	Total number of synchronous generators
m_{XY}	Modulation signal in the $X-Y$ reference frame at converter station
m_{abc}	Three-phase instantaneous modulation signal for converter
n	Total number of converter stations in MTDC grid
n_{ac}	Total number of buses in the AC network
v_{gabc}	Three-phase instantaneous voltage at the PCC of converter
v_{tabc}	Three-phase instantaneous voltage at converter AC terminal
β	Active power—DC voltage droop coefficient of converter
β_f	Active power—frequency droop coefficient of converter
δ	Rotor angle of synchronous generator
ω	Rotor angular speed of synchronous generator
ω_g	Angular frequency of the AC system at the PCC of converter
ω_s	Synchronous speed
ψ_{1d}	Sub-transient emf due to flux-linkage in d-axis damper of synchronous generator
ψ_{2q}	Sub-transient emf due to flux-linkage in q-axis damper of synchronous generator
Θ_g	Phase angle of AC system voltage phasor at the PCC of converter station

Subscripts:

i, j, k	ith, jth or kth element, parameter, quantity, or variable

Superscripts:

m	Parameters, quantities, or variables in the metallic return network of the MTDC grid
n	Parameters, quantities, or variables in the negative pole of the MTDC grid
p	Parameters, quantities or variables in the positive pole of the MTDC grid
$*$	Reference value
0	Nominal/pre disturbance value

CHAPTER 1

FUNDAMENTALS

1.1 Introduction

Commercial supply of electric power began in the late 1880s through electrification of the Wall Street area in New York City using direct current (DC) technology pioneered by Thomas Alva Edison. It was driven by the availability of DC generators and incandescent bulbs working with DC. Use of DC was the only option for electric supply until Nicola Tesla advocated for the use of alternating current (AC) form. Amidst fierce competition and lobbying for both DC and AC options, historically known as *war of currents* [1], AC started to win primarily due to more efficient power transmission enabled by use of transformers to step up or down voltage levels to reduce the power losses. As the need for long distance power transmission grew, the efficiency became a predominant consideration, which worked in favor of AC. For the first half of the twentieth century, AC transmission enjoyed unrivaled popularity and growth while DC was virtually ruled out for electric power transmission.

During the early 1950s, there was renewed interest in the use of DC technology primarily driven by the need for long distance cable transmission. It was realized that the power capacity of an AC cable reduces drastically due to excessive charging current even for moderate distances and voltage levels necessitating the use of DC

Multi-terminal Direct-Current Grids: Modeling, Analysis, and Control, First Edition.
Nilanjan Ray Chaudhuri, Balarko Chaudhuri, Rajat Majumder, and Amirnaser Yazdani.

cables where no such limitation exists. This led to the first DC cable link between mainland Sweden and Gotland island in 1953. Although DC reappeared in the scene in the context of cable transmission, it was soon realized that DC could be a cost-effective option even for overhead line transmission if the transmission distance is very high (beyond 1000 km) where AC transmission capacity is increasingly limited due to stability considerations.

Electric power generation and consumption continued to use AC, which meant converters were required at both ends to convert AC-to-DC and then DC-to-AC. At the beginning, these converters were based on mercury arc valves until the semiconductor switching devices like a thyrsitor was commercially available for high power applications. The converter technology evolved over time driving the costs down, which meant overhead DC transmission started to be cost-effective even for relatively small distances of the order of 700–900 km. This triggered a proliferation of long distance overhead DC links either embedded between two points within an AC system or interconnecting two separate AC systems. Alongside overhead DC lines, underground or subsea DC cables were also installed in different parts of the world. Until the late 1990s, high voltage direct current (HVDC) converter stations were built with either mercury-arc (before the seventies) or semiconductor switches, which could be turned on in a controllable way but relied on the polarity reversal of AC system voltage for turning off (or commutation). Over the years, the so-called line-commutated converter (LCC)-based HVDC technology got matured. Today, it constitutes the bulk of the installed and planned DC transmission capacity around the world.

It was only after 1997 that semiconductor switches with both controlled turn-on and turn-off capability, like an insulated gate bipolar transistor (IGBT) became commercially available at high power ratings. This enabled the use of voltage-sourced converter (VSC)-based HVDC technology, which offered significant advantages over its LCC counterpart. These include but are not limited to reliable operation with weak AC systems, low cost and footprint of converter stations, use of lighter, and stronger cables that makes VSC particularly attractive for offshore transmission. Despite the obvious potential and promise, the uptake of VSC technology was initially hindered by its limited power ratings (few hundred MWs) compared to LCC (up to 8000 MW). Rapid development in the VSC technology since has resulted in availability of relatively higher ratings (up to 1000 MW is under development now) for VSC-based HVDC links, but it is yet to catch up with the ratings offered by LCC.

Most of the HVDC links in operation today are connected between two points of a single AC system or two separate AC systems. These are commonly known as *point-to-point* HVDC links. There are only two exceptions around the world where the HVDC system has more than two points of connections to the AC system, which is referred to as *multi-terminal* direct current (MTDC) systems. Incidentally, both multi-terminal links in operation — Sardinia–Corsica–Italy link and Quebec–New England link — work with power flowing through the DC link from a generation center (e.g., hydro power from James Bay region in the north of Quebec province) to a main load center (e.g., Boston area and parts of New England) with another intermediate load center (e.g., Montreal region) on the way. However, unlike a meshed interconnected AC network, a truly meshed HVDC grid is yet to be realized in practice.

For overhead lines, HVDC is cost-effective only at large transmission distances (e.g., above 600 km) at which level meshed interconnections are not economically justifiable. For underground or subsea cable transmission, the distance beyond which DC technology is effective is much shorter. This has resulted in a number of point-to-point interconnections between AC systems separated by sea to allow exchange of cost-effective electricity.

With increased penetration of intermittent renewable energy sources (e.g., wind power), balancing the supply and demand is likely to be a major problem. To ensure reliable operation of the system, there is a growing need for meshed interconnection to effectively share the diverse portfolio of renewable energy resources and thereby increase operational flexibility. For instance, in Europe, the hydropower from Norway and solar power in Spain and Portugal could be utilized when the wind is not blowing in the UK or mainland Europe and vice versa. To enable such sharing of power and also harness remote offshore wind, there is a business case for setting up an European Offshore Supergrid [2, 3, 4, 5]. There are several visions for an offshore grid in Europe, some of which are shown in Fig. 1.1. One aspect in common with all such visions is that several DC links are connected at a single point forming a DC grid.

Because of the subsea transmission distances involved, the only viable option is to use DC, which essentially calls for an MTDC or meshed grid. In such a meshed DC grid, the power flow in the DC links would have to reverse frequently depending on the geographical distribution of renewable generation and the electricity price differential at a given point in time. The VSC technology allows such power flow reversals

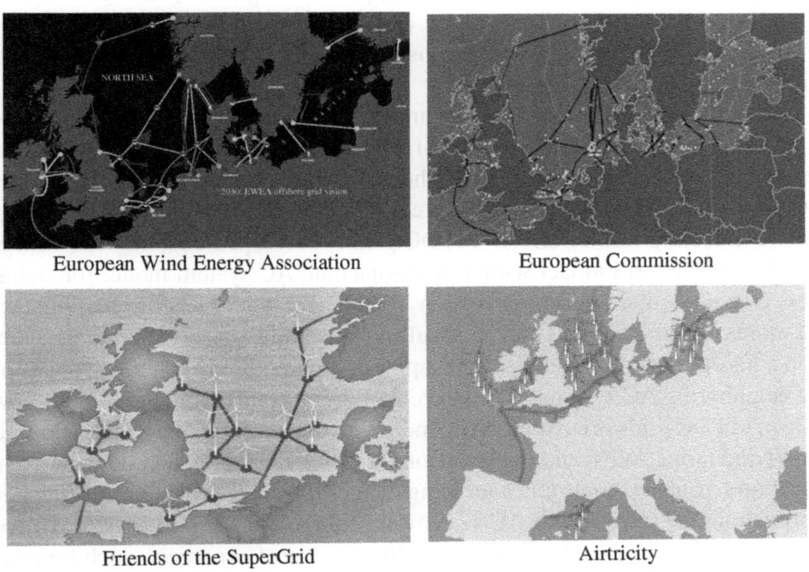

Figure 1.1 Visions for European Offshore Supergrid [2, 3, 4, 5].

without altering the DC voltage polarity and is therefore the only option for a meshed DC grid. LCC on the other hand relies on reversing voltage polarity to change the power flow direction which is not a problem for a point-to-point link but does not support a meshed grid operation without physical isolation. The situation is different from the two multi-terminal links in operation today where power flow is essentially unidirectional (from generation to load centers). Hence, they work with LCC technology, which was the only available option anyway at the time of installation of these systems.

The business case for meshed MTDC grids based on the VSC technology is getting stronger in Europe and elsewhere in the world. In China, the five-terminal Zhoushan project is expected to be operational in 2015 [6]. Despite tremendous potential, a VSC MTDC grid is yet to be realized in practice. There are several technical barriers that manufacturers, network planners, and operators are trying to resolve to facilitate the deployment of an MTDC grid. Protection and fault-current interruption in VSC MTDC grids are arguably the most challenging research and development problems that the manufacturers and academia are presently engaged in. As VSC MTDC grids are unprecedented with no operational experience, the network operators do not have much understanding of the interaction between an MTDC grid and the host AC systems and the overall stability. Moreover, there is lack of clarity about whether and how an MTDC grid could be operated and controlled to support the host AC systems. The prerequisite to studying the above in a systematic way is to develop a generic modeling and stability analysis framework for VSC MTDC grids that is compatible with those for conventional AC systems. This can then be used to analyze the overall stability of AC-MTDC grids and identify and resolve potential interactions. System support provisions (for instance frequency support) through an MTDC grid are beneficial for the network operation especially, considering the lower inertia of the turbine systems. However, they need to be exercised carefully to avoid adverse interactions leading to potential overall instability.

A comprehensive modeling, analysis, and control design framework aimed at evaluating the impact and potential benefits of DC grids on the surrounding AC network is the subject matter of this monograph. The main body of the monograph begins with an overview of the VSC systems which are the basic building blocks of MTDC grids. A generic modeling framework for MTDC grid is developed enabling easy integration of the MTDC grid model with a multi-machine AC system model for stability studies. One particular concern is how an MTDC grid would react to loss/outage of one or more converter stations and the resulting power imbalance. Sharing the burden of such a loss/power mismatch has to be appropriate in order to minimize the impact on the neighboring AC systems. The concept of droop control for autonomous power sharing in MTDC grids is covered. With increasing penetration of asynchronous generation (wind farms, WFs) and transmission (HVDC), reduction in system inertia and its effect on frequency control is a growing concern. Hence, provision for frequency support among the surrounding AC systems across the MTDC grid is critical. Modification of the autonomous power sharing (droop) control to exercise such frequency support across the MTDC grids is described. Toward the end of the monograph, a chapter is dedicated to review of the protection issues of MTDC grids including the present state of research in high power DC circuit breakers and fault blocking VSC systems. In addition, the challenges in MTDC grid protection are highlighted along with an overview of some of the protection strategies that could be used in future.

1.2 Rationale Behind MTDC Grids

In any power system, the objective is to ensure that the power generation matches the demand while considering the quality, security, reliability, and economics of the system. The rationale behind choosing an MTDC grid over multiple point-to-point DC links lies with the merits of the integrated operation of a power systems that can meet these requirements. Let us consider an example system shown in Fig. 1.2 where two options of interconnecting multiple offshore WFs to the onshore grid are shown. In one option individual WFs can be connected to the onshore grid via multiple point-to-point HVDC links as shown in Fig. 1.2(a). The other option considers integration of the WFs using a meshed MTDC grid. The onshore AC grids can be interconnected or might be isolated from each other (as shown in this figure). The advantages of the MTDC grid can be summarized as follows:

1. Improved reliability: A single-point failure in the point-to-point HVDC link results in loss-of-infeed into the AC grid which might have a significant impact on the system stability depending on the amount of loss. An MTDC grid can

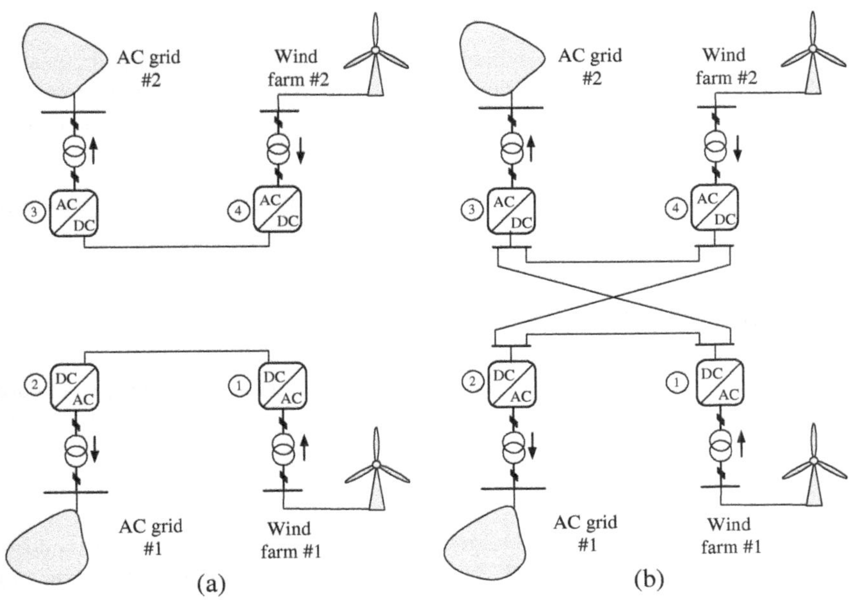

Figure 1.2 (a) Point-to-point HVDC links connecting wind farms to AC grids and (b) meshed MTDC grid connecting multiple wind farms to the AC grids.

reroute the power flow when a single-point failure occurs, thereby improving the reliability of the system. For example, the loss of cable connecting stations #3 and #4 will lead to loss-of-infeed from WF #2 for the point-to-point link, whereas the power can be rerouted in the meshed system.

2. Reduced capacity: The peak demand of different AC systems (e.g., AC grids #1 and #2) do not occur at the same time. Therefore, the peak demand of the MTDC system is much less than the sum of peak demands of multiple point-to-point HVDC links. Thus, the installed capacity of power generation can be reduced as well as the ratings of the expensive converter stations.

3. Reduced reserve capacity: The spinning reserve requirement in the AC systems can be reduced a lot when the DC grid connects the generation to different load centers.

4. Reduced curtailment from WFs: When multiple WFs are connected via multiple point-to-point HVDC links to different AC grids, the system operators might have to do wind curtailment[1] from individual farms to match the load profile of the corresponding AC systems. When the WFs are integrated through the MTDC grid, the curtailment can be reduced since the power exchange can happen between different areas.

5. Reduced variability in renewable generation: Since the MTDC grid can facilitate integration of multiple WFs, it can reduce the variability in the generation profile.

6. Ease of maintenance: The annual maintenance and preventive maintenance of the generators and the converter systems become much easier since power can be rerouted through the meshed grid in the event of shutdown of one or more converter stations.

7. Power exchange and trading: MTDC grid can facilitate energy trading between multiple regions like today's AC power systems.

1.3 Network Architectures of MTDC Grids

An MTDC grid can be realized by connecting the converter stations in series or parallel configurations or in combinations thereof. These architectures are dual in nature.

1.3.1 Series Architecture

In the series architecture of an MTDC grid, the converter stations have to be connected around a loop as shown in Fig. 1.3. The series connection of converter stations imposes circulation of the same DC current I_d around the network. One converter station

[1] when the power output of the WF is reduced even though wind energy is available

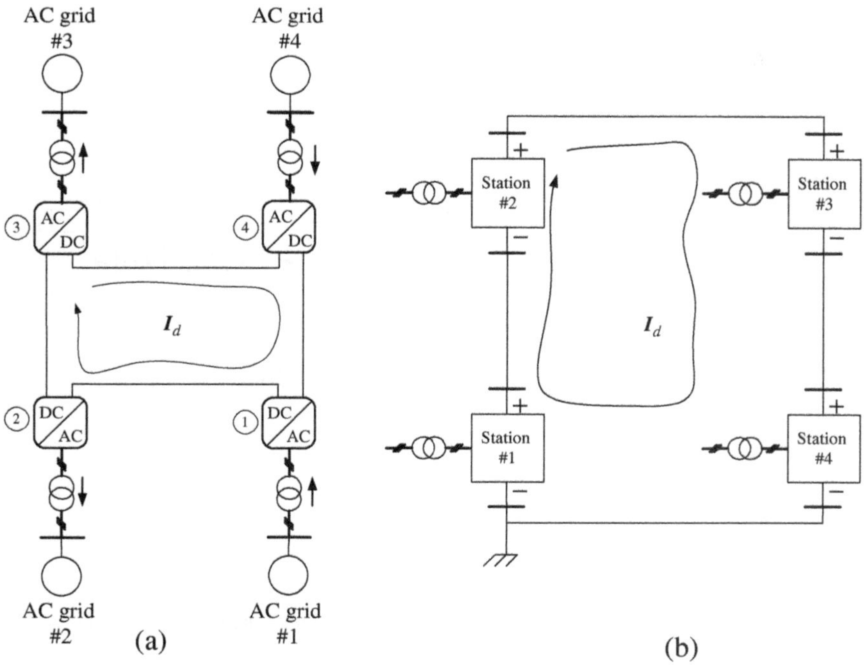

Figure 1.3 (a) MTDC network and (b) series architecture of converter stations.

is grounded and the voltage ratings of the converter stations are proportional to their individual power ratings. However, each converter transformer should be rated for the full DC system voltage insulation.

The series-connected scheme suffers from certain limitations. Different converter stations need different levels of insulation which makes the insulation coordination very challenging. Moreover, future expansion of an existing series-connected MTDC scheme is highly difficult as it needs complete redesign of insulation level. In addition, a permanent DC fault leads to the shutdown of the entire DC grid.

1.3.2 Parallel Architecture

In this architecture, the converter stations are connected in parallel, and therefore, they operate with the same voltage rating. Dual to the series counterpart, the current ratings of the converter stations are proportional to the power ratings. The MTDC grid can have either radial or meshed connection. A radial connected MTDC grid in parallel architecture is shown in Fig. 1.4 and a meshed connected grid is shown in Fig. 1.5.

Depending on the technology used (i.e., LCC or VSC), the control philosophies and corresponding challenges differ. Till date, only radial MTDC grids with parallel architecture based on LCC technology exist in practice. These schemes are described

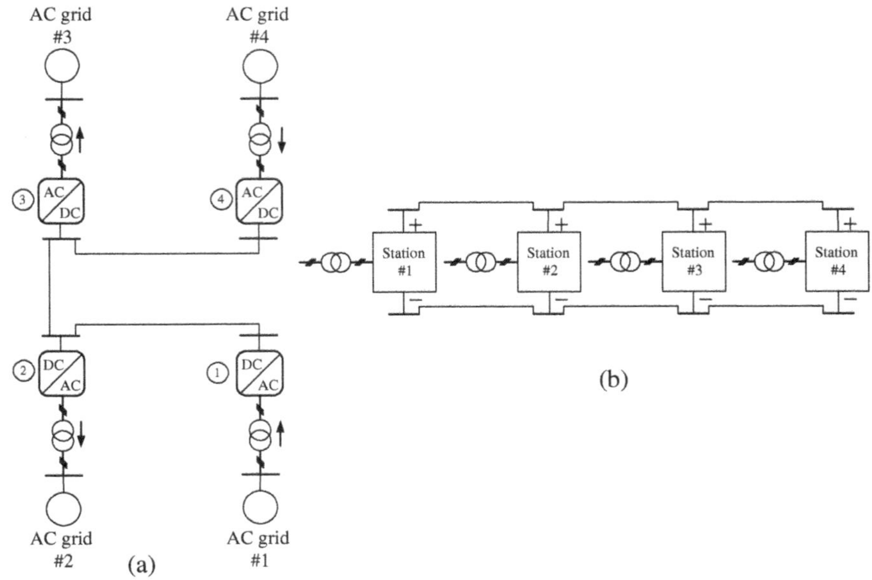

Figure 1.4 (a) Radial MTDC network and (b) parallel architecture of converter stations.

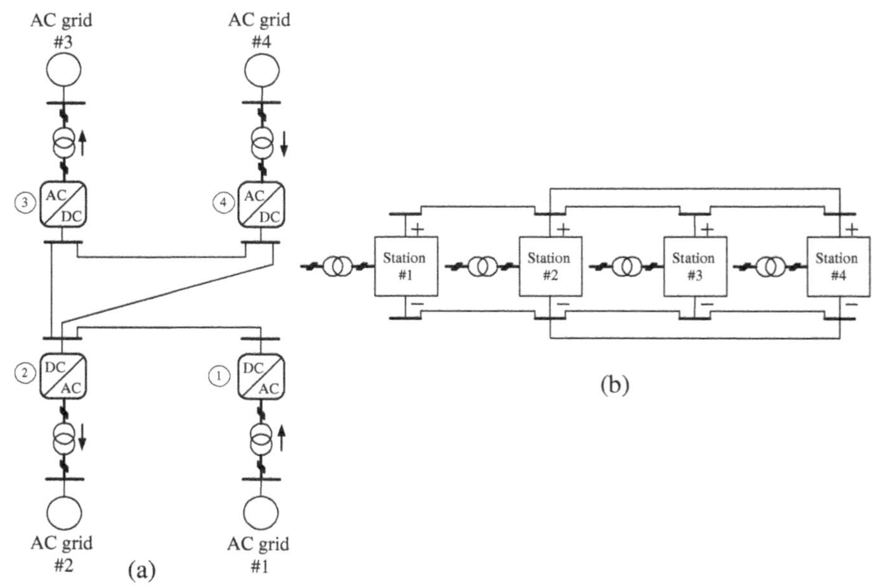

Figure 1.5 (a) Meshed MTDC network and (b) parallel architecture of converter stations.

in Section 1.4.1.2. Section 1.4 elaborates on the control philosophies and limitations of the parallel architecture for the LCC and VSC technologies. Irrespective of the type of converter used, it is widely accepted that the parallel connection is easier to control, has lower line losses, and is more flexible toward future expansion.

In the rest of this monograph, MTDC grid with parallel architecture will be considered.

1.4 Enabling Technologies and Components of MTDC Grids

The point-to-point HVDC links use either LCC or VSC technology for the converters. In principle, both LCC and VSC converter technologies can be used for realizing an MTDC grid, although VSC offers some advantages. In this section, the LCC and VSC technologies are briefly reviewed.

1.4.1 LCC Technology

The LCC technology is the most mature HVDC technology and is commonly used for bulk DC power transmission. Details of this technology can be found in many text books including [7, 8, 9, 10, 11]. The LCC converter stations use thyristors as switching devices that can be turned on using gate pulses when they are forward biased. However, they are naturally turned off only when the instantaneous AC commutating voltage becomes negative. Large smoothing reactors are connected to the DC side and fundamentally, the current in the DC side of these converters is unidirectional, i.e., the DC side behaves as a current source. Therefore, this technology is also called the *current-sourced converter (CSC) technology*. The direction of real power flow is reversed by reversing the polarity of the DC voltage. The converters consume reactive power under rectification as well as inversion mode of operation due to the inductive nature of the commutation circuit consisting converter transformer leakage reactance. The typical reactive power consumption is 50-60% of the real power at full-load condition. The consumption of reactive power changes with the loading condition.

In the AC side, converter transformers connect the terminals of the converter stations to the AC grid. Tap changing transformers are almost always used. Capacitor banks are used to meet the reactive power demand. Since the reactive power demand changes with loading, circuit breakers are used to switch the capacitor banks in and out. Fixed harmonic filters are connected to AC switchyard for harmonic cancelation and reactive power compensation. The footprint of an LCC HVDC converter station is large due to the requirement of large capacitor banks, harmonic filters, and AC switch gears. Both overhead lines and cables are used for DC power transmission. Most of the LCC cable transmission systems use mass impregnated cables. The polymeric cables cannot handle voltage polarity reversal.

For point-to-point HVDC links, the two converter stations work under different modes of operation. It is a common practice for the rectifier to control the DC line current and the inverter to control the DC voltage or commutation margin angle under nominal conditions. The margin angle control mode is less stable compared to the DC link voltage control mode. Mode shifts occur under off-nominal operating conditions by an arrangement called the *current margin control* [9, 12]. The tap changing transformers operate to bring the firing angles of the converter stations within the nominal operating range. Following a fault in the DC side, the fault current is extinguished by a technique called *forced retard* where the rectifier is pushed into the inversion zone. This generates a reverse voltage polarity at the rectifier side and takes the fault energy away from the DC side.

LCC converter stations face challenges when connected to weak AC systems due to reasons including high dynamic over voltage, voltage instability, large voltage flicker, and harmonic instability. Synchronous condensers or static VAr compensators (SVCs) are connected at the inverter bus to overcome some of these problems.

1.4.1.1 Control Modes in LCC-based MTDC Grid For the stable operation of the MTDC grid, the operating characteristics of the converter control should intersect at a common point. To achieve this, one converter station should maintain the voltage at its terminal and the others should control current injected/ drawn at their DC terminals. A rectifier or an inverter can be tasked with the DC voltage control. However, it is always more beneficial if a large rectifier terminal maintains DC link voltage while the rest of the stations control the current. This makes the system more stable and less dependent on fast communication. The major drawbacks of the LCC MTDC system include [12]:

- need for fast acting mechanical switches during power reversal.
- commutation failure at one inverter terminal drawing current from other terminals which can make the recovery more difficult.

1.4.1.2 Examples of Existing LCC MTDC Systems At present, there are two MTDC systems in the world that are operational: (1) the Sardinia-Corsica-Italy (SACOI) interconnection and (2) the Hydro-Quebec-New England interconnection. Both projects started as point-to-point links and later got extended to multi-terminal schemes. The LCC technology has been used in the converter stations, which are connected in parallel architecture.

The SACOI Interconnection [13, 14] The first phase of the SACOI interconnection was commissioned in 1967 by English Electric as a point-to-point 200 MW 200 kV monopole interconnection between San Dalmazio in Tuscany on the mainland Italy and Codrongianos on Sardinia. The converter stations were using mercury arc technology. The link was used to import power from Corsica to mainland Italy. The transmission system had three overhead line segments and two subsea cable sections. The overhead line segments were 22 km long on Italy, 156 km on Corsica, and 86 km over Sardinia, whereas the undersea cable segments between Italy and Corsica was 105 km long and 16 km in distance between Corsica and Sardinia, respectively.

A third converter station of 50 MW capacity was commissioned by CGE Alstom at Lucciana in Corsica in 1988 making it an MTDC station. This station used

Figure 1.6 The Sardinia-Corsica-Italy (SACOI) interconnection [14].

the thyristor technology. The Lucciana converter station was equipped with high speed changeover switches to facilitate bidirectional power flow from Corsica—although the power flow is unidirectional from Sardinia to Italy.

In 1992, the mercury-arc rectifier-based stations at Dalmazio and Codrongianos were decommissioned. Two new 300 MW, 200 kV stations based on thyristor technology were built in Suvereto and Codrongianos by Ansaldo A.S.I. (Italy) and General Electric (USA). Figure 1.6 [14] shows the geographic location of this radial MTDC interconnection.

The Hydro-Quebec-New-England Interconnection This is the only MTDC system involving more than three converter stations. The project was done in two phases. During Phase I, a bipolar point-to-point HVDC transmission corridor rated 690 MW was built in 1986 between Des Cantons located near Sherbrooke, Quebec, and Comerford, near Monroe, New Hampshire.

The Quebec–New England HVDC link was built during Phase II. Power generated at the La Grande II hydro power station in the James Bay area is converted into DC at the Radisson Converter Station and transmitted over the multi-terminal system to load centers in Montreal and Boston [15]. In this phase, three additional converter stations were installed and modifications were made to the existing scheme. In the north, the transmission system was extended from Des Canton over a distance of 1100 km to the Radisson terminal. In the south, an extension was done over a distance of 214 km from Comerford to a new 1800 MW converter terminal at Sandy Pond, Massachusetts.

In 1992, another terminal was placed in service on the MTDC system. This terminal is rated 2138 MW and located at Nicolet in the Montreal area [15]. Figure 1.7 shows the geographic location of this radial five-terminal MTDC interconnection. Table 1.1 summarizes the basic data of this MTDC system.

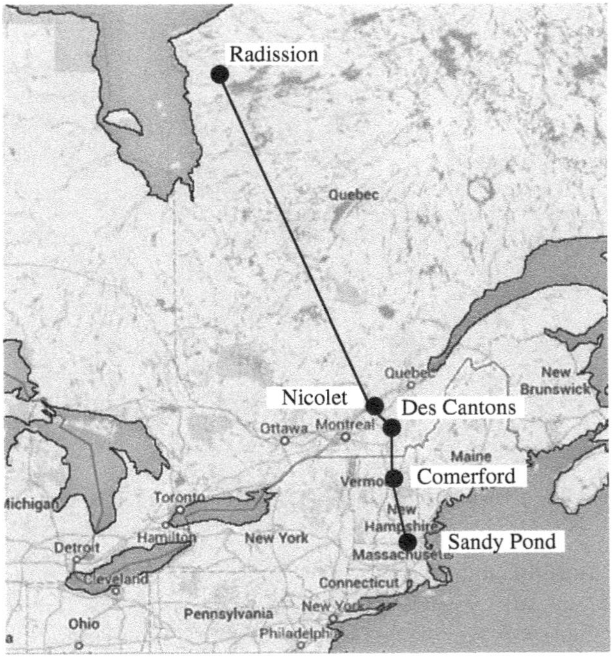

Figure 1.7 The Quebec–New England interconnection.

Table 1.1 Basic data of Quebec-New England MTDC grid [15]

Ratings	Radisson	Nicolet	Des Cantons	Comerford	Sandy Pond
Power, MW	2250	2138	690	690	1800
DC voltage, kV	450	450	450	450	450
max kV	500	475	463	458	460
DC current, A	2500	2250	773	768	2000
AC voltage, kV	315	230	230	230	345

1.4.2 VSC Technology

The VSC technology has been used for point-to-point HVDC transmission since the late nineties. This is still a developing technology unlike the LCC, which is mature and well-proven. The VSCs typically use IGBTs with antiparallel diodes. IGBTs offer both controlled turn-on and turn-off capability which can be exploited to derive an alternating voltage out of a fixed DC voltage maintained across the DC link capacitors. Both the magnitude and phase angle of the synthesized AC voltage can be controlled enabling independent control over active and reactive power exchange with the AC system at both ends. With advanced multi-level control of VSCs, much smoother AC voltage waveforms are synthesized using relatively low switching frequencies than what could be achieved with two- or three-level conversion using high frequency switching. This has drastically reduced the need for harmonic filtering and the power losses in the converters. Details of the VSC topologies and control are provided in Chapter 2. Due to significantly less (if not zero) filtering and reactive power requirement, the footprint (and hence the platform cost) of a VSC station is much less than an equivalent LCC station which makes the former particularly attractive for offshore installations.

Unlike an LCC HVDC link, the DC link voltage polarity of a VSC HVDC is fixed. Reversal of power flow is achieved through the reversal in the direction of current flow. Fixed DC side voltage polarity permits the use of polymeric (e.g., cross-linked polyethylene, XLPE) cables which are stronger and lighter and particularly suited for harsh subsea environment. Most of the VSC HVDC installations worldwide are driven by the need for subsea cable transmission. One exception is the Caprivi link [16], which is an overhead line. The motivation there is the strengths (short circuit ratios) of the terminal AC systems which are inadequate to support an LCC HVDC operation. Since the LCC systems rely on AC system voltage for turning off the thyristors, they can only operate if the AC system strengths are above a certain threshold. There is no such restriction for a VSC HVDC link which can work even with weak/isolated AC systems and even offer black start capability [10].

In case of a fault on the DC side, the antiparallel diodes of the IGBT modules act as an uncontrolled rectifier even if the IGBTs are blocked. Due to a very low inductance in the DC side, the rate of rise of short circuit current is extremely high — which is further aggravated by the DC link capacitor discharge. Since the converter is no longer controllable in this fault case (in contrast to LCC HVDC), circuit breakers on the AC side are typically opened to interrupt the fault current. In principle, circuit breakers on the DC side could be used as well but high power fast acting DC circuit breakers are yet to be commercially available [17]. Due to this vulnerability to the DC-side faults, the VSC technology is more suited for cable transmission rather than overhead lines which are more exposed to short circuits. Details of the fault response of the VSC and the protection issues are provided in Chapter 6.

VSC HVDC links are limited in terms of their rated capacities. This is due to the available voltage and current ratings of individual IGBTs (up to 5 kV, 2 kA) and also the voltage ratings (up to 320 kV) of the subsea polymeric cables. The largest VSC HVDC link commissioned until 2013 is the East-west interconnector project (EWIP) between Ireland and Wales [18] with a capacity of 500 MWs. The INELFE project between France and Spain which is to be commissioned in 2014 would use two VSC HVDC corridors each rated at 1000 MW. A list of the major VSC HVDC projects presently operational or is being planned/under construction is provided in Table 1.2.

Due to the late development and relatively low available capacity, the deployment of VSC technology has been limited to only a small fraction (< 10 GWs) of the overall HVDC installed capacity worldwide (over 100 GWs with an additional 200 GWs being planned in China alone [20]; majority of which are LCC HVDC). However, there is enormous potential for VSC HVDC especially, for offshore applications. Today,

Table 1.2 List of major VSC HVDC projects: present and future [19]

Project Name	Year	Supplier	MW	Voltage	Converter
Murray link	2002	ABB	220	±150 kV	3-level
Estlink	2006	ABB	350	±150 kV	2-level
BorWin1	2009	ABB	400	±150 kV	2-level
Trans Bay Cable	2010	Siemens	400	±200 kV	ML
Caprivi Overhead Link	2010	ABB	300	350 kV	2-level
East-West Interconnect	2013	ABB	500	±200 kV	2-level
BorWin2	2013	Siemens	800	±300 kV	ML
HelWin1	2014	Siemens	576	±250 kV	ML
DolWin1	2014	ABB	800	±320 kV	ML
INELFE (France to Spain)	2014	Siemens	2×1000	±320 kV	ML
SylWin1	2014	Siemens	864	±320 kV	ML
South-West Link	2014	Alstom	1440		ML
HelWin2	2015	Siemens	690	±320 kV	ML
Dolwin2	2015	ABB	900	±320 kV	ML

the VSC technology is, in principle, available for higher powers (up to 2000 MWs), limited mainly by the voltage ratings of XLPE DC cables.

For VSC HVDC system, there is more flexibility in control than an LCC HVDC. In a point-to-point VSC HVDC link, one converter station controls the DC link voltage, while the other usually sets the active power reference. In addition, each converter can independently control the AC side voltage or reactive power at either end. In the case of remote offshore WFs connected to the onshore network through a VSC HVDC link, the offshore converter could be operated as an ideal AC voltage-source, which would absorb all the incoming power from the WF and pass it on to the DC link [21].

Similar to the conventional AC networks, the power reversal in a particular VSC HVDC link within an MTDC grid is achieved simply by reversing the current through that link while maintaining the voltage polarity at both ends. Use of LCC technology would require reversal of voltage polarity to reverse the power flow direction in a particular link, which could affect the power flow direction in the other links as well. Thus, the VSC technology is the obvious and arguably the only choice for meshed DC grids. For the rest of this book, the scope and focus of discussion would be MTDC grids using the VSC technology.

1.5 Control Modes in MTDC Grid

Control of an MTDC grid is fundamentally different from that of a traditional AC grid that relies primarily on the control of generators (governors and exciters) to regulate the power flows and voltages. In an MTDC grid, the fast acting converters can control the power exchange (both active and reactive) with the AC systems and regulate the voltages on both AC and DC sides. The basic operating principle of a VSC system is based on a fixed voltage on the DC side. Hence, it is essential to firmly control the DC link voltage similar to controlling the frequency in AC systems. Frequency is a global variable with a common steady-state value everywhere while the DC voltage differs across the MTDC grid.

As a natural extension to the control of point-to-point VSC HVDC link, any one converter station within an MTDC grid could control the DC link voltage. The remaining converter stations would control the active power exchange with the respective AC systems. In addition, all the converters could control either the AC voltage or reactive power exchange. With one converter controlling the DC link voltage while the rest operating with a set active power reference there is a danger that following the outage of one converter, the converter under DC voltage control picks up the entire burden of the power imbalance. This issue is highlighted through case studies in Chapter 3. This is not desirable as the AC system connected to that particular converter station will experience a large change. Instead, it is desirable that following an outage of one or more converters, all the remaining ones should share the resulting power imbalance in a certain proportion. Thus, all the converter stations should operate in DC link voltage control mode rather than trying to follow a set active power references. However, set values of the DC link voltage references at all the converter stations could be conflicting unless they are modified properly depending on the reference and actual power and the DC link voltage.

The situation is similar to the parallel operation of generators in an AC system where power-frequency droop is used in the governor control loop to avoid potential conflict in frequencies. Along the same lines, active power-DC link voltage droop can be used to modify the reference values of DC link voltages at each converter station (under voltage control mode) as described in Chapter 4. Several mechanisms for controlling the DC link voltage in a VSC MTDC grid have been proposed in the literature. An overview is presented in Chapter 4. The droop control can be further modified to allow exchange of frequency support among the AC systems interconnected through the MTDC grid. This is particularly important as AC systems of future are going to have less effective inertia (e.g., due to high wind power penetration), which necessitates additional frequency support from other AC systems. Exchange of frequency support through an MTDC grid is covered in Chapter 5.

1.6 Challenges for MTDC Grids

Protection: A major challenge toward the reliable operation of an MTDC grid arises from the need for a very fast (within 1 ms) detection of a DC side fault and identification of the faulty cable. Locating a DC side fault within an MTDC networks is difficult. In AC systems, impedance relays are typically used to identify the location of the fault. As the resistance of the cables in a DC grid is negligible compared to the impedance of AC networks, it is virtually impossible for a relay to determine whether a DC fault is within its protective zone. Moreover, the protection system in a DC rid needs to act in a very short time (about 1ms), as the rate of rise of DC fault current is much higher compared to the AC fault current which is restricted by the circuit reactance. The DC grid protection system should be discriminative enough to distinguish between the AC and the DC side faults since the currents and the voltages in the DC system are also affected during an AC fault. Challenges and potential solutions toward designing appropriate protection systems for MTDC grids are discussed in Chapter 6.

Fault current interruption: Following a DC side fault within an MTDC grid, the fault current increases very fast. In order to limit the fault current within manageable levels, it should be interrupted within 2 ms. Existing point-to-point VSC HVDC links use the AC side circuit breakers to interrupt the fault current leading to loss of the whole link. Adopting a similar measure for an MTDC grid with large power transfer levels (potentially tens of GWs) would require opening of all the AC circuit breakers at interfaces with the AC systems which could be prohibitively disruptive in terms of loss of infeed. It is therefore, critical to isolate only the faulted cable and keep the remaining MTDC grid operational. This could be achieved primarily in two ways: (1) Using fast acting high power DC circuit breakers at both ends of each DC cable. Unfortunately, such DC breakers are yet to be commercially available. (2) Using fault-blocking converters where there is always a controllable switch (IGBT) in the fault current path; together with isolators which are simpler switches meant to open/close once the fault-blocking converters have ramped the fault current down to zero. Both the above-mentioned approaches rely on the presence of fast semiconductor switches in the normal current path which increases the on-state losses. Ongoing research and development is trying to strike a proper balance between efficient normal operation and fast and reliable fault current interruption as discussed in Chapter 6.

Power flow control: The basic challenges in controlling the MTDC grid have been described in Section 1.5. In addition to those, another key challenge is the control of power in a DC line. In traditional meshed AC grids, the power flow through lines can be controlled by changing the AC voltage magnitude and/or phase angle at either end of the line. Also, controllable elements like Flexible AC Transmission Systems (FACTS) devices can be used to change the line reactance to control the power flow. In DC lines, the line reactance is zero and only voltage magnitude can be altered at each end of the lines, i.e., at the converter DC terminals.

Therefore, it is not easy to control the line flow in a particular DC line without additional power electronics (DC FACTS) installed in series with the DC lines.

1.7 Configurations of MTDC Converter Stations

Possible configurations of the converter stations in an MTDC grid are described in this section. Figure 1.8 shows the converter stations in monopole configuration. For the

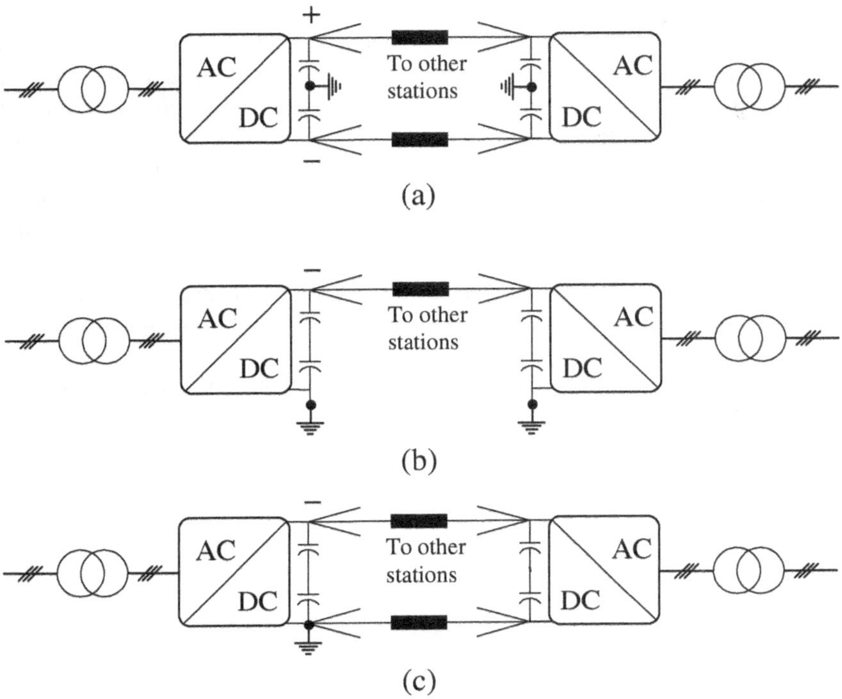

Figure 1.8 Configurations of converter stations. (a) Symmetric monopole, (b) asymmetric monopole with ground return, and (c) asymmetric monopole with metallic return.

sake of clarity, only two stations are illustrated — each of which could be connected to multiple converter stations. In this configuration, the converter station has only one independent converter bridge or group of bridges.

In the symmetric monopole configuration shown in Fig. 1.8(a), the center of the DC bus of each converter station is grounded and the converter stations are connected to two transmission lines at equal and opposite potentials. Note that this center point might be an example of the physical node defined as the *DC-side midpoint* in Section 2.2.

In the asymmetric monopole configuration, each converter station is connected to a transmission line (usually negative polarity for overhead lines) while the current returns through the ground (Fig. 1.8(b)) or metallic return path (Fig. 1.8(c)). For ground return configuration, each station is grounded whereas the metallic return needs only one grounding point. Metallic return is used under situations where ground return is not acceptable due to high earth resistivity, interference with underground metallic installations, or maritime regulations. The metallic return path has a very low insulation requirement. A negative polarity for the overhead transmission line is preferred over the positive polarity due to reduced corona losses and radio interference. The monopole configuration suffers from the reliability consideration if the outage of the converter or degradation of insulation in one line is treated probabilistically as (N-1).

The asymmetric monopole configuration can be commissioned as the first phase of a bipolar installation. A bipolar converter station has two independent converter bridges or group of bridges. Figure 1.9(a) shows the symmetric bipole configuration. One terminal of the DC bus of an independent group is grounded, and the other is connected to the transmission line with positive or negative polarity. Each converter pole can operate independently. Under normal conditions, the positive and the negative poles carry equal current and no current flows through the ground return path. For overhead lines, it also produces considerably less harmonic interference with nearby lines and telephonic interference compared to the asymmetric monopolar link.

Following the outage of a converter pole or outage of one of the transmission lines, the other pole can be overloaded based on the reserve capacity of the converters, transmission lines, ambient condition, and duration. The converters can be operated with a reserve capacity so that they can be overloaded by 200% for a short time.

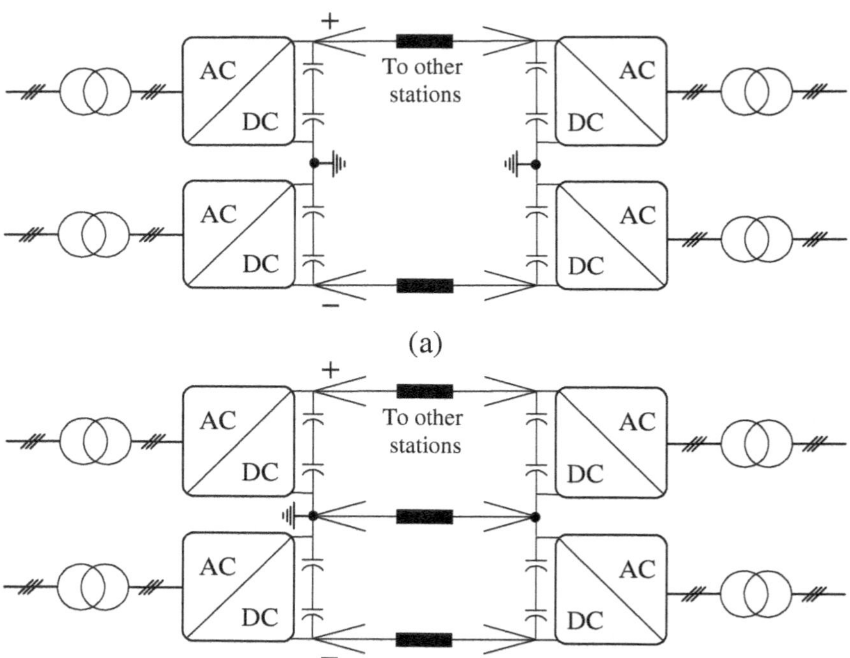

Figure 1.9 Configurations of converter stations. (a) Symmetric bipole and (b) asymmetric bipole with metallic return.

The ground return current under such conditions will be equal to the current flowing through the healthy transmission line.

When ground return is not viable, a metallic return network can be used as shown in Fig. 1.9(b). This configuration is called the *asymmetric bipole* with metallic return. The insulation requirement of the metallic return pole is minimal. For overhead transmission system, this line may act as shielding wire. From the reliability standpoint, if the outage of the converter or degradation of insulation in one line is treated probabilistically as (N-1), the asymmetric bipolar link is equivalent to the double circuit line in AC systems.

Another configuration called the *homopolar configuration* needs a continuous flow of ground current twice the rated current. This is not discussed here as it has not been adopted in any practical installation. Most point-to-point installations are based on bipole configuration and operate as monopole only under contingency situations or in the first phase of development.

Each converter pole within above-mentioned configurations consists of multiple valve groups connected in series on the DC side and in parallel on the AC side by means of converter transformers to match the voltage and current rating requirements.

1.8 Research Initiatives on MTDC Grids

Following a review of the possible MTDC grid architectures, converter technology, control modes, and converter station configurations, this section highlights the current research initiatives dedicated to MTDC grids. Recently, The International Council on Large Electric Systems (CIGRE) has initiated multiple research activities through its study committees. CIGRE study committee B4 on HVDC and Power Electronics has several working groups (WGs), some jointly with other study committees (like B5), which are currently active on studying different aspects of MTDC grids [22]. The aim of some of these WGs are summarized in the following:

B4-56: Guidelines for Preparation of Connection Agreements or Grid Codes for HVDC Grids The objective of this WG is to provide information for the HVDC grid designer, concerning the constraints within which the network should be designed, e.g., direct voltage range, temporary voltages and durations, power ramp rates, and faults current stresses. It would also provide information concerning information required by anyone providing new equipment for an existing HVDC grid in order that his terminal will integrate satisfactorily in the existing system, without any negative impact on the overall system performance.

B4-57 Guide for the Development of Models for HVDC Converters in an HVDC Grid The objective of this WG is to facilitate the development of generic models that will allow a high level study of the HVDC grid and AC network interactions to be performed. A EMTDC model will be used as an example of conversion to a "black box" model. The input/output signals for this model will be defined in collaboration with WG B4.56.

B4-58 Devices for Load Flow Control and Methodologies for Direct Voltage Control in a Meshed HVDC Grid The objectives of the WG are to (1) confirm the feasibility of power flow control in the HVDC grids to meeting changing import and export needs at the HVDC grid nodes, (2) identify methods and devices for controlling the power flow in HVDC grids, and (3) compare the merits of different methods and devices.

B4-60 Designing HVDC Grids for Optimal Reliability and Availability Performance The WG will develop the definitions of reliability and availability for HVDC grids, e.g., considering the ability of different parts of the system to transport power between each part, as well as the capability of the converter stations to provide the relevant power interface between the HVDC grid and the AC network. Issues to be considered will include mean time to failure, repair times, capacity loss, etc. Reference will be made to the criteria developed for AC networks, as well as those developed for point-to-point HVDC systems. All components of the HVDC grid will be considered, but unavailabilities in the AC system should be disregarded in this context. The WG will provide guidance in respect to the additional costs that may be incurred by adopting different reliability and availability criteria.

B4/B5-59 Control and Protection of HVDC Grids The objective of this WG is to provide guidelines for the control and protection system needed for an HVDC grid, with the emphasis on the protection of the HVDC grid and the elements within it, and not on the converter or AC side equipment. The WG will consider the entire HVDC grid protection system, e.g., the measurement, fault detection, and interruption devices. It will develop the requirements for the system in relation to availability and reliability including redundancy, requirements, speed, selectivity, and maintainability. All possible faults within the HVDC grid or affecting the AC grid, including DC line faults, "DC breaker" failures, and converter faults, will be considered.

JWG A3/B4.34 Technical requirements and specifications of state-of-the-art DC switching equipment This joint working group (JWG) will investigate technical requirements and the state-of-the-art specifications for "mechanical" based DC switching equipment in order to identify a guideline that may facilitate the development of efficient, reliable, and cost beneficial DC switching equipment applied to the future DC grids and systems. This way, this JWG will review the technical requirements of DC switching equipment for different applications such as MTDC systems and offshore WF connections. It will collect all available specifications of the state-of-the-art DC switching equipment used at transmission and distribution voltages. It will also investigate the technical capabilities and limitations of existing and projected switching equipment mainly with mechanical operating drives and then foresee the future capability of these DC switching equipments.

JWG B4/C1.65 Recommended voltages for HVDC Grids Prior CIGRE activities within B4 recognized that agreeing on a set of recommended DC voltage levels to be used in HVDC Grids could be very beneficial, for anyone planning HVDC projects, which might potentially become part of a future HVDC grid. However, while adopting the recommended DC voltages would facilitate future extensions toward HVDC grids, they would preclude the optimization of DC voltage levels in individual projects, thereby leading to potentially higher investment costs. Therefore, while a first proposal for recommended voltages for HVDC grids could be produced, it was recognized that before these values and guidelines for their use could be published outside CIGRE and used by, e.g., IEC as input to standardization activities, the views, and recommendations from System Development and Economics experts within CIGRE should be included. This WG is jointly organized with C1 for this purpose.

For further details about the scope and specific objectives of this WG, the readers should refer to the individual Terms of Reference documents available at [22].

1.9 Focus and Scope of the Monograph

The aim of this monograph is to facilitate understanding of the system level aspects of an MTDC grid and provide a systematic framework to study its impact on AC system operation. To that end, an overview of the VSC technology and its modeling for system level studies is provided in Chapter 2. A generic modeling framework for stability analysis of a combined AC-MTDC system is presented in Chapter 3. Autonomous power sharing among the converter stations within an MTDC grid and its implication on stability is discussed in Chapter 4 along with a case study for demonstration. In Chapter 5, exchange of frequency support between AC systems (including offshore WFs) connected through an MTDC grid and its impact on overall system stability is illustrated and a case study is presented. An overview of the technical challenges in MTDC grid protection and interruption of DC fault current using fault-blocking converters or DC circuit breakers and ongoing research and development is presented in Chapter 6. Some of the discussions in Chapters 3, 4, and 5 are based on recent research undertaken by the authors while Chapter 6 summarizes the recent and ongoing work by others.

The monograph is intended for the academic community with research and/or teaching interest in DC electric power transmission, graduate students focusing on DC grids, manufacturers of DC grid technologies, developers of offshore transmission network, and utilities/grid system operators who has or is planning to have VSC DC links connected to their system or contemplating being part of a DC grid in future.

CHAPTER 2

THE VOLTAGE-SOURCED CONVERTER (VSC)

2.1 Introduction

The voltage-sourced converter (VSC) would constitute the backbone of future MTDC grids. A VSC interfaces an AC grid with a DC grid and enables an efficient bidirectional trade of energy between the two. More specifically, (1) it may impress the voltage of the DC grid and then let the energy flow result based on the prevailing conditions or, alternatively, (2) operate off an impressed DC voltage (e.g., impressed by one or more other VSCs interfaced with the same DC grid). In addition, it can exchange (in either direction) a pre-specified amount of reactive power with the AC grid. The aforementioned functionalities are realized by a circuit comprising power diodes, semiconductor power switches with turn-off capability, inductors, and capacitors. As circuit elements, resistors are avoided, but they inevitably exist as parasitic elements.

Adopting a generalized approach, this chapter discusses the modeling, control, and simulation of the VSC.

Multi-terminal Direct-Current Grids: Modeling, Analysis, and Control. First Edition.
Nilanjan Ray Chaudhuri, Balarko Chaudhuri, Rajat Majumder, and Amirnaser Yazdani.

2.2 Ideal Voltage-Sourced Converter

Consider the system of Fig. 2.1 in which a five-terminal device, labeled as the *ideal VSC*, serves as an energy transfer medium between a DC grid and an AC grid. As Fig. 2.1 indicates, the two left-hand side terminals of the ideal VSC identify a port, referred in this book to as the *DC side*, whose voltage and current are denoted by, respectively, V_{dc} and I_{dc} (both functions of time, in general); a positive current is assumed to be one entering the positive terminal of the port. The three remaining terminals identify the *AC side* of the VSC and assume voltages v_{ta}, v_{tb}, and v_{tc}, relative to a potential reference node, 0 in Fig. 2.1. This potential reference node is referred to as the *DC-side midpoint* of the VSC and is a (physical or virtual) node whose voltage is the algebraic mean of the voltages of the DC-side positive and negative terminals (relative to any other node). In other words, relative to the DC-side midpoint, the positive and negative DC-side terminal voltages of the VSC are equal to within a sign. Typically, the DC-side midpoint of the VSC is not an accessible physical node, but it is a virtual node used to simplify the analysis. The currents i_{ta}, i_{tb}, and i_{tc} leave the corresponding AC-side terminals of the VSC and enter the AC network.

The following properties characterize the ideal VSC:

1. The AC-side voltages v_{ta}, v_{tb}, and v_{tc} constitute a balanced three-phase set of sinusoidal voltages whose magnitude and phase angle are controllable.

2. The ideal VSC is a network of passive, loss-less, memory-less, circuit elements, including the ideal fully controllable switch, the ideal diode, and the ideal transformer.

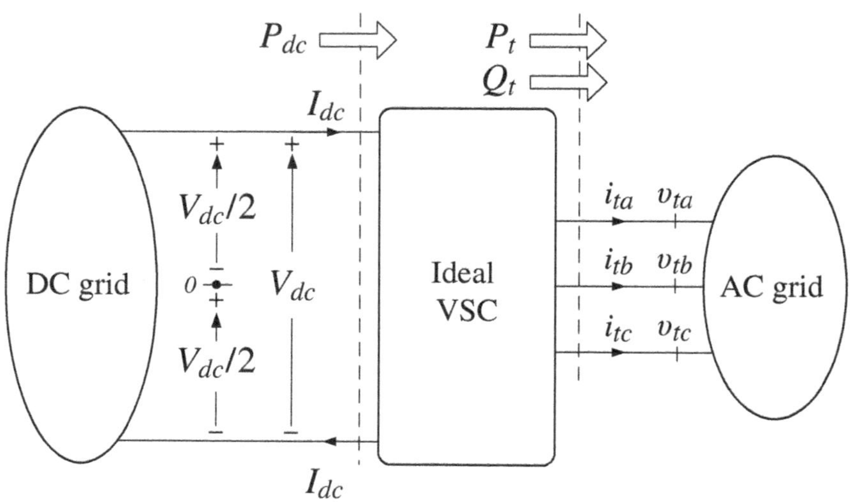

Figure 2.1 Ideal VSC interfacing a DC grid with an AC grid.

The first assumption above can be mathematically expressed as

$$v_{ta}(t) = V_t(t)\cos\left[\varepsilon(t)\right] \tag{2.1}$$

$$v_{tb}(t) = V_t(t)\cos\left[\varepsilon(t) - \frac{2\pi}{3}\right] \tag{2.2}$$

$$v_{tc}(t) = V_t(t)\cos\left[\varepsilon(t) - \frac{4\pi}{3}\right] \tag{2.3}$$

where $V_t(t)$ and $\varepsilon(t)$ are the (generally time-varying) magnitude and phase angle, respectively, of the balanced three-phase AC-side voltage. Commonly, $\varepsilon(t)$ is not directly controlled, but it is a by-product of the frequency, $\omega(t)$, as

$$\varepsilon(t) = \varepsilon_0 + \int_0^t \omega(\tau)d\tau \tag{2.4}$$

where ε_0 is the constant of integration (i.e., the initial phase angle). Thus, $\omega(t)$ is a stand-alone control variable.

Assuming a undistorted sinusoidal Thevenin voltage for the AC grid at the coupling point, a nonzero average energy transfer, i.e., a nonzero real-power flow, can exist only if the VSC operates at the same frequency as that of the AC grid, that is, if $\omega(t)$ is made equal to the power system frequency; the mechanism to ensure that this will be discussed in Section 2.4.2.5. Thus, if the power system frequency is constant at, say, ω_s, then, based on (2.1) through (2.4), the AC-side voltages assume the forms:

$$v_{ta}(t) = V_t\cos\left[\omega_s t + \varepsilon_0\right] \tag{2.5}$$

$$v_{tb}(t) = V_t\cos\left[\omega_s t + \varepsilon_0 - \frac{2\pi}{3}\right] \tag{2.6}$$

$$v_{tc}(t) = V_t\cos\left[\omega_s t + \varepsilon_0 - \frac{4\pi}{3}\right] \tag{2.7}$$

The second assumption implies that the instantaneous power that the VSC draws from the DC network, P_{dc}, equals the real power that the VSC delivers to the AC network, P_t:

$$\underbrace{V_{dc}I_{dc}}_{P_{dc}(t)} = \underbrace{v_{ta}i_{ta} + v_{tb}i_{tb} + v_{tc}i_{tc}}_{P_t(t)} \tag{2.8}$$

which can be reordered as

$$I_{dc} = \frac{v_{ta}i_{ta} + v_{tb}i_{tb} + v_{tc}i_{tc}}{V_{dc}} \tag{2.9}$$

Based on the forgoing discussion, the operation of the ideal VSC can be explained as follows: The control results in the generation of a balanced three-phase AC-side voltage, (2.5–2.7), whose magnitude V_t is also proportional to the DC-side voltage V_{dc}. The AC-side voltage, on the other hand, results in the generation of a corresponding three-phase AC-side current. Thus, the interaction of the AC-side voltage and current translates into a power flow, which, based on (2.9), translates into a corresponding amount of DC-side current I_{dc}. Therefore, proper operation of the ideal VSC requires that V_{dc} be fairly constant. This, in turn, means that the DC grid should exhibit a small Thevenin impedance to the DC-side port of the ideal VSC. On the other hand, since v_{ta}, v_{tb}, and v_{tc} do not depend on the AC-side currents, i_{ta}, i_{tb}, and i_{tc}, proper operation of the VSC requires the AC grid to exhibit, from any two of the coupling nodes, a large Thevenin impedance to the VSC. The aforementioned requirements are equivalent to stating that proper operation of the VSC requires the host DC and AC grids to be of the voltage source and current source natures, respectively. If the AC grid is linear, then i_{ta}, i_{tb}, and i_{tc} also constitute a balanced three-phase set of sinusoids and, therefore, P_t ($=P_{dc}$) is constant, which, if V_{dc} is also constant, implies a constant I_{dc}. The aforementioned discussion indicates that the ideal VSC behaves as a controllable three-phase voltage source from its AC side, and it acts as a dependent current source from its DC port. This point will be further discussed in Section 2.5.

A VSC is never directly interfaced with the AC and DC grids in the way shown in Fig. 2.1. Rather, an AC filter is placed between the AC-side terminals and the AC grid. Similarly, a DC filter is placed between the DC port of the VSC and the DC grid. There are two reasons for the insertion of these filters: First, one cannot count on, let alone control, the Thevenin impedance of the grids, to be able to ensure the current-source nature of the AC grid and the voltage-source nature of the DC grid. The second reason is that, as will be discussed in Section 2.3, a practical VSC employs switching techniques for AC voltage synthesis and, consequently, its AC-side terminal voltage is harmonically distorted. If the Thevenin impedance of the AC grid is small, the voltage harmonics result in large corresponding current harmonics, which do not contribute to the energy transfer process, but stress the VSC (and perhaps even the grid) and decrease the efficiency. Therefore, some impedance must be deliberately inserted between each AC-side terminal of the VSC and the corresponding phase of the AC grid, at the coupling point, to act as a buffer between the VSC and the Thevenin voltage of the AC grid and, thus, ensure an acceptably small current harmonic distortion. In a dual manner, an excessively large DC grid Thevenin impedance results in a poor DC-side voltage regulation and adversely affects the operation of the VSC. Thus, a DC-side filter ensures that the DC port of the VSC is voltage-supported, irrespective of the Thevenin impedance of the grid.

Figure 2.2 shows the same system as that of Fig. 2.1 with the VSC equipped with the AC- and DC-side filters; both filters are indicated to be of the LC type. The AC-side filter inductance, L, provides a large harmonic impedance between the VSC terminals, while the capacitance C exhibits a short path to the current harmonics and, therefore, prevents them from entering the AC grid. This, in turn, ensures low harmonic voltage distortion at the grid side. The capacitance also supports the fundamental component of the grid voltage at the coupling point, $v_{g,abc}$. Thus, effectively a deterministic inductance, i.e., L, ties each AC-side terminal of the VSC to the corresponding phase of a three-phase voltage source. Hence, L is also know as the *tie reactor* of the VSC. Depending on the design, C may be connected in parallel with, or even be entirely substituted by, one or more series LC circuits tuned to certain harmonic frequencies. R represents the ohmic power loss of the inductor.

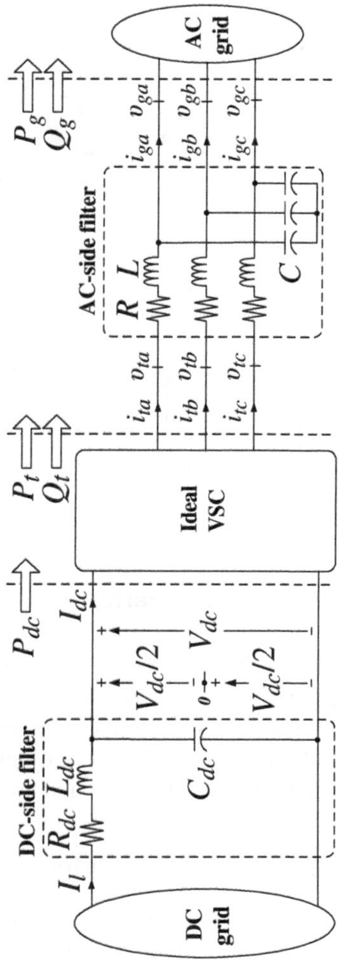

Figure 2.2 Ideal VSC equipped with DC-side and AC-side filters.

Similarly, the DC-side filter capacitor C_{dc}, while open for the average (DC) component of I_{dc}, must exhibit a small impedance to the harmonics of I_{dc}. To that end, the capacitance does not have to be large, since, typically, the harmonics are of high frequencies. However, an inductance, L_{dc}, may need to be included to ensure that the capacitor remains effective regardless of the impedance that the DC-grid exhibits to the VSC (R_{dc} represents the ohmic power loss of L_{dc}). On the other hand, the capacitance must be large to mitigate double-frequency ripples of V_{dc}, which develop due to grid voltage imbalances. Therefore, in practice, the capacitor consists of low-capacitance high-frequency capacitors connected in parallel with high-capacitance low-frequency capacitors. To enable large DC voltages, multiple such composite capacitors must be connected in series and parallel.

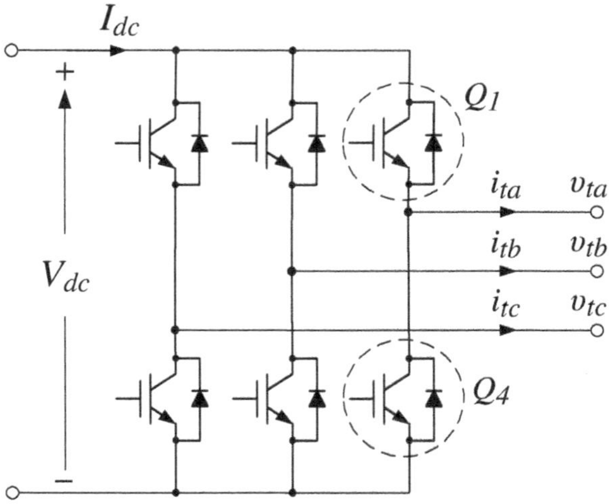

Figure 2.3 Simplified schematic diagram of the two-level VSC.

2.3 Practical Voltage-Sourced Converter

For obvious reasons, the ideal VSC described in Section 2.2 cannot exist in practice. However, it can be approached to varying degrees, depending on the expected performance and acceptable cost, by a variety of circuit configurations, which are invariably of the switched-mode type to be able to offer high efficiencies. This section introduces the most notable circuit configurations and switching strategies.

2.3.1 Two-Level Voltage-Sourced Converter

Figure 2.3 shows a simplified schematic diagram of the most basic (and perhaps the most widely used) of all such switched-mode circuits. The circuit, known as the *two-level VSC*, consists of three *half-bridge converters*, one per AC-side terminal, and the common DC sides of the half-bridge converters identify the DC port of the VSC. As Figure 2.3 indicates, each half-bridge converter consists of two fully controllable, reverse-conducting, semiconductor switches. The switches can be based on different technologies such as the insulated gate bipolar transistor (IGBT) and integrated gate-commutated thyristor (IGCT)[1] and are turned on and off in a complementary manner, that is, when the one switch is commanded to turn on, the gate command of the other switch must have been canceled. Consequently, at each switching instant, the terminal voltage of the half-bridge converter transitions from one of the two possible levels $-V_{dc}/2$ and $V_{dc}/2$ to the other one. Thus, if the aforementioned switching action is periodic, then the terminal voltage has a fundamental component, which

[1]Regardless of the technology, in this book, we will use the symbol of the IGBT to illustrate a fully controllable semiconductor switch.

can be made to have the desired frequency ω, as

$$\langle v_{ta}\rangle_1(t) = V_t(t)\cos\left[\varepsilon(t)\right] \tag{2.10}$$

$$\langle v_{tb}\rangle_1(t) = V_t(t)\cos\left[\varepsilon(t) - \frac{2\pi}{3}\right] \tag{2.11}$$

$$\langle v_{tc}\rangle_1(t) = V_t(t)\cos\left[\varepsilon(t) - \frac{4\pi}{3}\right] \tag{2.12}$$

where $\varepsilon(t)$ and $\omega(t)$ are related by (2.4).

The AC-side terminal voltage also includes harmonics, which are of odd orders if the switching action features *half-wave symmetry*. Thus, the number and moments of terminal voltage transitions (from $-V_{dc}/2$ to $V_{dc}/2$ and vice versa) in each half cycle may be controlled in such a way that (1) the magnitude of the fundamental component, $V_t(t)$ has the desired value, and (2) a number of low order harmonics are eliminated from the terminal voltage[2]. However, typically, this method of control is nonlinear, since the ratio $V_t(t)/V_{dc}$ is, in general, a highly nonlinear function of the switching instants.

The most frequently used switching strategy for the two-level VSC is the *suboscillation, carrier-based, pulse-width modulation (PWM)* strategy, in which the switching instants of a constituting half-bridge converter are determined by comparing a corresponding *modulating signal*, $m(t)$, with a high frequency periodic triangular *carrier signal*, as Figure 2.4 illustrates. A special case of the suboscillation, carrier-based, PWM strategy employs a sinusoidal modulating signal and, hence, is referred commonly to as the *sinusoidal pulse-width modulation (SPWM)* strategy. Figure 2.5 depicts waveforms typical of the SPWM switching strategy. Thus, in a three-phase VSC, there are three modulating signals, one per half-bridge converter, which are compared with a common carrier signal, and they constitute a balanced three-phase sinusoidal signal, i.e., they have the same magnitude, but are phase displaced by 120° relative to each other. Under the SPWM method, (1) the fundamental component of each AC-side terminal voltage is a magnified version of the corresponding modulating signal and (2) the frequencies of voltage harmonics are pushed up to multiples of the carrier signal frequency and, therefore, the AC-side filter inductance, L in Fig. 2.2, can be made smaller.

As mentioned earlier, based on the SPWM strategy, the fundamental components of the AC-side voltages are the amplified versions of their corresponding modulating signals, as

$$\langle v_{ta}\rangle_1(t) = \frac{1}{2}V_{dc}(t)m_a(t) \tag{2.13}$$

$$\langle v_{tb}\rangle_1(t) = \frac{1}{2}V_{dc}(t)m_b(t) \tag{2.14}$$

$$\langle v_{tc}\rangle_1(t) = \frac{1}{2}V_{dc}(t)m_c(t) \tag{2.15}$$

[2]This method of control is known as *selected-harmonic elimination (SHE)* switching strategy. No *triple-n harmonic* needs to be eliminated, however, if the AC network has three wires (such as shown in Fig. 2.1) and is also balanced.

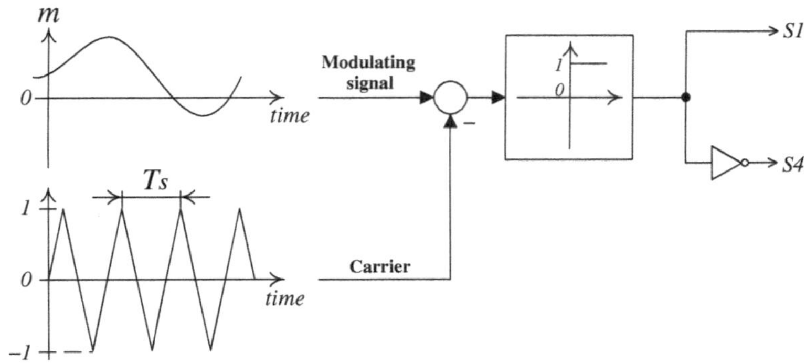

Figure 2.4 Signal flow diagram of the suboscillation, carrier-based, pulse-width modulation (PWM) switching strategy.

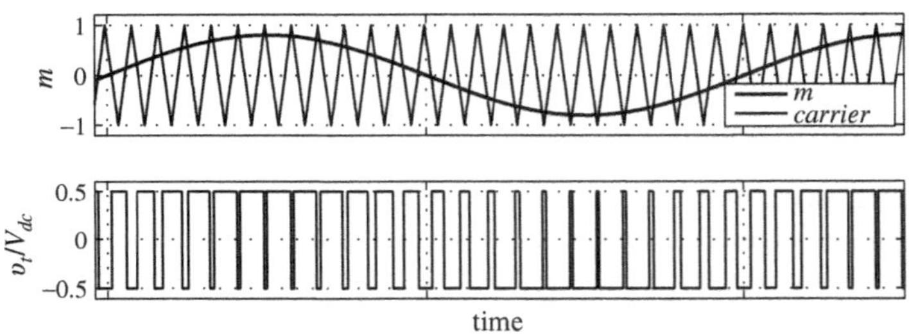

Figure 2.5 Waveforms of the modulating signal, carrier signal, and AC-side terminal voltage, based on the SPWM switching strategy.

where $m_a(t)$, $m_b(t)$, and $m_c(t)$ denote the modulating (control) signals, for phases a, b, and c, respectively. It then follows from (2.13) through (2.15) that, if

$$m_a(t) = M(t)\cos[\varepsilon(t)] \tag{2.16}$$

$$m_b(t) = M(t)\cos\left[\varepsilon(t) - \frac{2\pi}{3}\right] \tag{2.17}$$

$$m_c(t) = M(t)\cos\left[\varepsilon(t) - \frac{4\pi}{3}\right] \tag{2.18}$$

then,

$$\langle v_{ta}\rangle_1(t) = \frac{1}{2}V_{dc}(t)M(t)\cos[\varepsilon(t)] \tag{2.19}$$

$$\langle v_{tb}\rangle_1(t) = \frac{1}{2}V_{dc}(t)M(t)\cos\left[\varepsilon(t) - \frac{2\pi}{3}\right] \tag{2.20}$$

$$\langle v_{tc}\rangle_1(t) = \frac{1}{2}V_{dc}(t)M(t)\cos\left[\varepsilon(t) - \frac{4\pi}{3}\right] \tag{2.21}$$

where $M(t)$ is the magnitude of the modulating signals and an independent control variable. The phase angle $\varepsilon(t)$, however, is not a stand-alone control variable, but is calculated from the controllable frequency $\omega(t)$, based on (2.4). Thus, if the DC-side voltage is constant, then the fundamental components of the AC-side terminal voltages also constitute a balanced three-phase voltage. Further, if the magnitude and frequency of the modulating signals are also constants, say, at M and ω_s, respectively, then the fundamental AC-side voltages take the forms:

$$\langle v_{ta}\rangle_1(t) = V_t \cos\left[\omega_s t + \varepsilon_0\right] \tag{2.22}$$

$$\langle v_{tb}\rangle_1(t) = V_t \cos\left[\omega_s t + \varepsilon_0 - \frac{2\pi}{3}\right] \tag{2.23}$$

$$\langle v_{tc}\rangle_1(t) = V_t \cos\left[\omega_s t + \varepsilon_0 - \frac{4\pi}{3}\right] \tag{2.24}$$

where the constant $V_t = 0.5V_{dc}M$ is the magnitude of the AC-side three-phase voltage.

It should be pointed out that each switch of the two-level VSC must be able to withstand an off-state voltage of, at least, V_{dc}. Therefore, in order to enable the VSC to operate at higher DC and AC voltage levels, each switch may be implemented by series-connecting multiple individual switches, which are gate-commanded simultaneously. For a higher current-handling capability, multiple individual switches may need to be connected in parallel and gated in synchronism.

2.3.2 Three-Level Voltage-Sourced Converter

The terminal voltage of a VSC can more faithfully approximate a sinusoid, i.e., it can have a smaller harmonic distortion, if it is allowed to assume more levels than just two. For example, a three-level voltage waveform can be synthesized by an *H bridge*[3] (Fig. 2.6) in conjunction with the so-called *unipolar* SPWM switching strategy illustrated by the signal flow diagram of Fig. 2.7. Thus, as Fig. 2.8 illustrates, the AC-side terminal voltage v_t swings between the values 0 and V_{dc} during the positive half cycle of the sinusoidal modulating signal and between 0 and $-V_{dc}$ in the negative half cycle of the modulating signal. Then, as shown in Fig. 2.9, a three-phase three-level VSC may be realized by three H bridges, which are controlled by three corresponding modulating signals constituting a balance three-phase set of waveforms; the H bridges are connected in parallel from their DC ports to identify the DC port of the composite VSC, and their AC outputs are combined by three corresponding single-phase transformers whose grid-side windings are connected as a *wye* configuration.

A three-level voltage waveform can, however, be most economically synthesized by the configuration of Fig. 2.10, known as the three-phase *neutral-point clamped (NPC)* VSC[4]. Here, the voltage of an AC-side terminal swings between the values 0 and $V_{dc}/2$ during the positive half cycle of the modulating signal and between 0 and $-V_{dc}/2$ in the negative half cycle of the modulating signal, as shown in Fig. 2.11. In particular, the voltage level 0 calls for a connection between the AC-side terminal to

[3]Formally, the *full-bridge converter*.

[4]For a more elegant use of acronyms, in the remainder of this book, we use the term *NPC converter* instead of "NPC VSC."

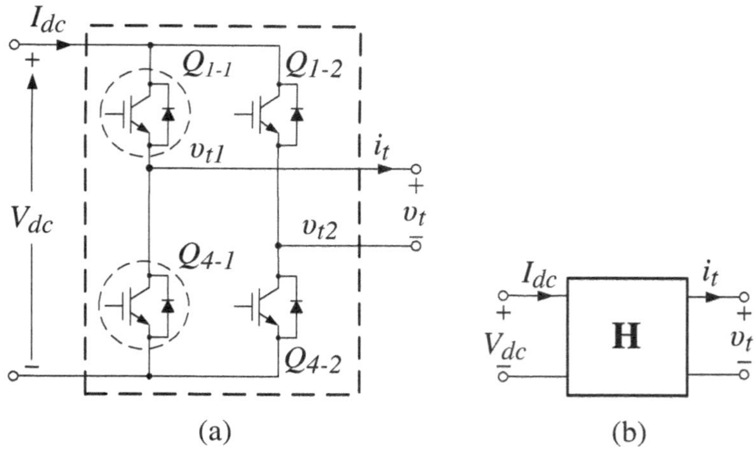

Figure 2.6 Schematic diagram of the H bridge.

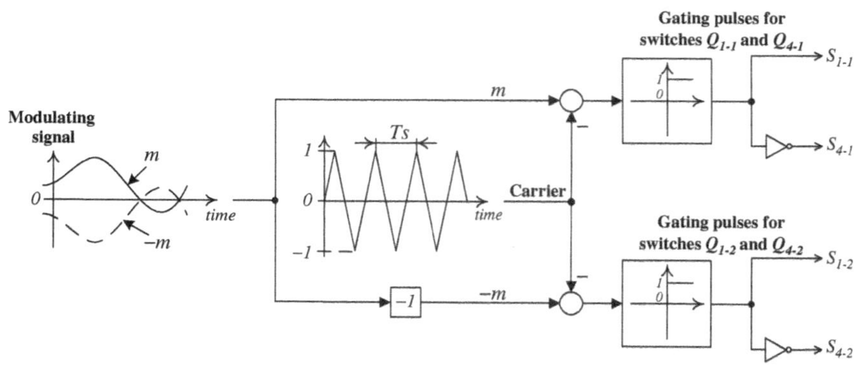

Figure 2.7 Signal flow diagram of the unipolar carrier-based PWM switching strategy for the H bridge.

the DC-side midpoint of the VSC (that is, to the potential reference). Consequently, in contrast to the two-level VSC, the DC-side midpoint in the NPC converter is a physical node. As indicated in Fig. 2.10, the DC-side midpoint is taken as the midpoint of a capacitive voltage divider that is connected in parallel with the DC port of the converter, and it consists of two nominally identical capacitors, $2C_{dc}$. Practically, these two capacitors act, effectively, also as the DC-side filter capacitor for the converter (Fig. 2.2). A three-level AC-side terminal voltage can be synthesized by the SPWM switching strategy of Fig. 2.12, which is a slightly different variation of the scheme of Fig. 2.7; here, the triangular carrier signal is allowed to swing between 0 and 1. Similar to the case of the two-level VSC, under the SPWM strategy, the magnitude and phase angle of the three-phase AC-side terminal voltage can be

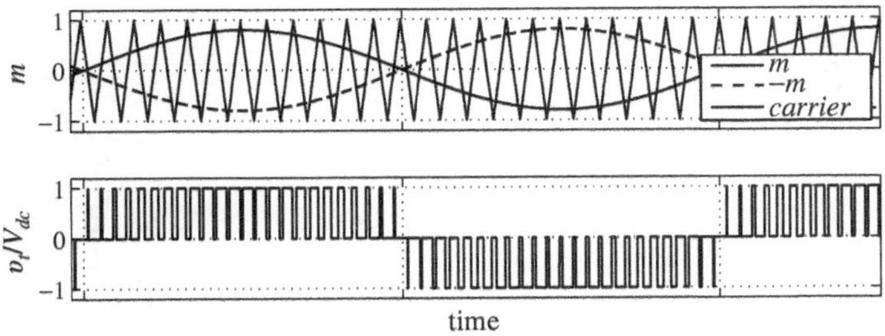

Figure 2.8 Waveforms of the modulating signal, carrier signal, and AC-side terminal voltage of the H bridge, based on the SPWM switching strategy for the H bridge.

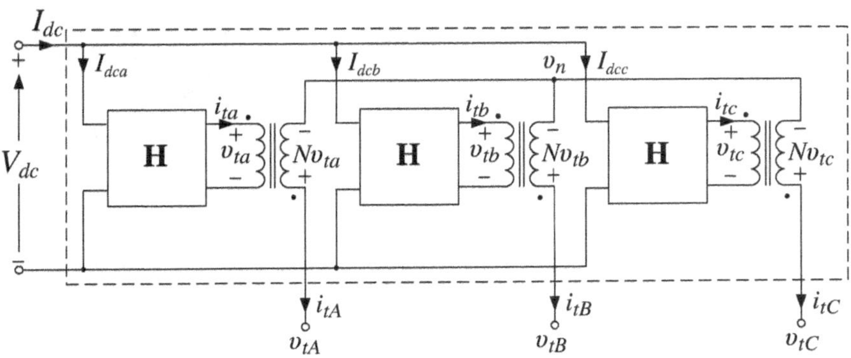

Figure 2.9 Schematic diagram of a three-phase three-level converter.

controlled, respectively, by those of a balanced three-phase modulating signal, and low frequency harmonics are absent from the voltage spectrum.

In a three-phase NPC converter (Fig. 2.10), if the circuit elements are identical and switching instants are precise, i.e., ideally, the midpoint current i_{np} has a zero DC component, while it pulsates, almost sinusoidally, with a frequency of 3ω (ω is the angular frequency of the modulating signal). However, such inevitable imperfections as unequal on-state voltage drops, switching delays, and so on, result in a nonzero DC component for i_{np} which, if not nullified, causes the two capacitor voltages to drift until the DC-side voltage drops entirely on only one capacitor. Thus, to maintain (the DC components of) the capacitor voltages, the DC component of the midpoint current is dynamically changed, and forced to zero in the steady state, by a feedback mechanism that compares the capacitor voltages with one another and adds a corrective DC offset to the modulating signals of the PWM switching scheme. To mitigate the triple-frequency ripples of the capacitor voltages, which, in turn, are due to the triple-frequency component of i_{np}, the capacitance of the two capacitors. i.e., $2C_{dc}$, must be adequately large. This also ensures that the effective DC-side filter capacitance, C_{dc}, is adequately large.

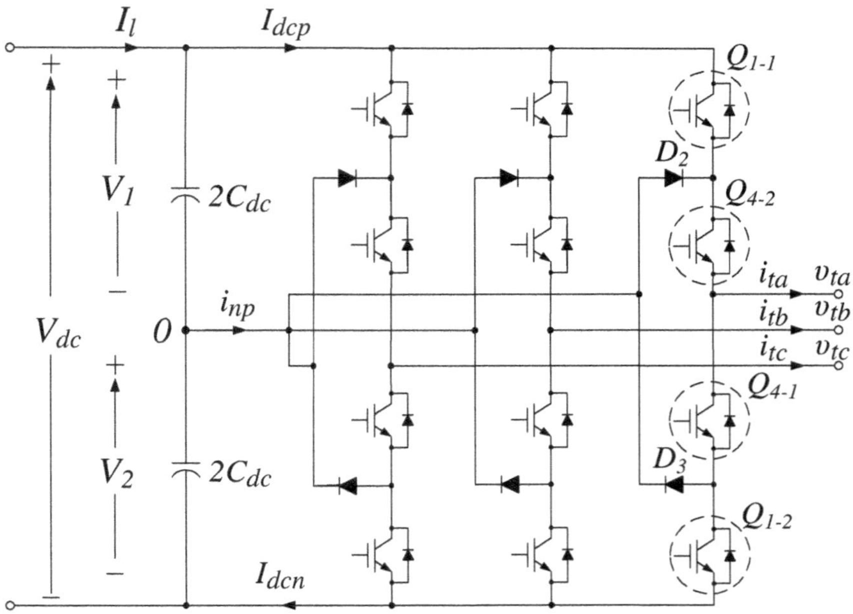

Figure 2.10 Schematic diagram of the neutral-point clamped (NPC) three-level VSC.

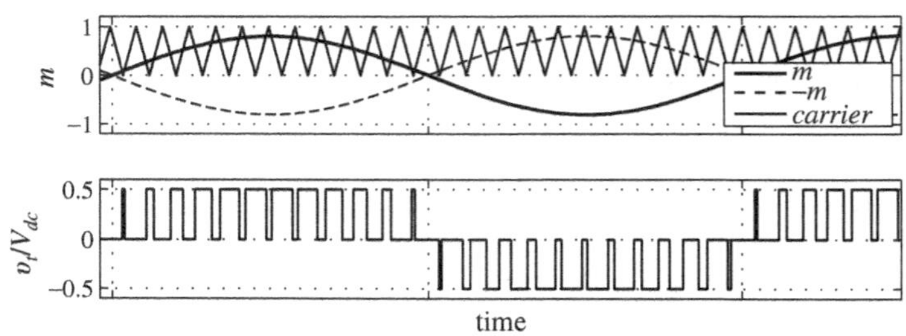

Figure 2.11 Waveforms of the modulating signal, carrier signal, and AC-side terminal voltage, based on the SPWM switching strategy for the NPC three-level VSC.

One technical advantage of the NPC converter over an equivalent two-level VSC, apart from its lower voltage harmonic content, is that the switches and diodes of the NPC converter are subjected to only half of the DC-side voltage. Therefore, the NPC converter can achieve twice the voltage rating of a two-level VSC that uses the same switches. Alternatively, with the same voltage rating as that of the two-level VSC, the NPC converter can employ power switches with half the voltage rating. This advantage also exists over the three-level converter of Fig. 2.9.

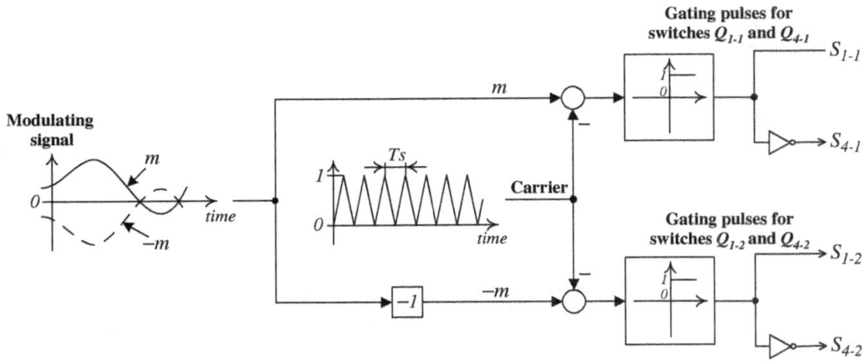

Figure 2.12 Signal flow diagram of the suboscillation carrier-based PWM technique for the NPC three-level VSC.

2.3.3 Multi-Level Voltage-Sourced Converter

Harmonic distortion of the AC-side terminal voltage of a VSC can be further reduced if the number of levels used to synthesize the waveform is increased beyond three. For example, the NPC converter of Fig. 2.10 can be evolved to a five-level, seven-level, or even nine-level *diode-clamped* VSC[5]; Fig. 2.13 shows the schematic diagram of a five-level diode-clamped VSC.

In recent years, however, a new breed of multi-level VSCs has found applications. This breed, known as the *modular multi-level converter (MMC)*, employs multiple *half-bridge* converters (Fig. 2.14) connected in series from their AC-side terminals, to synthesize a multi-level AC voltage waveform; Fig. 2.15 illustrates the configuration for one phase of a three-phase MMC.

Assume for the converter of Fig. 2.15 that the DC-side voltage, V_{dc}, is externally supported (for example, by a host DC grid) and that the DC-side capacitors of the constituting half-bridge converters, also known as the *sub-modules* of the MMC, are equally charged with a voltage of V_{dc}/n, where n is the number of series-connected half-bridge converters of one *arm*. Depending on the state of the corresponding switches, Q_1 and Q_4, the AC port of a half-bridge converter can be forced to assume a voltage of 0 (corresponding to an *off* converter) or V_{dc}/n (for an *on* converter). Therefore, the voltage across each arm, v_{arm1} or v_{arm4}, can be varied from 0 to $V_{dc}/2$, in steps of V_{dc}/n, by turning the half-bridge converters of the arm on or off.

Ideally, the AC-side terminal voltage of the MMC, v_t, is zero if the two arm voltages are equal at $V_{dc}/2$, and v_t can be varied in steps of V_{dc}/n, by varying v_{arm1} and v_{arm4} in a complementary fashion. Thus, v_t rises by one step if v_{arm1} is reduced by V_{dc}/n (i.e., if one of the half-bridge converters of arm 1 is turned off) while v_{arm4} is increased by V_{dc}/n (i.e., if one of the half-bridge converters of Arm 4 is turned on), and so on. Similarly, v_t drops by one step if v_{arm1} is increased by V_{dc}/n while v_{arm4} is

[5]It should be pointed out that the NPC converter (Fig. 2.10) is indeed a three-level diode-clamped VSC.

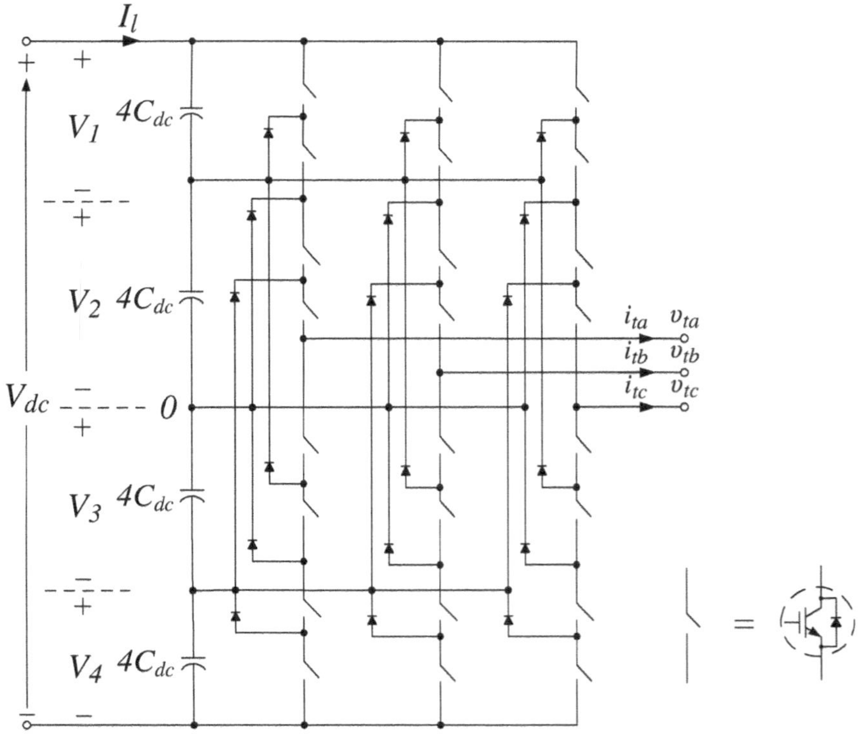

Figure 2.13 Schematic diagram of the five-level diode-clamped VSC.

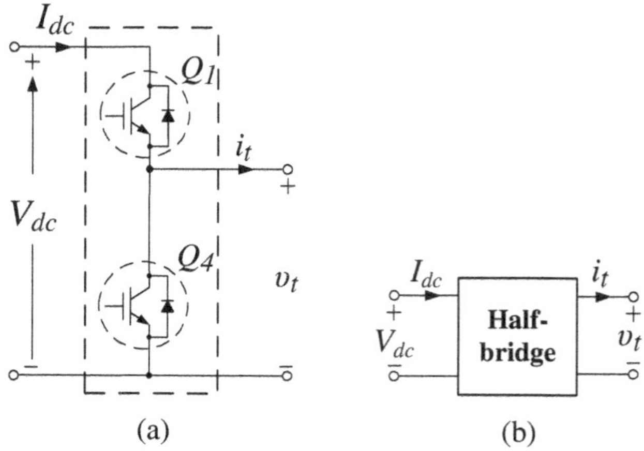

Figure 2.14 Schematic diagram of a half-bridge converter.

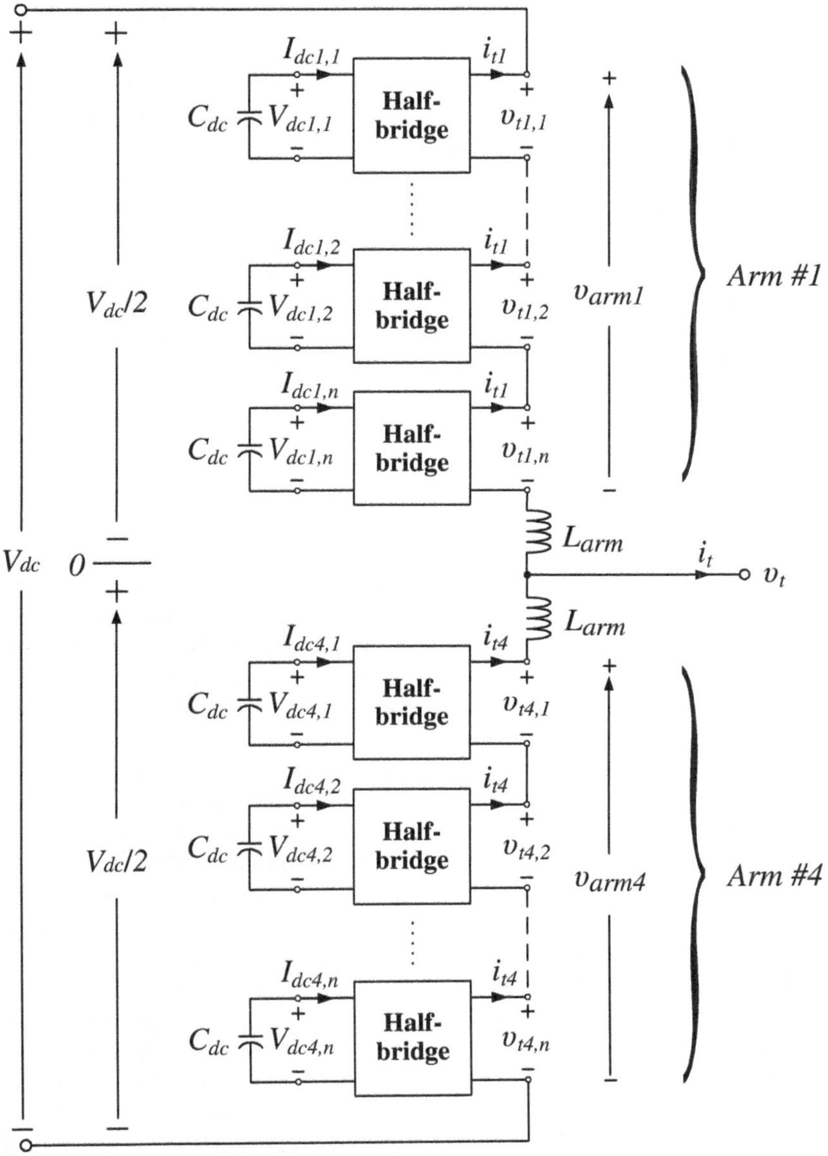

Figure 2.15 Schematic diagram showing one leg of a three-phase MMC.

decreased by V_{dc}/n, etc. Hence, the AC-side terminal voltage of the MMC has $n + 1$ steps, indicating that it can be made remarkably smooth with a sufficiently large number of half-bridge converters. Therefore, carrier-based PWM switching strategies are often not adopted for MMCs, and the switching power losses are almost nonexistent.

A salient merit of the MMC is its modularity and capability to bypass faulty half-bridge converters and continue to operate.

To absorb transient differences between V_{dc} and $v_{arm1} + v_{arm4}$ and, thus, to prevent the half-bridge converters from experiencing large transient currents, each arm is connected in series with a corresponding reactor, L_{arm}. To regulate the DC-side voltage of a half-bridge converter at V_{dc}/n, the switching scheme of the half-bridge converter must ensure that the average real-power flowing through the AC side of the half-bridge converter is zero, i.e., the respective DC-side capacitor is charged in a switching cycle for as long as it is discharged; this is achieved through a feedback mechanism. Moreover, since the reactors cannot absorb any long-term voltage mismatch, the control must also fine-tune v_{arm1} and v_{arm4}, through a feedback mechanism, such that the value $v_{arm1} + v_{arm4}$ is regulated at V_{dc}.

2.4 Control

2.4.1 Control of Real and Reactive Powers

Different control modes of a VSC and modeling of the corresponding controllers are described in this section. Section 3.2.3 will use the conclusions from this section to develop controller models for the converter stations in an MTDC grid. The control of a three-phase VSC boils, in essence, down to the control of the real and reactive power that the VSC exchanges with the host AC grid, P_g and Q_g (Fig. 2.2). Real-power control may be employed for power-flow control or, indirectly, for regulating the DC-side voltage of the VSC. Reactive-power control may be directly employed for ancillary services or, indirectly, for regulating the grid voltage magnitude at the coupling point. P_g and Q_g can be controlled based on either *voltage-mode control strategy* or *current-mode control strategy*, as discussed in the following.

In the voltage-mode control strategy, P_g and Q_g are controlled directly by the phase angle and magnitude of the terminal voltage of the VSC, $v_{t,abc}$, relative to those of the grid voltage $v_{g,abc}$. Thus, if the tie reactor of the VSC has a small R/L ratio, then the AC-side terminal powers P_t and Q_t are fairly precisely proportional to, respectively, the angle and magnitude of $v_{t,abc}$ (relative to those of $v_{g,abc}$). If the power losses of the tie reactor are small, then P_g equals the terminal real power P_t. In addition, Q_g equals the terminal reactive power Q_t minus the reactive power that the filter capacitance C generates. It is remembered from Sections 2.2 and 2.3 that the ability to control the magnitude and phase angle of the AC-side terminal voltage is the most essential capability of a VSC and its adopted switching strategy.

The voltage-mode control strategy is easy to implement, but it renders the VSC vulnerable to external AC-side faults. Thus, a drop in $v_{g,abc}$, for example due to a fault within the AC grid, results in corresponding drops in P_g and Q_g. Consequently, in an attempt to maintain P_g and Q_g, the control of the VSC increases the phase angle and magnitude of $v_{t,abc}$, with respect to those of $v_{g,abc}$. Due to the typically small values of L and R, this corrective action results almost certainly in a damaging large AC-side current. Another issue of the voltage-mode control strategy lies in the fact that power control through the control of phase angle and magnitude is based on a steady-state relationship (model) and, therefore, typically results in a fairly poor transient performance. Further, the control becomes sensitive and difficult to stabilize as the tie reactance L is reduced, for example, in a multi-level VSC. For these reasons, the current-mode control strategy is invariably preferred.

In the current-mode control strategy, P_g and Q_g are controlled by the AC-side current $i_{t,abc}$, with reference to the grid voltage $v_{g,abc}$ (with a discrepancy of the reactive power that the filter capacitance C generates). In turn, $i_{t,abc}$ is regulated through the AC-side terminal voltage of the VSC, $v_{t,abc}$. Therefore, the VSC is protected against over-currents and external faults since the magnitude of $i_{t,abc}$ will be limited if the magnitude of the reference current is constrained. This is the main reason for the invariable selection of the current-mode control strategy over its voltage-mode counterpart.

It should be pointed out that both the voltage-mode and current-mode control strategies require a measure of the grid voltage $v_{g,abc}$. The current-mode control strategy also needs a measure of the current $i_{t,abc}$.

2.4.2 Design and Implementation of Control

2.4.2.1 *Space Phasors* A three-wire three-phase AC system, such as the AC side of a three-phase VSC, is most elegantly represented, analyzed, and controlled through the concept of *space phasors*.

Let $f_a(t)$, $f_b(t)$, and $f_c(t)$ signify three signals of arbitrary waveforms that are required to only satisfy the relationship:

$$f_a(t) + f_b(t) + f_c(t) \equiv 0 \tag{2.25}$$

Then, their corresponding space phasor[6] is defined as

$$\begin{aligned}\overline{F}(t) &= F_\alpha(t) + jF_\beta(t)\\ &= \frac{2}{3}\left[e^{j0}f_a(t) + e^{j\frac{2\pi}{3}}f_b(t) + e^{j\frac{4\pi}{3}}f_c(t)\right]\end{aligned} \tag{2.26}$$

and is a complex-valued time function for which $F_\alpha(t)$ and $F_\beta(t)$ are the real and imaginary components, respectively. In terms of real-valued signals, one can write

$$\begin{bmatrix} F_\alpha(t) \\ F_\beta(t) \end{bmatrix} = \frac{2}{3}\mathbf{C}\begin{bmatrix} f_a(t) \\ f_b(t) \\ f_c(t) \end{bmatrix} \tag{2.27}$$

where

$$\mathbf{C} = \begin{bmatrix} 1 & -\frac{1}{2} & -\frac{1}{2} \\ 0 & \frac{\sqrt{3}}{2} & -\frac{\sqrt{3}}{2} \end{bmatrix} \tag{2.28}$$

[6] Also known as a *space vector*.

Conversely, if the space phasor is given, the three constituting signals can be retrieved as

$$f_a(t) = Re\left\{\overline{F}(t)e^{-j0}\right\} \tag{2.29}$$

$$f_b(t) = Re\left\{\overline{F}(t)e^{-j\frac{2\pi}{3}}\right\} \tag{2.30}$$

$$f_c(t) = Re\left\{\overline{F}(t)e^{-j\frac{4\pi}{3}}\right\} \tag{2.31}$$

where $Re\{\cdot\}$ means "real part of." In terms of the real-valued components F_α and F_β, the constituting signals are given by

$$\begin{aligned}\begin{bmatrix} f_a(t) \\ f_b(t) \\ f_c(t) \end{bmatrix} &= \begin{bmatrix} 1 & 0 \\ -\frac{1}{2} & \frac{\sqrt{3}}{2} \\ -\frac{1}{2} & -\frac{\sqrt{3}}{2} \end{bmatrix} \begin{bmatrix} F_\alpha(t) \\ F_\beta(t) \end{bmatrix} \\ &= \mathbf{C}^T \begin{bmatrix} F_\alpha(t) \\ F_\beta(t) \end{bmatrix}\end{aligned} \tag{2.32}$$

where $\mathbf{C}$ is defined by (2.28) and superscript T denotes matrix transposition.

The significance of transform (2.26) and its inverse, (2.29–2.31), can be best understood through the example of a balanced three-phase sinusoidal signal. To that end, consider the general case of a variable-amplitude, variable-frequency signal:

$$f_a(t) = F(t)\cos\left[\theta(t)\right] \tag{2.33}$$

$$f_b(t) = F(t)\cos\left[\theta(t) - \frac{2\pi}{3}\right] \tag{2.34}$$

$$f_c(t) = F(t)\cos\left[\theta(t) - \frac{4\pi}{3}\right] \tag{2.35}$$

where

$$\theta(t) = \theta_0 + \int_0^t \omega(\tau)d\tau \tag{2.36}$$

and $\theta(t)$, $\omega(t)$, and $F(t)$ are the instantaneous phase angle, frequency, and magnitude of the three-phase signal, respectively, and θ_0 is the initial phase angle. Then, based on (2.26), the space phasor of the three-phase signal is

$$\overline{F}(t) = F(t)e^{j\theta(t)} \tag{2.37}$$

Let us now graphically represent (2.37) by a phasor on the complex plane, as Fig. 2.16 shows; we have labeled the real and imaginary axes as α and β axes, respectively. Thus, the three-phase signal (2.33–2.35) is equivalent to a rotating phasor whose length equals the (instantaneous) peak value of the signal (i.e., $F(t)$), whose angular displacement relative to the α-axis equals the phase angle of the signal (i.e., $\theta(t)$), and

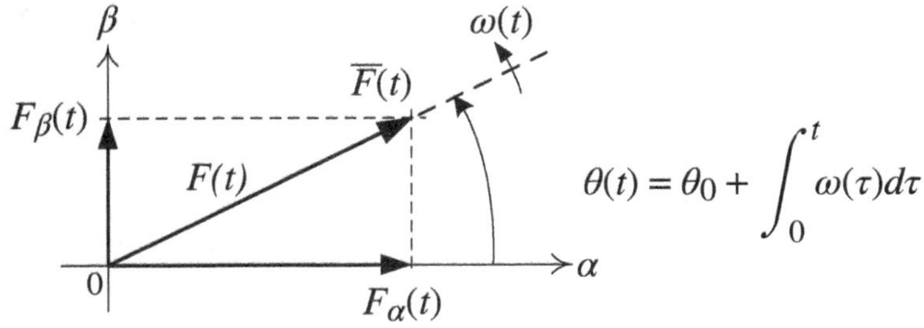

Figure 2.16 Graphical representation of the space phasor (Eq. 2.37).

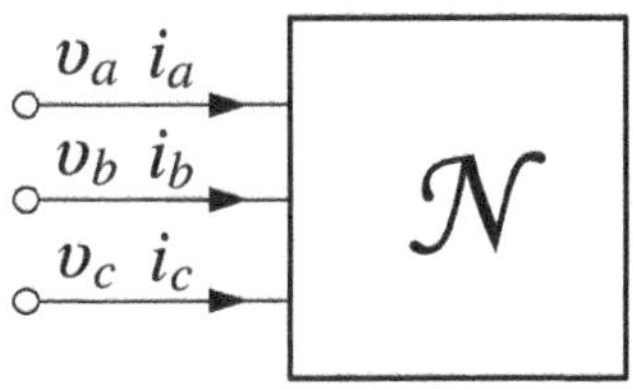

Figure 2.17 A three-wire three-phase network.

whose instantaneous angular velocity equals the instantaneous angular frequency of the signal (i.e., $\omega(t)$). Therefore, it follows from (2.29–2.31) that changes made to any of the attributes length, phase angle, and rotational speed of a space phasor are instantaneously reflected in, respectively, the peak value, phase angle, and frequency of the constituents $f_a(t)$, $f_b(t)$, and $f_c(t)$. This important property is the basis of the so-called *vectorial control* of power-electronic converters, in electric drive systems and utility applications, which will be detailed in Section 2.4.2.4.

We conclude this section by defining the instantaneous real and reactive powers, using space phasors. Consider the three-wire three-phase network of Figure 2.17 whose terminal voltages and currents are v_{abc} and i_{abc}, respectively; neither v_{abc} nor i_{abc} has to be balanced or even sinusoidal. The instantaneous power flowing into the network is given by

$$P(t) = v_a(t)i_a(t) + v_b(t)i_b(t) + v_c(t)i_c(t) \tag{2.38}$$

Using (2.29–2.31), v_{abc} and i_{abc} in (2.38) can be expressed in terms of their corresponding space phasors, to yield

$$\begin{aligned} P(t) = {} & Re\left\{\overline{V}(t)e^{j0}\right\} Re\left\{\overline{I}(t)e^{j0}\right\} \\ & + Re\left\{\overline{V}(t)e^{-j\frac{2\pi}{3}}\right\} Re\left\{\overline{I}(t)e^{-j\frac{2\pi}{3}}\right\} \\ & + Re\left\{\overline{V}(t)e^{-j\frac{4\pi}{3}}\right\} Re\left\{\overline{I}(t)e^{-j\frac{4\pi}{3}}\right\} \end{aligned} \tag{2.39}$$

Using the identity $Re\{\alpha\}Re\{\beta\} = (Re\{\alpha\beta\} + Re\{\alpha\beta^*\})/2$, one can rewrite (2.39) as

$$P(t) = \frac{Re\left\{\overline{V}(t)\overline{I}(t)\right\} + Re\left\{\overline{V}(t)\overline{I}^*(t)\right\}}{2} + \frac{Re\left\{\overline{V}(t)\overline{I}(t)e^{-j\frac{4\pi}{3}}\right\} + Re\left\{\overline{V}(t)\overline{I}^*(t)\right\}}{2} + \frac{Re\left\{\overline{V}(t)\overline{I}(t)e^{-j\frac{8\pi}{3}}\right\} + Re\left\{\overline{V}(t)\overline{I}^*(t)\right\}}{2} \tag{2.40}$$

where $*$ denotes "complex-conjugate of." Since $e^{j0} + e^{-j\frac{4\pi}{3}} + e^{-j\frac{8\pi}{3}} \equiv 0$, equation (2.40) is simplified to

$$P(t) = Re\left\{\frac{3}{2}\overline{V}(t)\overline{I}^*(t)\right\} \tag{2.41}$$

It should be noted that (2.41) is developed with no assumptions on the forms of v_{abc} and i_{abc}. These variables can have any arbitrary waveforms. However, if v_{abc} and i_{abc} are sinusoids of constant amplitudes and equal constant frequencies, (2.41) is reduced to the expression for the real power in the conventional phasor analysis. Therefore, one is motivated to also define instantaneous reactive power and instantaneous complex power, as

$$Q(t) = Im\left\{\frac{3}{2}\overline{V}(t)\overline{I}^*(t)\right\} \tag{2.42}$$

$$S(t) = P(t) + jQ(t) = \frac{3}{2}\overline{V}(t)\overline{I}^*(t) \tag{2.43}$$

where $Im\{\cdot\}$ denote the "imaginary part of." It can be verified that if v_{abc} and i_{abc} are constant-amplitude sinusoids of equal constant frequencies, (2.42) and (2.43) are identical to the expressions of reactive and complex powers in the conventional phasor analysis.

2.4.2.2 *Space-Phasor Representation of the AC Side* Now consider the system of Fig. 2.2, in which the AC-side current of the VSC is governed by the following dynamic equations:

$$L\frac{di_{ta}}{dt} = -Ri_{ta} + v_{ta} - v_{ga} \tag{2.44}$$

$$L\frac{di_{tb}}{dt} = -Ri_{tb} + v_{tb} - v_{gb} \tag{2.45}$$

$$L\frac{di_{tc}}{dt} = -Ri_{tc} + v_{tc} - v_{gc} \tag{2.46}$$

Multiplying the two sides of (2.44–2.46) by, respectively, $\frac{2}{3}e^{j0}$, $\frac{2}{3}e^{j\frac{2\pi}{3}}$, and $\frac{2}{3}e^{j\frac{4\pi}{3}}$, adding the resulting equations, and using (2.26), one obtains

$$L\frac{d\overline{I}_t}{dt} = -R\overline{I}_t + \overline{V}_t - \overline{V}_g \tag{2.47}$$

It is then understood from (2.47) that the space phasor of the AC-side current (and, therefore, the phase angle, frequency, and magnitude of the phase currents i_{ta}, i_{tb}, and i_{tc}) can be controlled by the space phasor of the VSC terminal voltage, while the grid voltage phasor acts as a disturbance input. Further, regarding a conventional phasor (used in sinusoidal steady-state analysis) as a special case of a space phasor[7], one can devise control strategies for the VSC. For example, if the VSC in the system of Fig. 2.2 is to only deliver reactive power to the host AC grid, then its terminal voltage phasor must be made aligned with the grid voltage phasor, such that the current phasor is perpendicular to the grid voltage phasor[8]; whether the current leads or lags the grid voltage phasor depends on the magnitude of the terminal voltage phasor, relative to that of the grid voltage phasor. The desired terminal voltage phasor, which is specified by the controller, is then transformed to three desired phase voltages, based on (2.29–2.31), which are synthesized by the VSC using a switching algorithm (Section 2.3). Care, however, should be taken in interpreting the results: The magnitude of a space phasor equals the peak value of its respective three-phase signal, whereas the traditional phasor uses rms characterization.

2.4.2.3 Current Control in the Stationary Frame Expressing each space phasor in (2.47) in terms of its α-axis (i.e., real) and β-axis (i.e., imaginary) components and splitting the resulting equation to its real and imaginary components, one finds

$$L\frac{dI_{t\alpha}}{dt} = -RI_{t\alpha} + V_{t\alpha} - V_{g\alpha} \tag{2.48}$$

$$L\frac{dI_{t\beta}}{dt} = -RI_{t\beta} + V_{t\beta} - V_{g\beta} \tag{2.49}$$

Therefore, the control of the VSC boils down to controlling two subsystems; and the two subsystems are decoupled if the grid voltage components $V_{g\alpha}$ and $V_{g\beta}$ are independent, as in a stiff grid. Thus, two independent control loops, respectively, determine the desired terminal voltage components, $V^*_{t\alpha}$ and $V^*_{t\beta}$, based on the control objectives, for example, regulation of I_α and I_β. Then, using (2.29–2.31), the desired three-phase signals v^*_{ta}, v^*_{tb}, and v^*_{tc} are calculated from the space phasor $\overline{V}^*_t = V^*_{t\alpha} + jV^*_{t\beta}$ and fed for synthesis to the switching scheme of the VSC. The method of control is known as control in the *stationary frame* or *$\alpha\beta$ frame*. It then follows from substituting for $\overline{V}_g = V_{g\alpha} + jV_{g\beta}$ and $\overline{I}_t = I_{t\alpha} + jI_{t\beta}$ in (2.41) and (2.42)

[7] See [23] for a proof.

[8] The phase angle and/or magnitude of a space phasor can be changed by multiplying the space phasor by a complex-valued function, $ke^{j\rho}$, in a phase-shifter/scaler signal processing block; k and ρ are the scale factor and phase shift, respectively [23].

that

$$P_g(t) = \frac{3}{2}\left(V_{g\alpha}I_{t\alpha} + V_{g\beta}I_{t\beta}\right) \tag{2.50}$$

$$Q_g(t) = \frac{3}{2}\left(-V_{g\alpha}I_{t\beta} + V_{g\beta}I_{t\alpha}\right) + Q_c(t) \tag{2.51}$$

where $Q_c(t)$ is the instantaneous reactive power generated by the bank of filter capacitors (Fig. 2.2)[9]. Thus, knowing the grid voltage components $V_{g\alpha}$ and $V_{g\beta}$, one can control P_g and Q_g by the controlling the VSC AC-side current components $I_{t\alpha}$ and $I_{t\beta}$.

The α- and β-axis components of a space phasor corresponding to set of three time-varying signals are time-varying themselves. In particular, in power system applications, such that those discussed in this book, the α- and β-axis components are sinusoids, as (2.37) indicates. Consequently, to ensure small errors and high performance, the compensators and filters of the control scheme must be designed such that the control loops feature sufficiently large gain crossover frequencies, corresponding to adequately wide closed-loop bandwidths; and the design is typically onerous. The control design and implementation tasks are considerably simpler if the signals are DC in nature (except in transients). This objective can be fulfilled if the control is performed in an orthogonal *rotating frame* whose angular velocity equals the angular frequency of the grid voltage. This topic is further discussed in Section 3.2.6 where integration of space-phasor-based converter and controller models with phasor-based host AC systems is described.

2.4.2.4 Current Control in a Rotating Frame Consider a variable-amplitude, variable-frequency, balanced, three-phase grid voltage, as

$$v_{ga}(t) = V_g(t)\cos\left[\theta_g(t)\right] \tag{2.52}$$

$$v_{gb}(t) = V_g(t)\cos\left[\theta_g(t) - \frac{2\pi}{3}\right] \tag{2.53}$$

$$v_{gc}(t) = V_g(t)\cos\left[\theta_g(t) - \frac{4\pi}{3}\right] \tag{2.54}$$

where

$$\theta_g(t) = \theta_{g0} + \int_0^t \omega_g(\tau)d\tau \tag{2.55}$$

and $\theta_g(t)$, $\omega_g(t)$, and $V_g(t)$ are the instantaneous phase angle, frequency, and magnitude of the grid voltage, respectively, and θ_{g0} is the initial phase angle. Then, based on (2.26), the corresponding space phasor is

$$\overline{V}_g(t) = V_g(t)e^{j\theta_g(t)} \tag{2.56}$$

[9]We are ignoring transient excursions in the magnitude and phase angle of $\overline{V}_g(t)$. Otherwise, the power exchanged with the capacitor would also have a real component, $P_C(t)$.

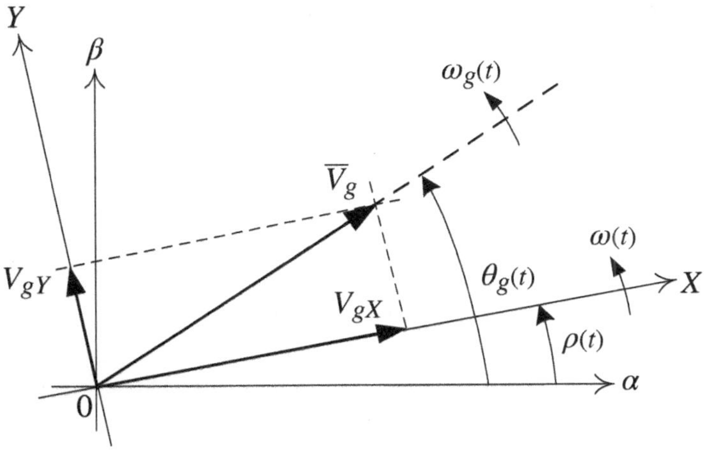

Figure 2.18 A space phasor and a rotating frame.

which is a phasor that rotates with an instantaneous angular velocity of $\omega_g(t)$, as shown in Fig. 2.18. Now consider another Cartesian frame of coordinates, which, hereafter, is referred to as an XY frame[10], whose X-axis makes an angle of $\rho(t)$ with respect to the α-axis of the stationary frame. Thus, the frame rotates around the origin at an instantaneous angular velocity of

$$\omega(t) = \frac{d\rho}{dt} \tag{2.57}$$

Commonly, however, $\omega(t)$, rather than ρ, is a control variable, whereas $\rho(t)$ is a by-product. Thus, one can calculate ρ from $\omega(t)$, as

$$\rho(t) = \rho_0 + \int_0^t \omega(\tau)d\tau \tag{2.58}$$

where ρ_0 is the initial value of ρ. Then, if the signal magnitude, V_g, is constant, the components of $\overline{V}_g(t)$ along the X- and Y-axes, i.e., V_{gX} and V_{gY}, become time-invariant if and only if the XY-frame rotates at the same angular velocity[11] as that of $\overline{V}_g(t)$, that is, if $\omega(t) = \omega_g(t)$; in the same frame, the other waveforms also correspond to time-invariant phasors based on the fact that they have the same angular frequency, in the steady state, as that of the grid voltage.

The expression of a space phasor, $\overline{F}(t) = F_\alpha(t) + jF_\beta(t)$, in the XY-frame of Fig. 2.18 is mathematically formulated as

$$F_X(t) + jF_Y(t) = \overline{F}(t)e^{-j\rho(t)} \tag{2.59}$$

[10]It is most commonly known as a *dq* frame. In this book, however, we have chosen the name *XY* frame to circumvent potential confusions with a similar concept employed for the analysis of synchronous machines.

[11]And, therefore, in the same direction.

which we refer to as the $\alpha\beta$- to XY-frame transformation. In terms of real-valued signals, one can write

$$\begin{bmatrix} F_X(t) \\ F_Y(t) \end{bmatrix} = \mathbf{R}\left[\rho(t)\right] \begin{bmatrix} F_\alpha(t) \\ F_\beta(t) \end{bmatrix} \tag{2.60}$$

where

$$\mathbf{R}\left[\rho(t)\right] = \begin{bmatrix} \cos\rho(t) & \sin\rho(t) \\ -\sin\rho(t) & \cos\rho(t) \end{bmatrix} \tag{2.61}$$

Conversely, the XY- to $\alpha\beta$-frame transformation is defined by

$$\overline{F}(t) = \left(F_X(t) + jF_Y(t)\right) e^{j\rho(t)} \tag{2.62}$$

which, in terms of real-valued signals, can be rewritten as

$$\begin{aligned} \begin{bmatrix} F_\alpha(t) \\ F_\beta(t) \end{bmatrix} &= \mathbf{R}^{-1}\left[\rho(t)\right] \begin{bmatrix} F_X(t) \\ F_Y(t) \end{bmatrix} \\ &= \mathbf{R}\left[-\rho(t)\right] \begin{bmatrix} F_X(t) \\ F_Y(t) \end{bmatrix} \end{aligned} \tag{2.63}$$

where

$$\mathbf{R}^{-1}\left[\rho(t)\right] = \mathbf{R}\left[-\rho(t)\right] = \begin{bmatrix} \cos\rho(t) & -\sin\rho(t) \\ \sin\rho(t) & \cos\rho(t) \end{bmatrix} \tag{2.64}$$

It can also be verified that

$$\mathbf{R}^{-1}\left[\rho(t)\right] = \mathbf{R}^{T}\left[\rho(t)\right] \tag{2.65}$$

It is also useful to derive an expression for the derivative of a space phasor, in terms of its XY-frame expression:

$$\frac{d\overline{F}}{dt} = \left(\frac{dF_X}{dt} + j\frac{dF_Y}{dt}\right) e^{j\rho} + \omega\left(-F_Y + jF_X\right) e^{j\rho} \tag{2.66}$$

where $\omega = d\rho/dt$ (2.57). The AC side of the system of Fig. 2.2 can now be expressed in an XY-frame, if each space phasor in the model of (2.47) is expressed in terms of the corresponding XY-frame components, based on (2.62) and (2.66). The result is

$$L\frac{dI_{tX}}{dt} = -RI_{tX} + L\omega I_{tY} + V_{tX} - V_{gX} \tag{2.67}$$

$$L\frac{dI_{tY}}{dt} = -RI_{tY} - L\omega I_{tX} + V_{tY} - V_{gY} \tag{2.68}$$

It then follows from replacing $\overline{V}_g$ and $\overline{I}_t$ in (2.41) and (2.42) with, respectively, $\left(V_{gX}+jV_{gY}\right)e^{j\rho}$ and $\left(I_{tX}+jI_{tY}\right)e^{j\rho}$ that

$$P_g(t)=\frac{3}{2}\left(V_{gX}I_{tX}+V_{gY}I_{tY}\right) \tag{2.69}$$

$$Q_g(t)=\frac{3}{2}\left(-V_{gX}I_{tY}+V_{gY}I_{tX}\right)+Q_c(t) \tag{2.70}$$

where $Q_c(t)$ is the instantaneous reactive power generated by the bank of filter capacitors (Fig. 2.2). Therefore, knowing the grid voltage components V_{gX} and V_{gY}, one can control P_g and Q_g by the controlling the VSC AC-side current components I_{tX} and I_{tY}.

Equations (2.67) and (2.66) represent a two-input/two-output dynamic system (Fig. 2.19) for which V_{tX}, V_{tY}, and ρ are the control inputs, I_{tX} and I_{tY} are the outputs, and V_{gX} and V_{gY} are the disturbance inputs. The control inputs, V_{tX} and V_{tY} are, in turn, produced through a switching strategy, from two other control variables. For example, based on the SPWM switching strategy (Section 2.3.1), V_{tX} and V_{tY} are proportional to m_X and m_Y, i.e., the X- and Y-axis components of the three-phase modulating signal; the system of Fig. 2.19 assumes the SPWM switching strategy. In practice, due to the harmonic distortion of the terminal voltage $v_{t,abc}$, V_{tX} and V_{tY} include time-varying components in addition to the DC components $(V_{dc}/2)m_X$ and $(V_{dc}/2)m_Y$. These components, not shown in Fig. 2.19, act as disturbance signals

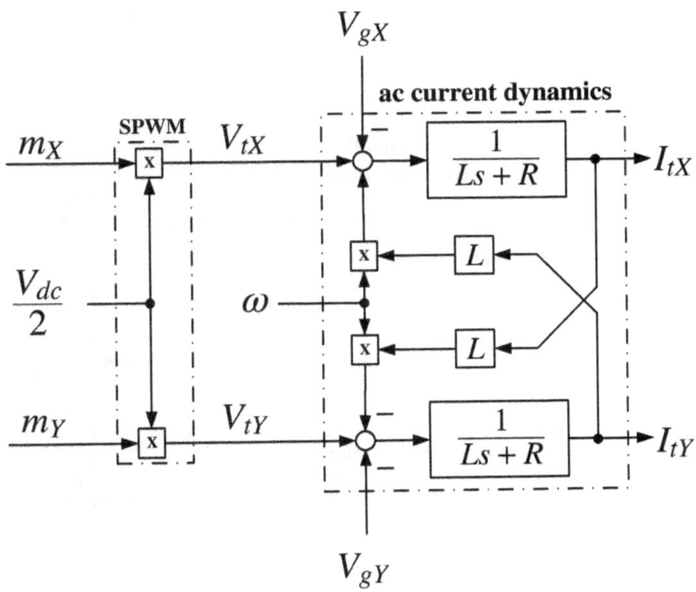

Figure 2.19 Block diagram of the control plant describing the dynamics of the AC-side current in a general XY-frame.

and result in corresponding time-varying components in I_{tX} and I_{tY}, which, in turn, translate into harmonic distortion of the AC-side current $i_{t,abc}$. The magnitudes of the time-varying components of I_{tX} and I_{tY} depend on the magnitudes and frequencies of the harmonics, relative to the corner frequency of the filter $1/(Ls+R)$, i.e., R/L; the filter corresponds to the tie reactor of the VSC. Therefore, for a sinusoidal AC current, one must either minimize the harmonic distortion of $v_{t,abc}$ by, for example, multi-level AC voltage synthesis techniques or increase the harmonic frequencies of $v_{t,abc}$ by, for example, the SPWM switching strategy.

For the system of Fig. 2.19, let us assume that the control objective is to regulate I_{tX} and I_{tY} at two respective setpoints I^*_{tX} and I^*_{tY}, by the control inputs V_{tX} and V_{tY}; to ensure DC signals, ω will be regulated at the angular frequency of the grid voltage, ω_g, through a mechanism that we will shortly describe. The control, however, is not straightforward as I_{tX} and I_{tY} are coupled due to the terms $\omega(t)I_{tX}$ and $\omega(t)I_{tY}$. Moreover, in practice, the grid voltage $v_{g,abc}$ depends on the current contribution of the VSC, $i_{t,abc}$, due, most importantly, to nonzero Thevenin impedance of the host AC grid. Consequently, V_{gX} and V_{gY} depend on I_{tX} and I_{tY} and, as such, may not be considered as disturbance signals. Further, the dependence of $v_{g,abc}$ on $i_{t,abc}$ translates into the dependence of ω on I_{tX} and I_{tY}, for the reasons discussed in Section 2.4.2.5. Furthermore, ω_g (and therefore ω, which is made to follow ω_g) depends on I_{tX} and I_{tY}, if the VSC is large relative to the rating of the host AC grid.

Figure 2.20 shows the block diagram of a control scheme that largely (almost entirely) overcomes the aforementioned couplings and nonlinearities associated with the control plant of Fig. 2.19. Thus, the control scheme may be considered to consist of two sub-controllers, the X- and Y-axis controllers. For the X-axis controller, a

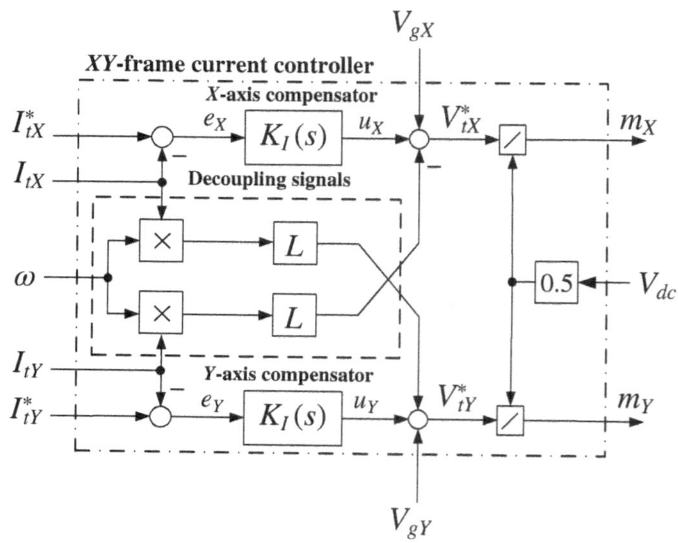

Figure 2.20 Block diagram of the current-control scheme for control plant of Fig. 2.19.

corresponding compensator, $K_I(s)$, processes the error signal $e_X = I^*_{tX} - I_{tX}$ and generates the control signal u_X. Then, the desired value of the X-axis AC-side terminal voltage, V^*_{tX}, is produced by supplementing the compensator output u_X with two signals, $-I_{tY}\omega$ and V_{gX}; the former signal decouples I_{tX} from I_{tY}, whereas the latter acts as a feedforward signal and mitigates the dependance of I_{tX} on V_{gX} in Fig. 2.19. Similarly, in the Y-channel controller, a corresponding compensator processes the error signal $e_Y = I^*_{tY} - I_{tY}$ and generates the control signal u_Y. Then, the desired value of the Y-axis AC-side terminal voltage, V^*_{tY}, is produced by supplementing u_Y with two signals, $I_{tX}\omega$ and V_{gY}, which decouple I_{tY} from I_{tX} and V_{gY}. Finally, if the SPWM switching strategy is employed, the X- and Y-axis components of the three-phase modulating signal, m_X and m_Y, are produced by dividing the desired AC-side terminal voltage components V^*_{tX} and V^*_{tY} by the gain of the VSC. The gain, $V_{dc}/2$, can either be measured, thus acting as a feedforward signal, or be considered a constant equalling the nominal steady-state value of the VSC DC-side voltage. The control requires signal transformations from the *abc* frame to the *XY* (for feedback and feed-forward signals) and from the *XY*- frame to the *abc*-frame (for the production of the modulating signals m_a, m_b, and m_c from the controller outputs m_X and m_Y), which take place in the transformation blocks of Fig. 2.21. The transformations require angle $\rho(t)$, which is calculated from $\omega(t)$, based on (2.58).

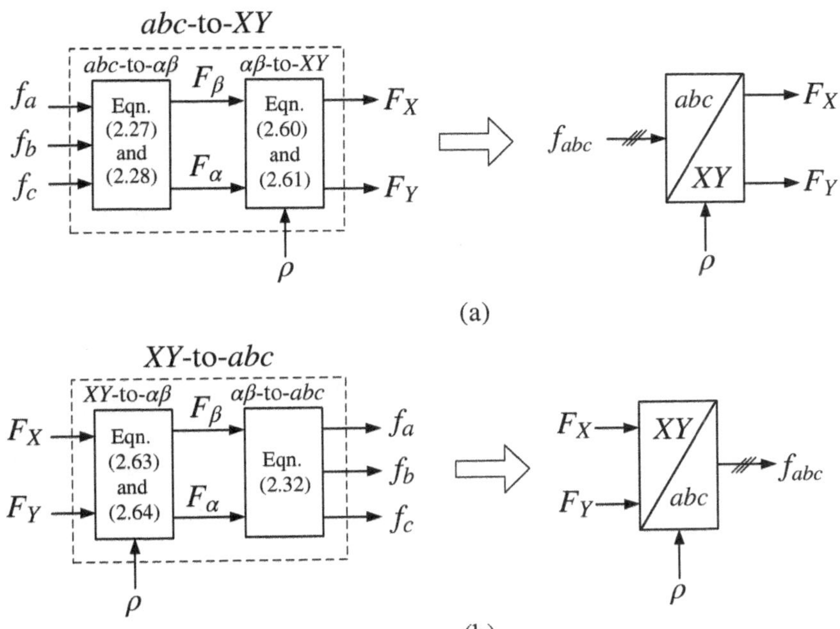

Figure 2.21 Block diagrams of (a) *abc*- to *XY*-frame signal transformation, and (b) *XY*- to *abc*-frame signal transformation.

We shall remember that the feedback and feedforward signals in the control scheme of Fig. 2.20 are measures of the actual corresponding variables in the plant of Fig. 2.19. However, we use the same symbol for both a variable and its measure, as we assume that the measurement dynamics are fast, or, alternatively, the voltage and current transducers offer sufficiently wide-band, low pass, frequency responses, such that the measure of a variable equals the variable itself over the frequency range of interest.

Figure 2.22 illustrates the current-control process, based on the control scheme of Figure 2.19 and control plant of Fig. 2.20. It can be readily verified that the control is equivalent to two independent current-control loops, as shown in Fig. 2.23. One loop regulates I_{tX} at setpoint I^*_{tX}, whereas the other loop regulates I_{tY} at I^*_{tY}. Moreover, as Fig. 2.23 indicates, the compensator outputs u_X and u_Y equal, respectively, the X- and Y-axis components of the voltage that is required to be dropped across the RL tie reactor. It is further noted that the control plant is the same for both control loops, $1/(Ls + R)$, and, therefore, the compensators can have the same transfer function, $K_I(s)$, as also indicated in the block diagram of Figure 2.20.

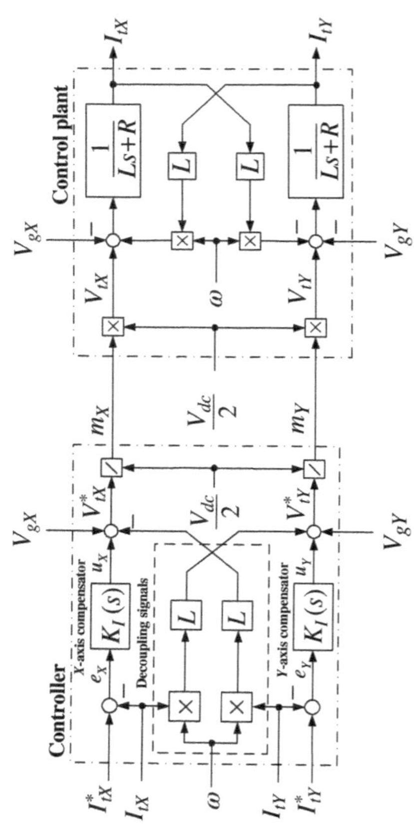

Figure 2.22 Block representation of the current-control process in an XY-frame.

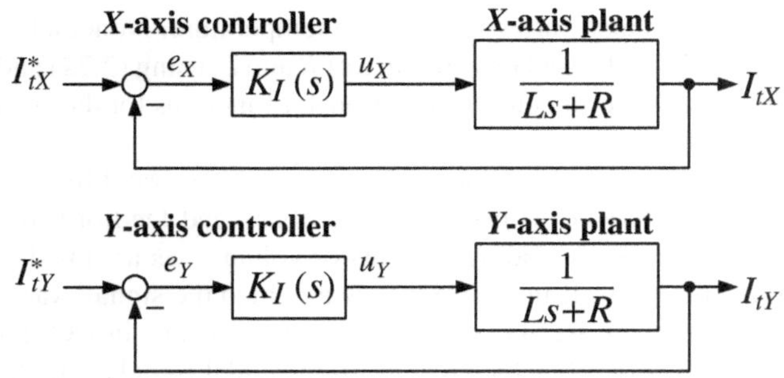

Figure 2.23 Block diagrams of the decoupled X- and Y-axis current-control loops equivalent to the current-control process illustrated in Fig. 2.22.

Considering the DC nature of the signals and first-order transfer function of the control plants in the control loops of Fig. 2.22(b), the simplest compensator that ensures regulation with zero steady-state error is a proportional-integral (PI) compensator, as

$$K_I(s) = \frac{k_p s + k_i}{s} \tag{2.71}$$

where k_p and k_i are the proportional and integral gains, respectively. A particularly interesting set of gains is

$$k_p = \frac{L}{\tau_i} \tag{2.72}$$

$$k_i = \frac{R}{\tau_i} \tag{2.73}$$

for which the closed-loop transfer functions assume the following first-order forms:

$$\frac{I_{tX}(s)}{I^*_{tX}(s)} = \frac{I_{tY}(s)}{I^*_{tY}(s)} = \frac{1}{\tau_i s + 1} \tag{2.74}$$

where τ_i is the desired time-constant of the closed-loop step response or, alternatively, reciprocal of the -3 dB bandwidth of the closed-loop transfer function.

The choice of compensator parameters based on (2.72) and (2.73) corresponds to canceling the plant pole, $-R/L$, by the compensator zero, $-k_i/k_p$, such that the transfer function of the open-loop gain is $\ell(s) = 1/(\tau_i s)$, thus resulting in a first-order closed-loop transfer function (2.74). A first-order closed-loop transfer function, on the other hand, is desirable as its response exhibits no overshoots. Therefore, to protect the VSC, it is sufficient to place suitable upper limits on I^*_{tX} and I^*_{tY}. The pole-zero cancelation, however, cannot be perfect, mainly, due to the fact that R is a poorly known parameter which also varies widely with temperature. Nonetheless, compensator parameters calculated based on estimates of R and L, using (2.72) and (2.73), result in predominantly first-order closed-loop transfer functions for the current-control loops of Fig. 2.22(b).

It should be noted that the block diagrams of Fig. 2.22 are valid for any XY-frame, i.e., irrespective of the $\omega(t)$, of course, if the feedback and feedforward signals are faithful replica of their corresponding currents or voltages, as assumed. The impact of ω, however, manifests itself in the frequency at which the signals vary in a steady state: Two extreme choices are (1) $\omega(t) = \omega_g(t)$, which results in DC signals as discussed earlier, and (2) $\omega(t) = 0$, which results in sinusoidal signals with the frequency ω_g. The control in the $\alpha\beta$ frame is a special case of this type of control, in which, as discussed earlier, the compensators have to process time-varying error signals. Moreover, the transducers producing feedback or feedforward signals must adhere to more stringent bandwidth requirements to be able to produce faithful measures of the electrical variables.

We conclude this section by pointing out to that fact that, regardless of the control strategy, i.e., current-mode or voltage-mode, a VSC interfaced with an AC grid is vulnerable to DC-side short circuits, due to the conduction of the antiparallel diodes of the switches, irrespective of its power circuit configuration, and this is considered the most important shortcoming of the VSC technology for power system applications. The response of two-level converters and MMC following DC-side faults will be analyzed in Sections 6.3.1 and 6.3.2, respectively. The concept of "Fault-Blocking Converters" will also be introduced in Section 6.4.

2.4.2.5 Phase-Locked Loop As discussed in Section 2.4.2.4, the most salient advantage of control in an XY frame, i.e., the involvement of DC signals, manifests itself if the angular velocity of the XY frame, $\omega(t)$, equals the angular frequency of the grid voltage phasor, $\omega_g(t)$ (Fig. 2.18). One way to achieve this goal is to force the angle $\rho(t)$ to track the angle $\theta_g(t)$, as explained in the following.

Consider the scenario illustrated in Fig. 2.18; the XY frame lags the grid voltage phasor $\overline{V}_g$, that is, $\rho(t)$ is smaller than $\theta_g(t)$. Therefore, the angular velocity of the XY frame, $\omega(t)$, must be increased, in order for the XY frame to reach $\overline{V}_g$. Alternatively, if the frame led $\overline{V}_g$, then $\omega(t)$ would need to be reduced, to allow $\overline{V}_g$ to overtake the XY frame. That the frame lags or leads the grid voltage phasor can be identified by the sign of V_{gY}, the Y-axis component of $\overline{V}_g$; a positive V_{gY} signifies a lagging XY frame (the scenario illustrated in Fig. 2.18), calling for an increased ω, whereas a negative V_{gY} represents a leading XY frame and indicates the need for a reduced ω. Once, ρ equals θ_g, $V_{gY} = 0$ and ω must be kept constant.

Figure 2.24 illustrates the feedback loop known as a *phase-locked loop (PLL)* that regulates $\rho(t)$ at $\theta_g(t)$, based on the strategy described earlier. Thus, first, the grid voltage is transformed to the XY frame, using angle ρ, and then V_{gY} is processed by a compensator, $H(s)$. The output of the compensator, ω, is then fed to a voltage-controlled oscillator (VCO) which produces ρ, a periodic sawtooth signal swinging between zero and 2π with a frequency of $\omega/(2\pi)$. In practice, the frequency of the

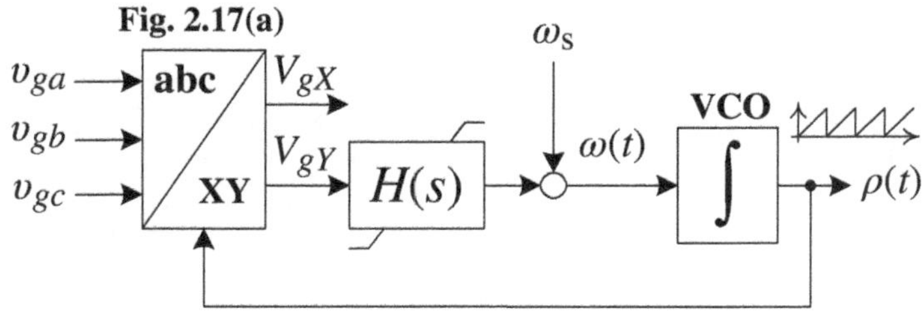

Figure 2.24 Block diagram of a PLL.

grid voltage varies in a small vicinity of a nominal value, ω_s. For example, $\omega_s = 377$ rad/s in a 60-Hz power system. Therefore, ω is also sufficient to be variable over a fairly narrow range about ω_s. Therefore, to circumvent the need for large compensator output variations, for example, in start-up or other severe transient regimes, the output of $H(s)$ is constrained to a small positive upper limit and a small (in absolute values) negative lower limit, and then offset by a constant value of ω_s, to produce ω (Fig. 2.24). Hence, the XY frame is guaranteed to rotate at an angular velocity of about ω_s, even if the output of $H(s)$ is saturated. Although ineffectual for small-signal disturbances, the aforementioned technique remarkably improves large-signal performance of the PLL.

To characterize and design $H(s)$, one must develop a control block diagram for the PLL of Fig. 2.24. To that end, let us first express $\overline{V}_g(t) = V_g(t)e^{j\theta_g(t)}$, (2.56), in the XY frame, using (2.59):

$$\begin{aligned} V_{gX}(t) + jV_{gY}(t) &= V_g(t)e^{j\theta_g(t)}e^{-j\rho(t)} \\ &= V_g(t)e^{j(\theta_g(t)-\rho(t))} \end{aligned} \tag{2.75}$$

Using Euler's identity $e^{j(.)} = \cos(.) + j\sin(.)$, one finds

$$V_{gY}(t) = V_g(t)\sin\left(\theta_g(t) - \rho(t)\right) \tag{2.76}$$

which, assuming that the feedback loop is stable and has kept $\rho(t)$ close to $\theta_g(t)$, can be approximated as:

$$V_{gY}(t) \approx V_g(t)\underbrace{\left(\theta_g(t) - \rho(t)\right)}_{e(t)} \tag{2.77}$$

It then follows from (2.77), in conjunction with the block diagram of Fig. 2.24, that the PLL can be represented by a unity-feedback control loop, as shown in Fig. 2.25, for which $\theta_g(t)$ and $\rho(t)$ are the reference and output signals, respectively. Thus, $e(t) = \theta_g - \rho$ is the error signal, $V_g(t)$, i.e., the instantaneous peak of the grid (line-to-neutral) voltage, is a time-varying gain, and the VCO, an integrator, is the control plant. The objective of this control loop is to tune the compensator $H(s)$ such that $\rho(t)$ is tightly and rapidly regulated at $\theta_g(t)$.

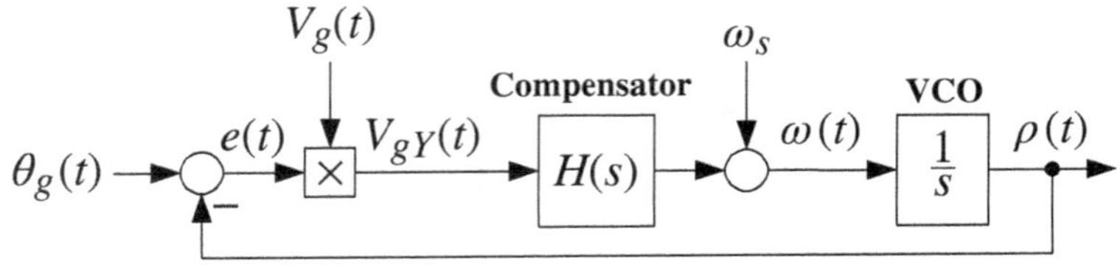

Figure 2.25 Control block diagram for the PLL of Fig. 2.24.

The compensator of the PLL, $H(s)$, must ensure sufficient phase margin, gain margin, and gain crossover frequency, for closed-loop stability, robustness to parameter variations, and disturbance rejection. Since the command signal $\theta_g(t)$ is more or less a ramp function of time[12], $H(s)$ must have at least one pole at $s = 0$, to ensure that the overall open-loop gain includes at least two integrators (including the integral action of the VCO), i.e., the control loop is of the so-called "type 2." This also renders constant disturbances ineffectual; these are the offset of ω_s added to the output of $H(s)$, the initial phase angle of the grid voltage, and the initial value of ρ. In its simplest form, $H(s)$ can be of the PI type. However, a more sophisticated filter is often preferred. For example, the grid voltage imbalance, however small, causes V_{gY} to pulsate at a frequency of $2\omega_g$. The pulsations, in turn, result in pulsations in both ω and ρ[13]. Hence, $H(s)$ may also be delegated the task of filtering of the double-frequency pulsations of V_{gY}, for ripple-free ω and ρ. In practice, under normal conditions, $V_g(t)$ may be considered constant at its nominal value and, therefore, acts as a pure gain.

Equations (2.52) through (2.54) assume that, as with any other node voltage, v_{ga}, v_{gb}, and v_{gc} are measured with reference to the DC-side midpoint of the VSC; and the DC-side midpoint may not be accessible or even be a physical node. However, it can be readily shown that the expression for the corresponding space phasor, (2.56), remains the same even if v_{ga}, v_{gb}, and v_{gc} are measured with reference to an arbitrary node whose voltage with reference to the DC-side midpoint is, for example, v_x. Thus, using (2.26) and the identity $e^{j0} + e^{j\frac{2\pi}{3}} + e^{j\frac{4\pi}{3}} \equiv 0$, one can verify that the space phasor corresponding to the three constituents $v_{ga} + v_x$, $v_{gb} + v_x$, and $v_{gc} + v_x$ has the same expression as that of (2.56). A convenient such measurement reference node is one of the three coupling nodes, for example, the one corresponding to phase c, as indicated in Fig. 2.26.

Once the PLL locks itself to the grid voltage and reaches a steady state, the $\rho(t) = \theta_g(t)$, $\omega(t) = \omega_g(t)$, $V_{gY} = 0$, and $V_{gX} = V_g(t)$. These relationships also hold in transients, albeit approximately. That is, $\rho(t) \approx \theta_g(t)$, $\omega(t) \approx \omega_g(t)$, $V_{gY} \approx 0$, and $V_{gX} \approx V_g(t)$, unless the disturbances are so large that the output of $H(s)$ is saturated (for example, due to an external fault within the AC grid). Therefore, in subsequent chapters, we shall denote a PLL by the symbol of Fig. 2.27. In addition,

[12] If $\omega_g(t)$ is constant at its nominal value, ω_s, then, based on (2.55), $\theta_g(t) = \omega_s t + \theta_{g0}$, which is a ramp function of time with a constant component.

[13] The impact of these pulsations is discussed in [23].

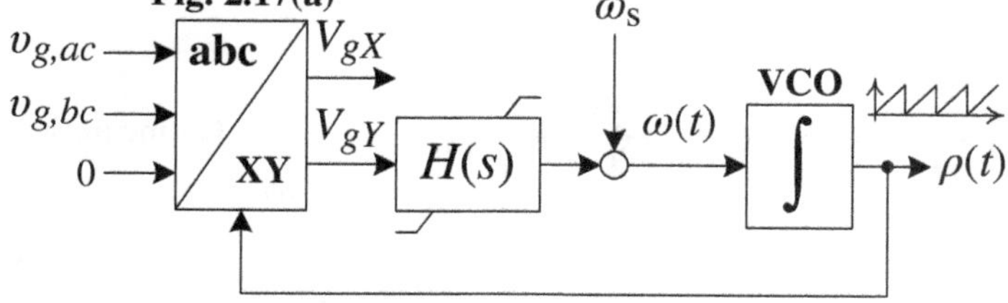

Figure 2.26 The same PLL as that of Fig. 2.24, but using the line voltages $v_{g,ac} = v_{ga} - v_{gc}$ and $v_{g,bc} = v_{gb} - v_{gc}$.

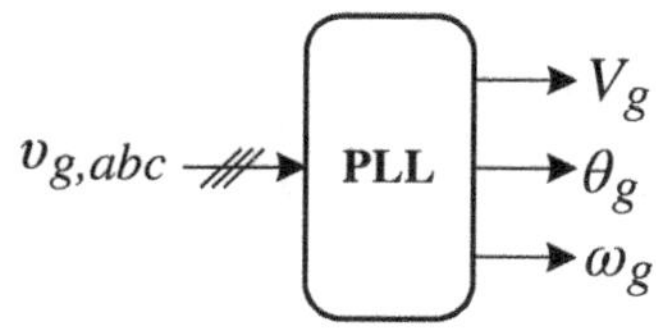

Figure 2.27 A simplified symbol for the PLL, based on its steady-state performance.

the current-control scheme of Fig. 2.20 can be simplified as shown in Fig. 2.28. The current-control scheme can be further simplified by replacing the signals $\omega_g(t)$ and $V_g(t)$ with their respective nominal values, in view of the fact that, in practice, the two aforementioned quantities vary over relatively narrow ranges around their respective nominal values and may be considered constant for control purposes.

However, perhaps, the most important and desirable result of the synchronization process enabled by the PLL is that the real- and reactive-power outputs of the VSC become proportional to and can therefore be independently controlled by the direct- and quadrature-axis components of the VSC AC-side current, i.e., I_{tX} and I_{tY}, respectively. This can be verified by substituting for $V_{gY} \approx 0$ in (2.69) and (2.70) to arrive at

$$P_g(t) = \frac{3}{2}V_g(t)I_{tX}(t) \tag{2.78}$$

$$\begin{aligned} Q_g(t) &= -\frac{3}{2}V_g(t)I_{tY}(t) + \underbrace{\frac{3}{2}C\omega_g(t)V_g^2(t)}_{Q_c(t)} \\ &= -\frac{3}{2}V_g(t)\left[I_{tY}(t) + C\omega_g(t)V_g\right] \end{aligned} \tag{2.79}$$

Hence, based on (2.78), to regulate P_g, one can compute the VSC X-axis current setpoint, I_{tX}^*, from the real-power setpoint, P_g^*. Similarly, to regulate Q_g, the setpoint for the Y-axis component of the VSC current, I_{tY}^*, can be calculated from the reactive-power setpoint, Q_g^*, based on (2.79), where, depending on the required precision, V_g and ω_g may either be received from the PLL or be approximated by their nominal values. Alternatively, for outmost precision, I_{tX}^* and I_{tX}^* may be determined through

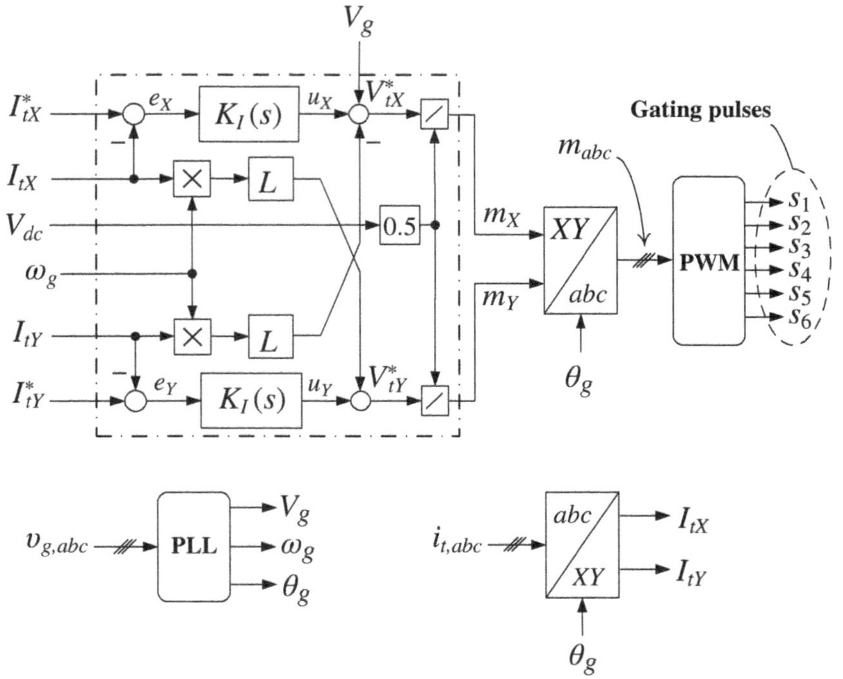

Figure 2.28 Block diagram of the current-control scheme (Fig. 2.20) with more details shown in regards to implementation.

two corresponding feedback mechanisms, as illustrated in Fig. 2.29; in their simplest forms, compensators $K_p(s)$ and $K_Q(s)$ can be of the PI type. The current setpoints I_{tX}^* and I_{tY}^* are received by the current-control scheme of the VSC (Fig. 2.28). They are bounded by judiciously chosen lower and upper limits, to ensure protection of the VSC.

2.4.3 Control of the DC-Side Voltage

As mentioned earlier, in the system of Fig. 2.2, a VSC may be employed for power-flow control, in which case the objective is to regulate P_g at a respective setpoint, by I_{tX}, based on (2.78), or through a closed-loop mechanism as shown in Fig. 2.29. Alternatively, however, P_g may be controlled as an intermediate variable, as before by I_{tX}, for the ultimate goal of regulating the DC-side voltage V_{dc}, at its respective setpoint V_{dc}^*.

The basis of controlling V_{dc} by P_g is the following power-balance equation:

$$\frac{d}{dt}W_{C_{dc}} = V_{dc}I_l - P_{dc} \tag{2.80}$$

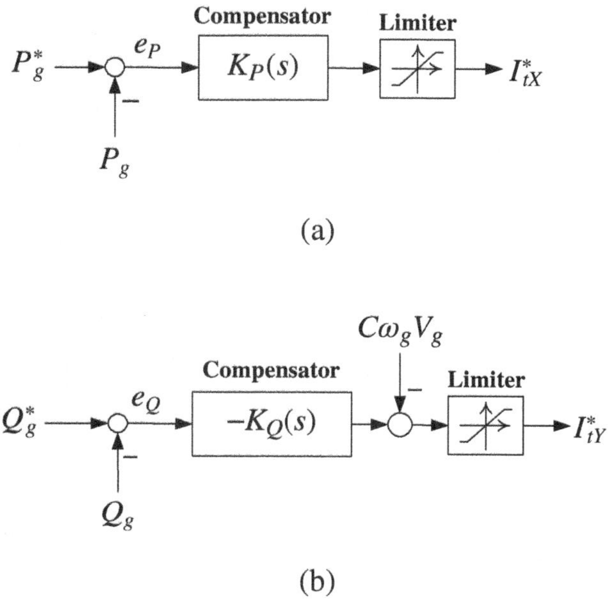

Figure 2.29 Closed-loop generation of the AC-side current components of the VSC for regulating (a) real-power output, P_g, and (b) reactive-power output, Q_g.

where W_C, equalling $\frac{1}{2}C_{dc}V_{dc}^2$, signifies the energy stored in the DC-side capacitance C_{dc}, and P_{dc} is the power entering the DC side of the VSC. Ignoring the power losses of the VSC, P_{dc} equals P_t, i.e., the power that leaves the AC-side terminals of the VSC. In turn, $P_t \approx P_g$, in view of the fact that the power loss of inductor L is small (i.e., R is small) and assuming that the instantaneous powers exchanged with L and C are negligible. Hence, $P_{dc} \approx P_g$ and, using (2.78), one can rewrite (2.80) as

$$\left(\frac{C_{dc}}{2}\right)\frac{dV_{dc}^2}{dt} = V_{dc}I_l - \underbrace{\frac{3}{2}V_g I_{tX}}_{P_g} \tag{2.81}$$

Figure 2.30 shows the block diagram of the process of regulating V_{dc}, based on the model of (2.81). As Fig. 2.30 illustrates, a compensator, $K_{Vdc}(s)$, processes the error between the square of the measure of V_{dc}, i.e., V_{dc}^2, and its corresponding setpoint, $(V_{dc}^*)^2$, in order to determine one component of the setpoint I_{tX}^*, i.e., u_{vdc}. The other component of I_{tX}^*, i.e., u_f, is determined through a feed-forward compensation strategy whose objective is to minimize the impact of the current I_l (Fig. 2.2) on the regulation process (2.81). In a special case where I_l is independent of V_{dc}, it acts as a disturbance input and its effect would be rejected by the closed-loop mechanism, without a need for feed-forward compensation. However, I_l almost always depends on V_{dc}, among other variables, in a substantial hard-to-characterize way, hence the

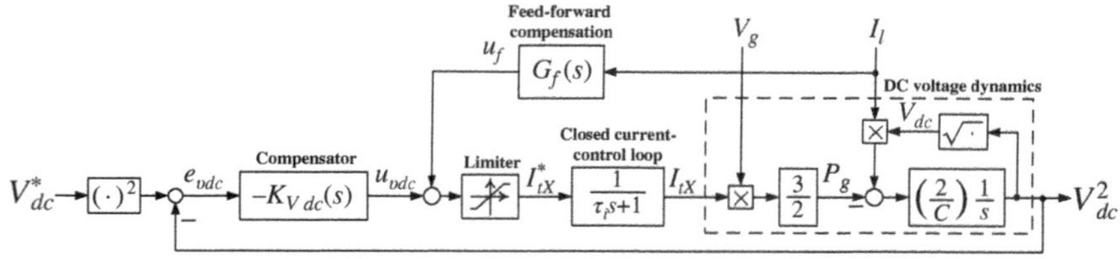

Figure 2.30 Block diagram illustrating the process of DC-side voltage regulation.

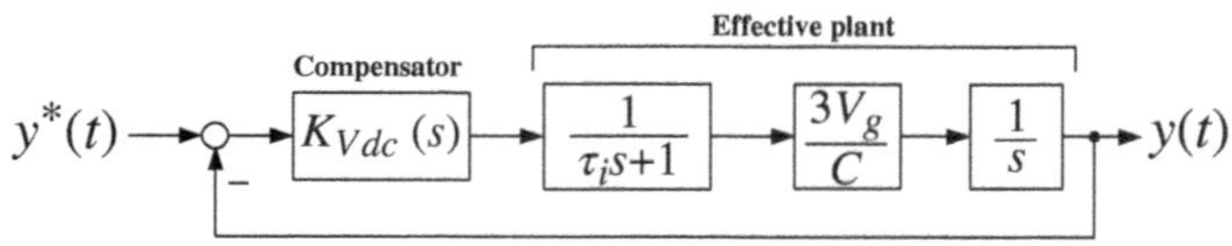

Figure 2.31 Block diagram of the effective DC-side voltage regulation loop.

feed-forward compensation. In its simplest form, the transfer function relating the measure of I_l to the feed-forward signal u_f (Fig. 2.30) can be the pure gain

$$G_f(s) = \frac{2V_{dc}}{3V_g} \tag{2.82}$$

for which V_{dc} and V_g are either dynamic measures or nominal values of the corresponding voltages. Then, I_{tX}^* is passed through a hard limiter and delivered to the current-control scheme of the VSC.

It should be noted that u_f, the feed-forward component of I_{tX}^*, is communicated through the X-axis current-control loop and, consequently, is delayed. Therefore, for an effective feed-forward compensation, one must ensure a fast current-control loop, corresponding to a small time-constant τ_i. Assuming this to be the case, the DC-side voltage regulation loop is equivalent to the control loop of Fig. 2.31. As with any typical control loop, the objective here is to design the compensator $K_{V\,dc}(s)$ such that the closed loop is stable, robust to uncertainties of the parameter (such as that of V_g which acts as a gain), and sufficiently fast. For a feedback loop of the form illustrated in Fig. 2.31, i.e., one whose plant consists of a first-order lag in series with an integrator, the simplest compensator is a PI transfer function whose gains are determined through the method of *symmetrical optimum* [24].

2.4.4 Control of the AC Grid Voltage

Irrespective of whether P_g is controlled for the objective of power-flow control or for regulating the DC-side voltage, the reactive-power output of the VSC, Q_g, can be controlled for reactive-power flow regulation, by direct assignment of I_{tY}^* or through the feedback loop of Fig. 2.29(b), or, alternatively, it can be controlled, as an intermediate variable, to regulate V_g, i.e., the magnitude of the grid voltage at the coupling point. This can be achieved by the feedback loop of Fig. 2.32.

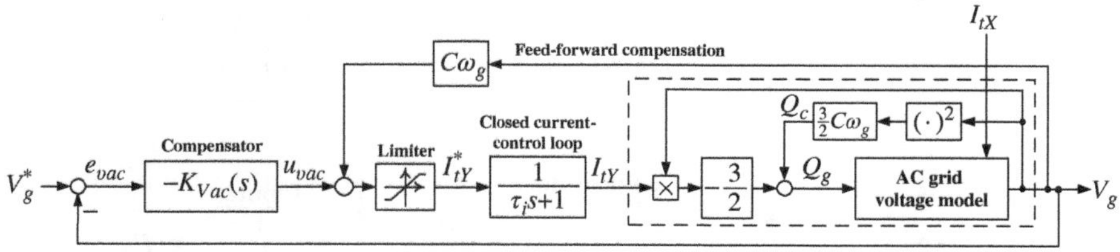

Figure 2.32 Block diagram illustrating the process of AC voltage regulation at the coupling point.

In the feedback loop of Fig. 2.32, a compensator, $K_{Vac}(s)$, compares the error between the measure of V_g and its setpoint, V_g^*, and determines one component of the VSC Y-axis current setpoint, while the feed-forward signal $C\omega_g V_g$ constitutes the other component of I_{tY}^*. The feed-forward signal is to compensate for Q_c, the reactive-power contribution of the filter capacitors (Fig. 2.2). It is straightforward to show that, assuming a purely inductive AC grid, V_g is incrementally proportional to $-I_{tY}$, with the constant of proportionality being equal to the fundamental-frequency Thevenin reactance seen from the point of coupling [23][14]. In practice, due to the resistance of the lines, V_g depends also on P_g and, therefore, on I_{tX}, as Fig. 2.32 indicates. The dependance is, however, negligible in high voltage grids.

2.4.5 Multi-unit Control of DC Grid Voltage and/or AC Grid Voltage

In a multi-terminal system, more than one VSCs may be given the task of regulating the DC grid voltage. Thus, depending on the resistances of the lines connecting the DC ports of the VSCs, the process can entail undesirable power circulations among the VSCs, or even instabilities, even if the VSCs are assigned (nominally) equal DC-side voltage setpoints.

One way to avoid the aforementioned conflicts is to modify the voltage setpoints based on the amount of power that the respective VSCs channel. Thus, as Fig. 2.33(a) illustrates, the DC-side voltage setpoint, $(V_{dc}^*)^2$, can be varied in dynamic proportion to the output real power of the VSC, P_g, which can be readily calculated from measures of the grid voltage and VSC AC-side current, based on (2.78); $\beta_P(s)$ is a filter with a positive DC gain (i.e., $\beta_P(0) > 0$). Consequently, if the regulation of V_{dc}^2 at $(V_{dc}^*)^2$ requires, for example, that an excessive amount of power be exported to the AC grid because, for example, the other VSCs have stiffly set the DC grid voltage at a larger value, then $(V_{dc}^*)^2$ is made larger by $\beta_P(s)P_g$ and the condition is relaxed. On the other hand, if the DC grid voltage is stiffly smaller than the setpoint, such that the VSC must import a large amount of power from the AC grid (i.e., P_g is negative with

[14]The box labeled as "AC Grid Voltage Model" in Fig. 2.32 indicates that V_g depends on Q_g, to remind one of the general understanding that the AC voltage is dependent on the reactive power. However, based on (2.79), it is understood that Q_g is, in turn, related to I_{tY}, as also illustrated by Fig. 2.32

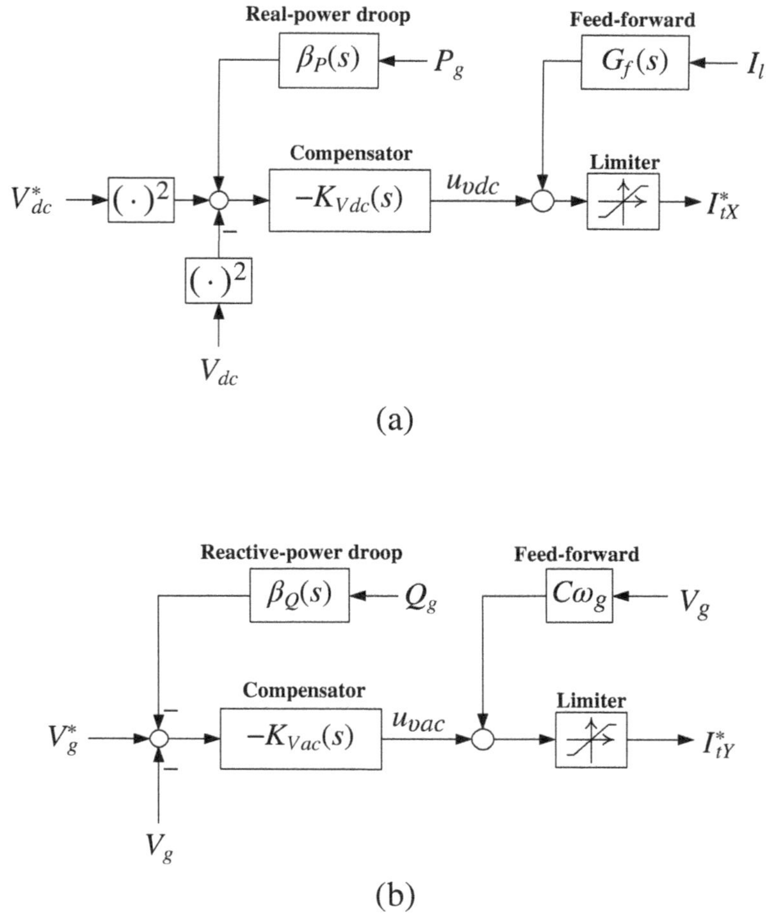

Figure 2.33 Multi-VSC regulation of the DC and AC grid voltages by drooping the setpoints: (a) against the real-power output of the host VSC and (b) against the reactive-power output of the host VSC.

a large absolute value), in order to increase V_{dc}^2 toward $(V_{dc}^*)^2$, then $(V_{dc}^*)^2$ is made smaller by $\beta_P(s)|P_g|$ to relax the condition. Eventually, contingent upon the stability of the overall system, an equilibrium is reached, in terms of the real power that each VSC channels and the DC-side voltage that each VSC assumes.

The aforementioned issue also arises if more than one VSCs attempt to regulate the AC grid voltage, depending on the reactances between the lines connecting the AC coupling points. Similarly, as shown in Fig. 2.33(b), the AC voltage setpoint, V_g^*, can be changed in dynamic proportion to the output reactive power, Q_g, which can be calculated from measures of the grid voltage and AC-side current of the VSC, based

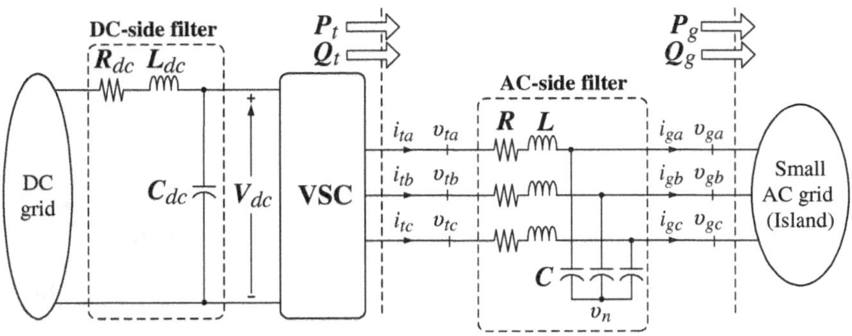

Figure 2.34 VSC energizing a small AC grid.

on (2.79), in order to enable stable multi-VSC regulation of the AC grid voltage; the filter $\beta_Q(s)$ has a positive DC gain, similar to its real-power counterpart, $\beta_P(s)$.

Each of the two correction mechanisms described earlier is known as a *droop mechanism*. Thus, V_{dc}^* and V_g^* are said to be *drooped* against the real power and the reactive power, respectively. Usually, $\beta_P(s)$ and $\beta_Q(s)$ are chosen as first-order filters of the forms:

$$\beta_P(s) = \frac{\beta_P}{\tau_P s + 1} \tag{2.83}$$

$$\beta_Q(s) = \frac{\beta_Q}{\tau_Q s + 1} \tag{2.84}$$

for which the DC gains β_P and β_Q are known as the *droop gains* or *droop coefficients*. These gains and the time constants τ_P and τ_Q are design choices. The magnitude of β_P is in a trade-off with the closeness of V_{dc} to V_{dc}^*, and the magnitude of β_Q is in a trade-off with the closeness of V_g to V_g^*. In other words, the droop mechanism results in steady-state errors, in the interest of cooperation of multiple VSCs. Further details of droop control will be analyzed in Chapter 4.

2.4.6 Control of Islands

In some applications, one or more VSCs may be required to energize, and regulate the voltage magnitude and frequency of, a relatively small AC grid. Such an AC grid hosts and exchanges energy with small-scale generators, for example, wind turbines in an off-shore wind farm, as well as regional loads. Thus, the loads and generators are synchronized to the AC grid voltage. Hereafter, to circumvent confusions with the AC grid mentioned in the previous sections, the small AC grid is hereafter referred to as an *island*. Figure 2.34 illustrates the scenario.

The task of regulating the voltage magnitude and frequency of the island can be fulfilled through different methods. However, the control we present here is based on the basic controls discussed in the previous sections. Thus, it is assumed that the VSC

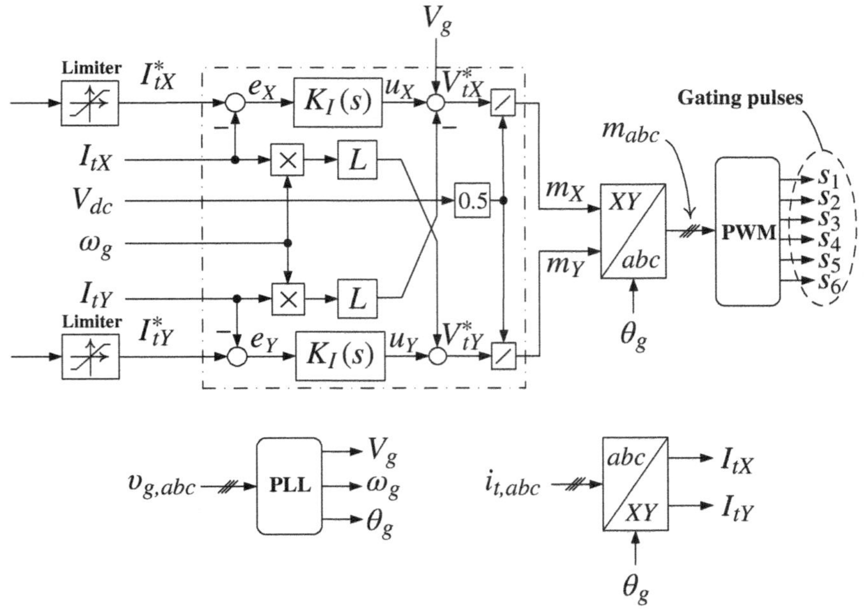

Figure 2.35 Current-control scheme for the VSC of Fig. 2.34.

is voltage-supported by the DC grid and current-controlled by the control scheme of Fig. 2.28, repeated here as Fig. 2.35 for the ease of reference. Hence, the dynamics of the grid voltage at the coupling point, $v_{g,abc}$, is described by

$$C\frac{d}{dt}\left(v_{ga} - v_n\right) = i_{ta} - i_{ga} \tag{2.85}$$

$$C\frac{d}{dt}\left(v_{gb} - v_n\right) = i_{tb} - i_{gb} \tag{2.86}$$

$$C\frac{d}{dt}\left(v_{gc} - v_n\right) = i_{tc} - i_{gc} \tag{2.87}$$

where v_n is the voltage of the common node of the AC-side filter capacitors, with reference to the DC-side midpoint of the VSC.

Multiplying both sides of (2.85–2.87) by, respectively, $\frac{2}{3}e^{j0}$, $\frac{2}{3}e^{j\frac{2\pi}{3}}$, and $\frac{2}{3}e^{j\frac{4\pi}{3}}$, adding the resulting equations, and using (2.26), one finds the following space-phasor equation:

$$C\frac{d\overline{V_g}}{dt} = \overline{I_t} - \overline{I_g} \tag{2.88}$$

in which v_n does not appear due to the fact that $e^{j0} + e^{j\frac{2\pi}{3}} + e^{j\frac{4\pi}{3}} \equiv 0$.

Next, it is assumed that the PLL has perfectly aligned the X-axis of the XY frame with the grid voltage space phasor and, therefore, $\overline{V_g} = V_g e^{j\theta_g}$, where V_g is the

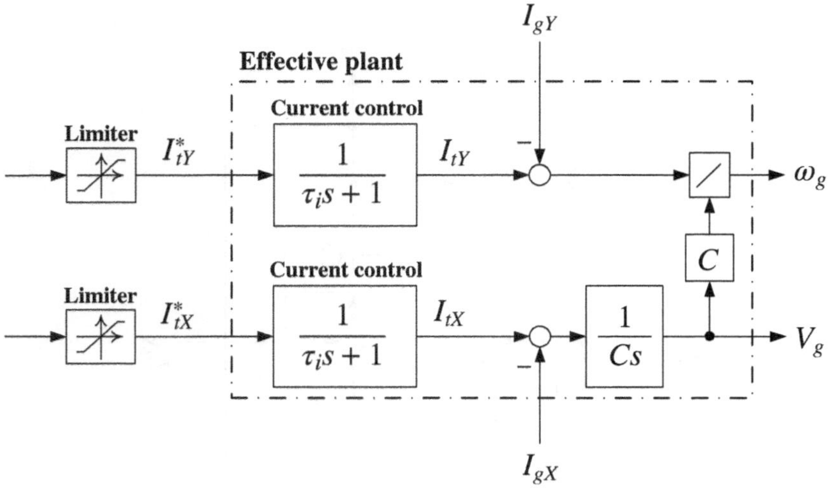

Figure 2.36 Control model for the magnitude and frequency of the island voltage.

instantaneous magnitude of the line-to-neutral grid voltage. Thus, substituting for $\overline{V_g} = V_g e^{j\theta_g}$, $\overline{I_t} = (I_{tX} + jI_{tY})\, e^{j\theta_g}$, and $\overline{I_g} = (I_{gX} + jI_{gY})\, e^{j\theta_g}$ in (2.88), performing the differentiations, and splitting the resulting equation into its real and imaginary components, we obtain

$$\frac{dV_g}{dt} = \left(\frac{1}{C}\right)(I_{tX} - I_{gX}) \tag{2.89}$$

and

$$0 = -C\omega_g V_g + I_{tY} - I_{gY} \tag{2.90}$$

which can be rewritten as

$$\omega_g = \left(\frac{1}{C}\right)\frac{I_{tY} - I_{gY}}{V_g} \tag{2.91}$$

where

$$\frac{d\theta_g}{dt} = \omega_g \tag{2.92}$$

Equations (2.89) and (2.91) indicate that the magnitude and frequency of $v_{g,abc}$ can be controlled by the converter current components I_{tX} and I_{tY}, which, in turn, are controlled by their respective setpoints I^*_{tX} and I^*_{tY}, based on (2.74).

Figure 2.37 shows a block diagram of the control scheme. Thus, two compensators, $K_{Vac}(s)$ and $K_{\omega}(s)$, process the corresponding error signals, $e_{vac} = V^*_g - V_g$ and $e_{\omega} = \omega^*_g - \omega_g$, and produce the setpoints I^*_{tX} and I^*_{tY}, respectively. The compensators are, however, difficult to design since the current components I_{gX} and I_{gY}

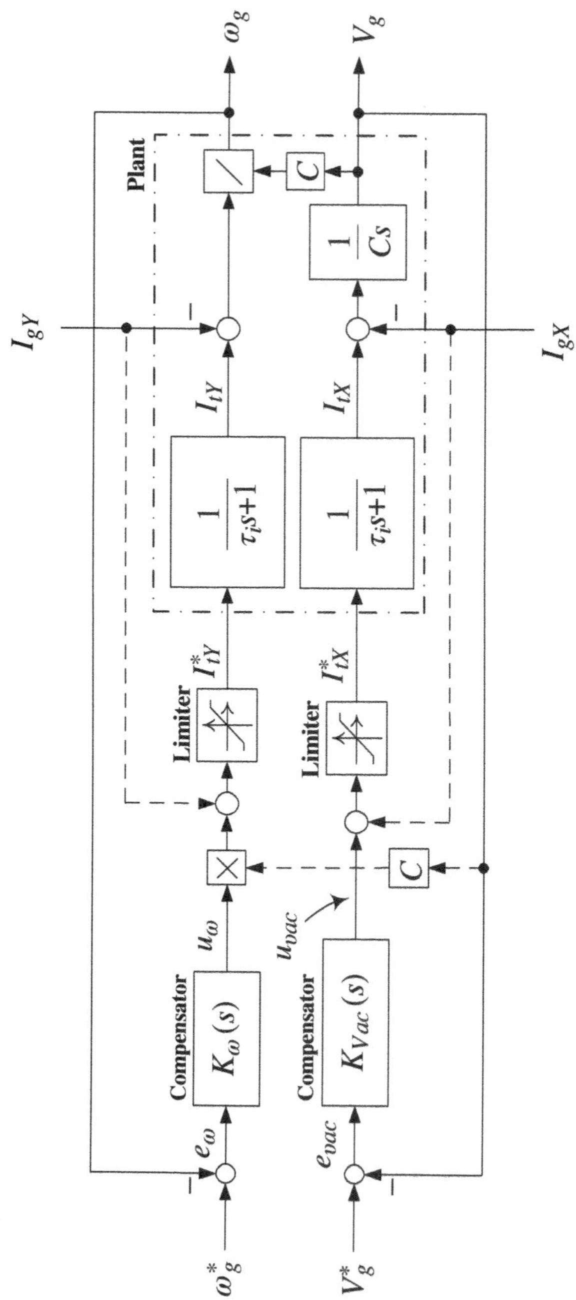

Figure 2.37 Voltage magnitude and frequency control loops.

depend on both ω_g and V_g, in an uncertain manner. Therefore, as Fig. 2.37 shows, the compensator outputs are augmented with measures of I_{gX} and I_{gY}, as feedforward signals. Figure 2.37 also shows that the output of $K_\omega(s)$ may be multiplied by CV_g to counteract the impact of V_g on ω_g (Fig. 2.36). However, assuming that V_g is regulated about a small neighborhood of its nominal value, the feedforward signal CV_g may be replaced with a constant gain, as a factor of $K_\omega(s)$. In order for the aforementioned feedforward compensation strategies to be effective, τ_i must be sufficiently small, corresponding to adequately fast X- and Y-axis closed current-control loops. In their simplest forms, $K_{Vac}(s)$ and $K_\omega(s)$ can be of the PI form. $K_{Vac}(s)$, in particular, can be tuned through the method of symmetrical optimum [24]. To enable multiple VSCs to concurrently regulate the voltage magnitude and frequency of an island, the setpoints V_g^* and ω_g^* of each VSC can be drooped against the reactive and real powers that the VSC exchanges with the grid, Q_g and P_g in Fig. 2.34, as

$$V_g^* = V_{gn} - \beta_Q(s)Q_g \tag{2.93}$$

and

$$\omega_g^* = \omega_s - \beta_P(s)P_g \tag{2.94}$$

where V_{gn} and ω_s are the nominal frequency and magnitude of the island voltage, respectively. Filters $\beta_P(s)$ and $\beta_Q(s)$ are described in Section 2.4.5.

2.5 Simulation

In general, simulation studies constitute the first and most important step in evaluation of a design and its impact on the host AC and DC grids. Depending on the available simulation platforms or the type of study to be conducted, a system of VSCs can be modeled in different ways, e.g., using the mathematical relationships and signal processing blocks in MATLAB®/SIMULINK® [25] software environment, etc. However, a circuit-based simulation model, developed on a specialized software platform, such as EMTDC/PSCAD [26], PSIM, or PSPICE, offers a more tangible perspective of the operation and performance of the system and its transients. Moreover, a circuit-based simulation model can be more conveniently expanded to study more complex systems which typically include transmission lines, electric machines, loads of different characteristics, transformers, switching devices and relays, and so on. To enable convenient modeling of the control, circuit-based simulation platforms offer libraries of various signal processing building blocks and, in some software packages, such as EMTDC/PSCAD, even the capability to link user-developed control and signal processing codes. The subsequent sections discuss the development of circuit-based simulation models for the VSCs in a multi-terminal AC/DC system.

To model a VSC, one major task is to model of switches of the VSC; the model of this specific component effectively determines the trade-off between the accuracy and speed of simulation of the model. Some software packages (e.g., PSPICE) offer sophisticated models for transistors and diodes (based on generic mathematical equations whose parameters may be specified by device manufacturers) and can therefore capture such transient phenomena as diode reverse recovery and transistor tailing current. However, some others represent a diode or a transistor by a nonlinear resistor. Thus, the type of study and transients to be captured, size and complexity of the overall simulation model, and tolerable simulation runtime are among the factors that determine the level of complexity required for the switches of a VSC. For example, detailed device models are suitable for studying fast overvoltage transients, device thermal states, switching power losses, etc., while they significantly increase the runtime of the simulation. By contrast, simplified device models cannot capture the aforementioned transients, but adequately capture transients relevant to power system studies, with a reasonably low runtime. A circuit-based simulation model of the VSC in which the switches and their switching scheme are included is referred to as a *topological model*, a *detailed model*, or a *switched model* and is generally considered the closest, in terms of accuracy, to a hardware prototype.

As the size and complexity of the study system grow, the simulation runtime based on a switched model can quickly become prohibitively large, even if the crudest models are adopted for the switches, especially if a multitude of VSCs coexist within the study system. The computational burden becomes particularly onerous when multiple VSCs connected to a large AC grid comprising many generators and transmission lines are studied. For such cases, the switched VSC models can be replaced by models in which no switching device scheme is included, but the AC-side terminal voltages (or currents) of the VSC are produced by dependent voltage sources. Throughout the rest of the book, this model will be referred as the *averaged model.* This averaged model will be used in Chapter 3 to develop the modeling framework for AC-MTDC grids.

Figure 2.38 illustrates a circuit-based VSC model in which the AC-side terminal voltages are produced by three dependent voltage sources. The positive terminals of the voltage sources correspond to the AC-side terminals of the VSC, while their negative terminals meet at a common node, "0," as the virtual DC-side midpoint of the VSC (Figs 2.1 and 2.2); to circumvent convergence difficulties, it is sometimes necessary to connect the node "0" to the circuit ground, through a large resistor, as shown in Fig. 2.38, especially if nowhere else within the AC grid is connected to the ground. In the model of Fig. 2.38, each voltage source establishes a voltage (with reference to node "0") that is equal in value to the fundamental component of the corresponding AC-side terminal voltage of the VSC. Thus, the control signals of the voltage sources producing the voltages v_{ta}, v_{tb}, and v_{tc} are obtained from XY- to abc-frame transformation of the signals V_{tX}^* and V_{tY}^*, i.e., the XY-frame components of the desired voltage phasor. In turn, V_{tX}^* and V_{tY}^* are obtained from the XY-frame components of the modulating signal, m_X and m_Y, which are produced by the control scheme of the VSC (for example, by the scheme of Fig. 2.28); the angle for the XY- to abc-frame transformation, θ_g, is the phase angle of the grid voltage phasor, which is obtained from the PLL of the control scheme (Section 2.4.2.5). The gain k_{PWM}, also commonly known as the *converter gain*, depends on the switching strategy. For example, for a two-level VSC (Fig. 2.3.1), k_{PWM} is 0.5 under the SPWM strategy,

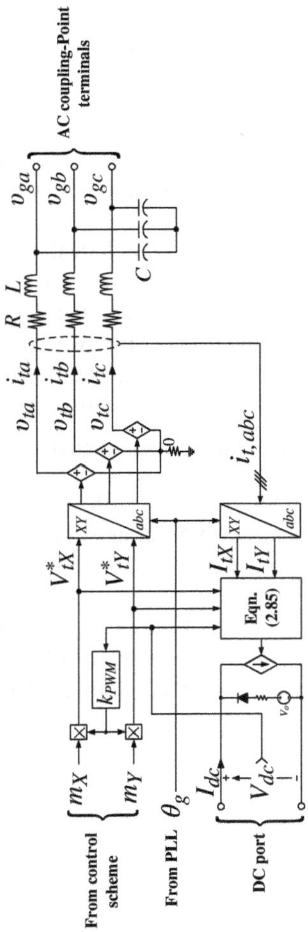

Figure 2.38 A circuit-based averaged model of the VSC based on dependent voltage and current sources.

0.577 (= $1/\sqrt{3}$) under the SPWM with third-harmonic injection, and 0.577 under the *space-vector modulation (SVM)* strategy. If the impact DC-side voltage on the AC-side terminal voltage is deemed to be negligible, for example, because V_{dc} is known to be constant or tightly regulated, the signals V_{tX}^* and V_{tY}^* may be directly obtained from the control scheme.

Figure 2.38 also indicates that the DC-side current of the VSC, I_{dc}, is produced by a dependent current source. Thus, the value of the current is determined by (2.9). Rewriting the numerator of (2.9) in terms of the XY-frame components of $v_{t,abc}$ and $i_{t,abc}$, one finds

$$I_{dc} = \frac{\frac{3}{2}\left(V_{tX}I_{tX} + V_{tY}I_{tY}\right)}{V_{dc} + \epsilon} \tag{2.95}$$

in which the relatively small constant ϵ is deliberately introduced to prevent division by zero which would otherwise occur when V_{dc} is zero, e.g., at the start of the simulation study. The signals I_{tX} and I_{tY} are obtained from an *abc*- to *XY*-frame signal transformation, as Fig. 2.38 illustrates.

The averaged model of Fig. 2.38 does not include the switches or switching scheme. Therefore, it runs considerably faster than its switched counterpart. However, it does not predict the voltage/current harmonics (generated through the PWM process) and their impact on the control and host grids. Moreover, assuming a SPWM switching strategy, it is valid for frequencies up to about one-third of the switching frequency (carrier signal frequency) and only if the VSC in the actual system or in the switched model operates in its linear modulation region, that is, if $\sqrt{V_{tX}^{*\,2} + V_{tY}^{*\,2}} \leq 0.5V_{dc}$ (the condition is modified to $\sqrt{V_{tX}^{*\,2} + V_{tY}^{*\,2}} \leq 0.575V_{dc}$ if the SPWM strategy with third-order harmonic injection or the SVM strategy is assumed). The averaged model, however, can be expanded to also include AC-side voltage harmonics, as illustrated by the following example.

Consider the three-level waveform of Fig. 2.39, representative of the AC-side terminal voltage of, for example, the NPC converter of Fig. 2.10, produced through the so-called *single-pulse modulation* switching strategy. The terminal voltage, v_{ta}, may be expressed as

$$v_{ta}(\varepsilon) = \underbrace{k\cos\alpha\cos\varepsilon}_{\langle v_{ta}\rangle_1} +$$

$$\underbrace{k\left(\frac{-1}{3}\cos 3\alpha\cos 3\varepsilon + \frac{1}{9}\cos 9\alpha\cos 9\varepsilon + \frac{-1}{15}\cos 15\alpha\cos 15\varepsilon + \ldots\right)}_{\text{triple-n harmonics}} +$$

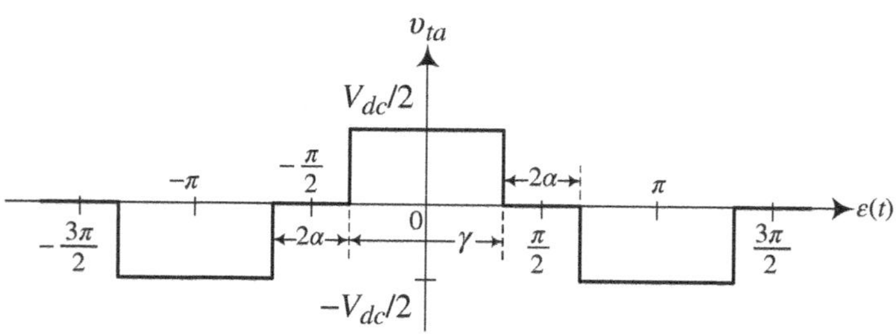

Figure 2.39 Waveform of a three-level AC-side terminal voltage produced by an NPC converter through single-pulse modulation strategy.

$$k\left(\frac{1}{5}\cos 5\alpha \cos 5\varepsilon + \frac{-1}{7}\cos 7\alpha \cos 7\varepsilon + \frac{-1}{11}\cos 11\alpha \cos 11\varepsilon + \ldots\right) + \tag{2.96}$$

where

$$\begin{aligned} k &= \left(\frac{V_{dc}}{2}\right)\left(\frac{4}{\pi}\right) \\ &= \left(\frac{2}{\pi}\right) V_{dc} \end{aligned} \tag{2.97}$$

and

$$\varepsilon(t) = \theta_g(t) + \varepsilon_0 \tag{2.98}$$

where θ_g is the phase angle of the grid voltage phasor provided by the PLL, and ε_0 is the phase angle of the fundamental component of v_{ta}, relative to θ_g. Thus, v_{ta} consists of a fundamental component, $\langle v_{ta} \rangle_1$, superimposed with odd-order harmonics grouped as triple-n harmonics and the other harmonics. Assuming a balanced three-phase operation, the phase voltages v_{tb} and v_{tc} can be found by replacing $\varepsilon(t)$ in (2.96) with $\varepsilon(t) - 2\pi/3$ and $\varepsilon(t) - 4\pi/3$, respectively:

$$v_{tb}(\varepsilon) = v_{ta}\left(\varepsilon - \frac{2\pi}{3}\right) \tag{2.99}$$

$$v_{tc}(\varepsilon) = v_{ta}\left(\varepsilon - \frac{4\pi}{3}\right) \tag{2.100}$$

Based on (2.26), one finds the space phasor corresponding to v_{ta}, v_{tb}, and v_{tc} as

$$\begin{aligned} \overline{V}_t = & \underbrace{(k\cos\alpha)\, e^{j\varepsilon}}_{\text{fundamental}} + \\ & \underbrace{\left(\frac{-k}{7}\cos 7\alpha\right) e^{j7\varepsilon} + \left(\frac{k}{13}\cos 13\alpha\right) e^{j13\varepsilon} + \left(\frac{-k}{19}\cos 19\alpha\right) e^{j19\varepsilon} + \ldots +}_{\text{positive-sequence harmonics}} \\ & \underbrace{\left(\frac{k}{5}\cos 5\alpha\right) e^{-j5\varepsilon} + \left(\frac{-k}{11}\cos 11\alpha\right) e^{-j11\varepsilon} + \left(\frac{k}{17}\cos 17\alpha\right) e^{-j17\varepsilon} + \ldots}_{\text{negative-sequence harmonics}} \end{aligned} \tag{2.101}$$

Thus, $\overline{V}_t$ is composed of individual space phasors, each corresponding to a harmonic component, including the fundamental component, of the three-phase voltage $v_{t,abc}$, with the exception of the triple-n harmonics which are common for the phase voltages v_{ta}, v_{tb}, and v_{tc}. The space phasors corresponding to harmonics of order $6n + 1$ (n a positive integer) rotates in the same direction as the space phasor corresponding to the fundamental component, whereas the space phasors corresponding to harmonics of order $6n - 1$ rotate in the opposite direction. Hence, relative to the fundamental component, a harmonic component belonging to the former group is referred to as a *positive-sequence harmonic*, whereas a member of the latter group is referred to as a *negative-sequence harmonic*. A triple-n harmonic is referred to as a *zero-sequence harmonic*.

Replacing ε with $\theta_g + \varepsilon_0$, based on (2.98), one can rewrite (2.101) as

$$\overline{V}_t = \left[V_{tX,1} + jV_{tY,1}\right] e^{j\theta_g} + \left[V_{tX,5} + jV_{tY,5}\right] e^{-j5\theta_g} + \left[V_{tX,7} + jV_{tY,7}\right] e^{j7\theta_g} + \ldots$$

where

$$\begin{aligned}
V_{tX,1} + jV_{tY,1} &= k\cos(\alpha)e^{j\varepsilon_0} \\
V_{tX,5} + jV_{tY,5} &= \frac{k}{5}\cos(5\alpha)e^{-j5\varepsilon_0} \\
V_{tX,7} + jV_{tY,7} &= \frac{-k}{7}\cos(7\alpha)e^{j7\varepsilon_0} \\
V_{tX,11} + jV_{tY,11} &= \frac{-k}{11}\cos(11\alpha)e^{-j11\varepsilon_0} \\
V_{tX,13} + jV_{tY,13} &= \frac{k}{13}\cos(13\alpha)e^{j13\varepsilon_0} \\
V_{tX,17} + jV_{tY,17} &= \frac{k}{17}\cos(17\alpha)e^{-j17\varepsilon_0} \\
V_{tX,19} + jV_{tY,19} &= \frac{-k}{19}\cos(19\alpha)e^{j19\varepsilon_0} \\
&\;\;\vdots
\end{aligned} \tag{2.102}$$

in which the numerical part of a subscript denotes the harmonic order.

Equation (2.102) indicates that each harmonic component of $v_{t,abc}$ can be produced by a corresponding *XY*- to *abc*-frame transformation, from the corresponding *X*- and *Y*-axis components given by (2.102); the phase angle for the *XY*- to *abc*-frame transformation is the harmonic order times θ_g (for a positive-sequence harmonic) and the harmonic order times $-\theta_g$ (for a negative-sequence harmonic). Figure 2.40 illustrates the concept for a simulation that is intended to include the 5th, 7th, and 11th harmonics.

Equation (2.102) indicates that the magnitude of the fundamental component of $v_{t,abc}$ is proportional to $\cos\alpha$, which varies from 0 to 1 as the pulse width γ (Fig. 2.39) is varied from 0 to π. Therefore, $\cos\alpha$ qualifies to be defined as the magnitude of the modulating signals, M:

$$M = \cos\alpha \tag{2.103}$$

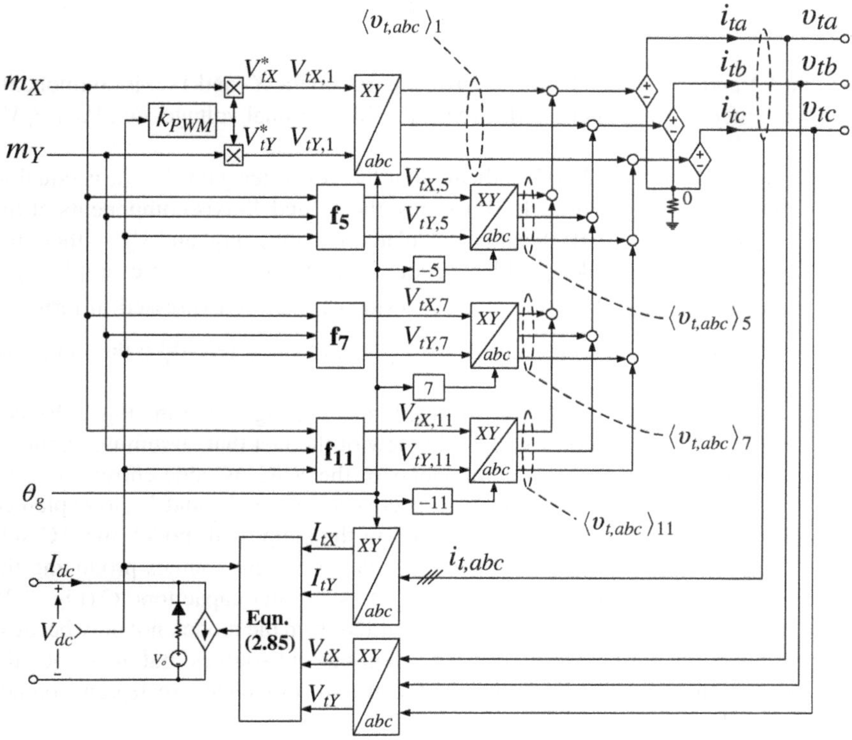

Figure 2.40 Circuit-based averaged model of Fig. 2.38 extended to include AC-side terminal voltage harmonics.

It then follows from (2.102) and (2.103) that

$$V_{tX,1} = kM\cos\varepsilon_0 \tag{2.104}$$

$$V_{tY,1} = kM\sin\varepsilon_0 \tag{2.105}$$

which, based on (2.97), can be rewritten as

$$V_{tX,1} = \underbrace{\frac{2}{\pi}V_{dc}m_X}_{V^*_{tX}} \tag{2.106}$$

$$V_{tY,1} = \underbrace{\frac{2}{\pi}V_{dc}m_Y}_{V^*_{tY}} \tag{2.107}$$

where

$$m_X = M\cos\varepsilon_0 \tag{2.108}$$

$$m_Y = M \sin \varepsilon_0 \tag{2.109}$$

Note that V_{tx}^* and V_{tY}^*, issued by the control, are indeed the X- and Y-axis components of the fundamental component of the VSC AC-side terminal voltage, i.e., $V_{tX,1} = V_{tx}^*$ and $V_{tY,1} = V_{tY}^*$.

Equations (2.106) and (2.107) indicate that the converter gain, k_{PWM}, is equal to $2/\pi$, in this example. As Fig. 2.40 illustrates, the X- and Y-axis components of the harmonic space phasors (2.102) can be calculated from m_X, m_Y, and V_{dc}, either analytically or by means of look-up tables (relationships $\mathbf{f_5}$, $\mathbf{f_7}$, etc.). For example, $V_{tX,5}$ $(= \frac{k}{5}\cos 5\alpha \cos 5\varepsilon_0)$ and $V_{tY,5}$ $(= -\frac{k}{5}\cos 5\alpha \sin 5\varepsilon_0)$ can be expressed in terms of $\cos\alpha$ $(= M)$, $sin\alpha$ $(= \sqrt{1-M^2})$, $\cos\varepsilon_0$ $(= \frac{V_{tX,1}}{kM})$, and $\sin\varepsilon_0$ $(= \frac{V_{tY,1}}{kM})$; the impact of V_{dc} is embedded in k.

For a current-controlled VSC, the averaged model of Fig. 2.38 (and Fig. 2.40) can be further simplified to that of Fig. 2.41, in view of the fact that, assuming an unsaturated control, the X- and Y-axis components of the VSC AC-side current, i.e., I_{tX} and I_{tY}, rapidly and precisely track their respective setpoints, I_{tX}^* and I_{tY}^*, as explained in Section 2.4.2.4. Thus, as Fig. 2.41 shows, in the simplified model, the AC-side voltage sources are substituted by three controllable current sources producing the AC-side currents i_{ta}, i_{tb}, and i_{tc}, and only the AC-side filter capacitors, C (Fig. 2.2), are retained; the three-phase tie reactors are omitted from the model, not only because their currents are now dictated by the dependent current sources, but also since the series connection of a series RL branch and a current source can create convergence problems for the simulation software.

In the model of Fig. 2.41, the control signals for the AC-side current sources are calculated through an XY- to abc-frame transformation of the filtered measures of the setpoints I_{tX}^* and I_{tY}^*, with the (first-order) filters emulating the dynamics of the closed current-control loops (2.74). These setpoints are determined based on the desired real- and reactive-power outputs of the VSC, as explained in Section 2.4.2.4, or they may be issued by the control schemes for DC- and/or AC-side voltage regulation (Sections 2.4.3, 2.4.4, and 2.4.5). The DC side of the model remains a dependent current source, as in the model of Fig. 2.38, whose value is, as before, calculated from (2.95). Note that the model of Fig. 2.41 assumes, as an approximation, that the AC-side terminal voltage of the VSC is equal to the grid voltage, i.e., $V_{tX} \approx V_{gX}$ and $V_{tY} \approx V_{gY}$.

Similar to the model of Fig. 2.38, the accuracy of the model of Fig. 2.41 is compromised if the constraint $\sqrt{V_{tX}^{*\,2} + V_{tY}^{*\,2}} \leq 0.5V_{dc}$ (for the SPWM strategy) or $\sqrt{V_{tX}^{*\,2} + V_{tY}^{*\,2}} \leq 0.575V_{dc}$ (for the SPWM strategy with third-order harmonic injection or for the SVM strategy) is not satisfied. Further, in the actual VSC or its switched model, the current-control scheme measures, estimates, or assumes the grid voltage components V_{gX} and V_{gY}, and controls the AC-side terminal voltage components V_{tX} and V_{tY} in such a way that the voltage across the filter reactor, L (Fig. 2.2), is always regulated in spite of the disturbances; it is the proper implementation of this process that makes the filter reactor act as a three-phase current source with controllable magnitude and phase angle. Depending on the structure, implementation, and parameters of the current-control scheme, the assumption of the current-source nature of the VSC and its tie reactor may or may not be valid if the grid voltage is unbalanced, sagged, swelled, etc. Therefore, to correctly interpret the results, one should be cognizant of the conditions under which the aforementioned composition of components acts as a current source. Finally, the model of Fig. 2.41 underestimates the power

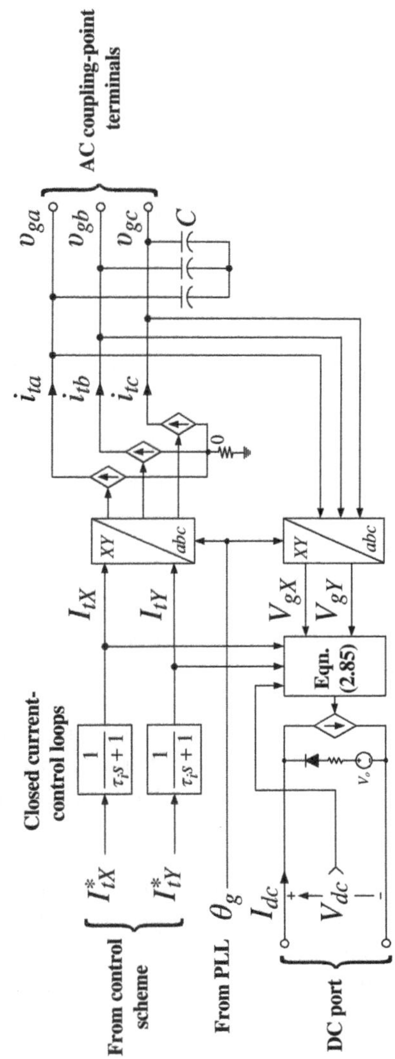

Figure 2.41 A simplified circuit-based averaged model of the VSC in which the AC-side currents are produced by controllable current sources.

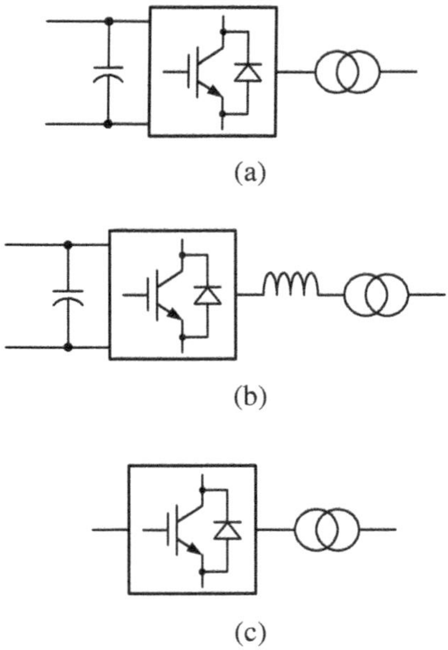

(a)–(c) labels: (a) (b) (c)

Figure 2.42 (a)–(c) Circuit symbols for the VSC.

loss of the VSC and its tie reactor, due to the omission of the resistance R (Fig. 2.2). However, the introduced error is typically negligible, since even the switched model may not be able to accurately predict the power losses if it employs a crude model for the switches of the VSC.

We conclude this section by making a remark about proper initialization of the simplified models of Figs 2.38, 2.40, and 2.41. When a switched model is adopted for a VSC (Fig. 2.2) in a circuit-based simulation, at the system start-up, the DC-side capacitor, C_{dc}, is charged by either the DC gird or the AC grid (through the antiparallel diodes of the VSC switches). However, the switches are absent from the simplified models of Figs 2.38, 2.40, and 2.41. Therefore, a mechanism is needed to ensure that the DC-side capacitor is adequately charged and, more importantly, whose voltage can never be negative. As Figs 2.38, 2.40, and 2.41 indicate, this goal can be achieved by, for example, connecting a series branch consisting of a DC voltage source, a resistor, and a diode in parallel with the DC-side terminals of the VSC model; the DC source voltage should be set to about the peak value of the AC grid line-to-line voltage. Subsequent to the start-up process and during normal operation, the DC-side voltage exceeds the peak value of the grid line-to-line voltage, due to the boosting property of the VSC, and, therefore, the diode turns off.

2.6 Symbols of the VSC

Depending on the context, the remainder of this book may employ any of the three symbols of Fig. 2.42 to graphically represent a VSC, irrespective of such details as the circuit configuration and technology of the constituting semiconductor switches.

CHAPTER 3

MODELING, ANALYSIS, AND SIMULATION OF AC–MTDC GRIDS

3.1 Introduction

It is envisaged that a subsea MTDC grid based on the VSC technology would be built around the North Sea to tap the rich wind resource of the region and also interconnect the UK and Nordic pool with continental Europe [27]. Sections 1.1 and 1.2 described the business case and rationale behind such an MTDC grid. It was also mentioned in Sections 1.4.1 and 1.4.2 that the VSC technology is the obvious choice over the LCC technology for this application.

For system planning and operation, both steady-state and dynamic simulation of the MTDC grid in conjunction with the surrounding AC network is essential. From the network operators' point of view, the possibility of an enormous loss of in-feed (tens of gigawatts) due to a single-point failure in such VSC based MTDC grid is one of the major concerns. Following DC side faults, isolating only the faulty component (a converter or a cable) of the MTDC grid is a challenge. Besides protection and DC breaker development issues, there are primary control problems such as autonomous sharing of power imbalance among the converters following a converter or cable outage. Moreover, the MTDC grid could be required to provide AC system support such as allowing exchange of frequency reserves. To study these issues, interaction

Multi-terminal Direct-Current Grids: Modeling, Analysis, and Control. First Edition.
Nilanjan Ray Chaudhuri, Balarko Chaudhuri, Rajat Majumder, and Amirnaser Yazdani.

between multi-machine AC systems and an MTDC grid along with the impact on the overall stability of the combined AC–MTDC system needs investigation, which is the subject matter of this chapter.

A lot of research attention has been focused on MTDC grids based on the LCC technology. Modeling of such systems and representation in stability program are well reported [28, 29]. However, VSC-based MTDC has received relatively less attention until the recent past when modeling [30, 31], control strategies [32, 33, 34], and protection [35] were studied.

The focus of this chapter is to present a comprehensive modeling and analysis of an MTDC grid interconnecting the AC networks. The following topics are covered: First, an asymmetric bipole MTDC grid model with the provision of metallic return network is described. The averaged model of the converters described in Chapter 2 is modified to integrate the MTDC grid with the AC grid. Different control modes of operation are covered, followed by the modeling of the DC network. Next, the dynamic model of the AC grid is briefly discussed. A steady-state model of the combined AC–MTDC grid is presented, and the load flow analysis methodology of such networks is described. Also, small-signal stability analysis of the AC–MTDC grid is performed to characterize different modes using participation factors and analyze the interaction of the MTDC network with the AC system. The approach for nonlinear time-domain simulation for transient stability analysis is presented. Finally, three case studies are presented to demonstrate the validation of the proposed averaged model in MATLAB®/SIMULINK® [25] against a switched model in EMTDC/PSCAD [26] software enabling easy integration of the MTDC grid with the multi-machine AC system models for stability studies.

3.2 MTDC Grid Model

Different configurations of MTDC converter stations were shown in Section 1.7. From a modeling standpoint, the asymmetric bipole configuration with a metallic return network (Fig. 1.9(b)) is the most generic one, and except the symmetrical monopole (Fig. 1.8(a)), the rest are special cases of this configuration. From an N-1 reliability consideration, such a configuration has the highest availability. Figure 3.1 shows further details of two converter stations in an MTDC grid connected in this configuration. In this book, this particular MTDC converter configuration is considered. Fundamentals of modeling of VSC were described in Chapter 2. Based on the conclusions drawn in Chapter 2, the converter modeling and control is described, followed by the generic model of the DC transmission network in the subsequent sections.

3.2.1 Modeling Assumptions

As described before, the motivation behind developing the model of the MTDC grid which can be integrated to surrounding AC grid model is to carry out *dynamic*

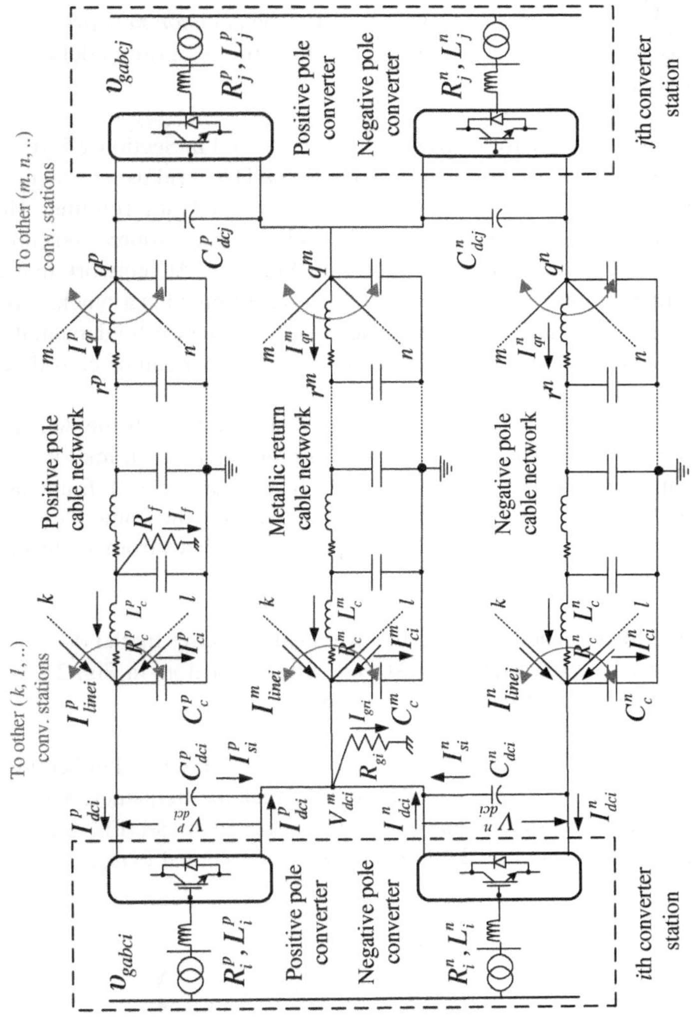

Figure 3.1 The *i*th and the *j*th converter stations in the MTDC grid connected in asymmetric bipole configuration with metallic return network. ©[36], 2011.

simulation studies for planning and operation. To that end, the following assumptions are made:

1. AC grid model: For stability studies, the positive sequence fundamental frequency phasor model of the AC system is traditionally used. Other than the dynamic components (such as generators, exciters, etc.), the AC transmission network is represented by algebraic power balance equations. The AC network voltages and currents are expressed in a reference frame *Re–Im* rotating

with synchronous speed ω_s. This is called the *common network frame of reference* as shown in Fig. 3.2. Further details of the AC grid model is described in Section 3.3.

2. Modeling adequacy for converters: As mentioned in Section 2.5, dynamic simulation of a large AC grid connected to the MTDC grid using a switched model is unrealistic because of a prohibitively large simulation runtime. Therefore, a circuit-based averaged model using dependent voltage source and current source (Fig. 2.38) can be used for this purpose. Since the AC network interfacing the AC coupling point terminals in Fig. 2.38 is represented in a synchronously rotating phasor reference frame, only the positive sequence fundamental frequency component of the voltage generated by the converter can be considered.

3. Reference frame for converter model: To integrate with the AC network expressed in a synchronously rotating common reference frame, it is more convenient to express the circuit model of Fig. 2.38 in the X–Y frame as shown in Fig. 2.19. The ith converter station is modeled using the frame X_i–Y_i rotating at a speed ω_{gi}, as shown in Fig. 3.2. The speed ω_{gi} deviates from ω_s during dynamic conditions.

4. AC side filter capacitor: Since the averaged converter model does not consider any harmonic AC voltage, the AC side filter capacitors in Fig. 2.2 are not considered. This leads to the simplification $i_{t,abc} = i_{g,abc}$.

5. PLL dynamics: The phase-locked loop dynamics described in Section 2.4.2.5 is neglected assuming it to be fast enough to track the frequency. The PLL will be represented by its steady-state performance-based model (Fig. 2.27). This leads to simplifications in Fig. 2.18, where ω becomes equal to ω_g.

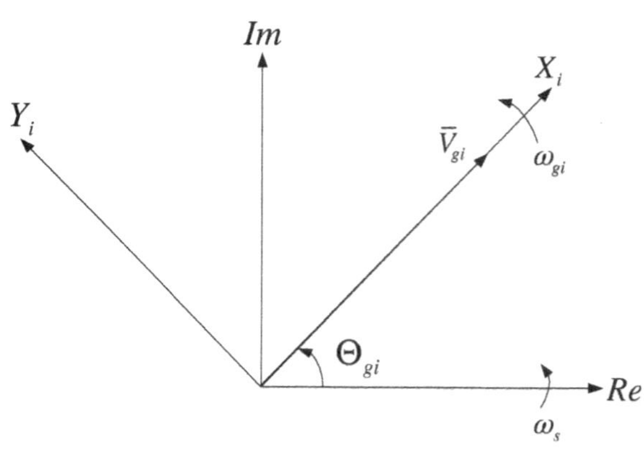

Figure 3.2 The common network reference frame *Re*–*Im* and the reference frame X_i–Y_i for modeling the ith converter station.

6. Interface of converter and AC system: The bus voltages in the AC network are expressed as phasors which differ from the the space phasors used to express the converter terminal quantities, Section 2.4.2.1. The positive sequence fundamental frequency voltage space phasor is different in magnitude from its phasor counterpart, which will be handled using a per-unit system described in Section 3.2.6.1. Additionally, the phase angle of the space phasor is expressed with respect to a stationary frame of reference (Figs 2.16 and 2.18), whereas the phasors are expressed in a synchronously rotating frame (in this case the *Re–Im* frame shown in Fig. 3.2).

3.2.2 Converter Model

The single line diagram of the converter in the positive pole of the *i*th converter station is shown along with the control loops in Fig. 3.3. All the parameters and variables are referred to the converter side of the transformer shown in Fig. 3.1. The parameters of the *i*th converter station are denoted with a subscript *i*, and the positive (negative) pole is denoted by a superscript *p* (*n*). The convention of the direction of current I^n_{dci} and voltage V^n_{dci} is chosen in such a manner that under normal conditions they are equal and opposite in sign as compared to I^p_{dci} and V^p_{dci}, respectively.

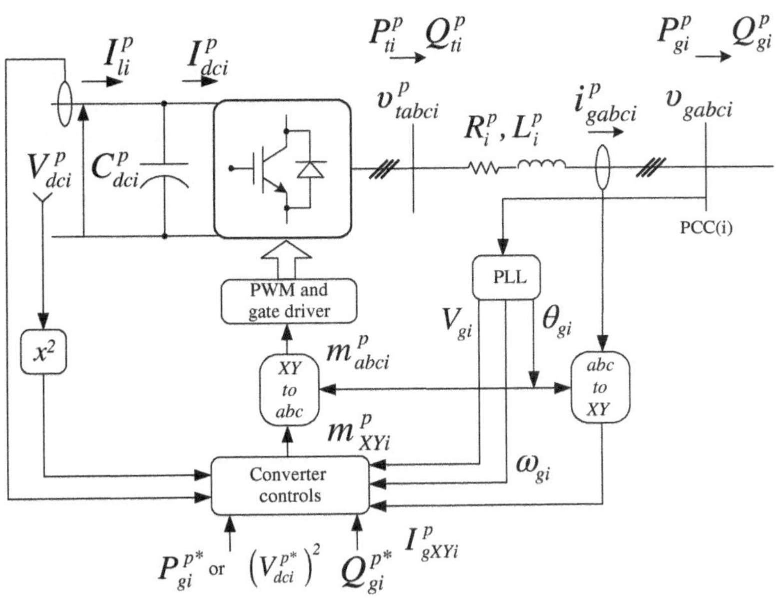

Figure 3.3 Single line diagram of the converter in the positive pole of the *i*th converter station and its controls. ©[36], 2011.

From Fig. 3.3, the KVL equation of the AC side of the converter in *abc*-frame can be written as

$$L_i^p \frac{di_{gabci}^p}{dt} = -R_i^p i_{gabci}^p + v_{tabci}^p - v_{gabci} \tag{3.1}$$

where R_i^p, L_i^p are the aggregated resistance and inductance of the converter transformer and phase reactors. This equation can be expressed in a space phasor (or space vector) form defined in Section 2.4.2.2 as

$$L_i^p \frac{d\overline{I}_{gi}^p}{dt} = -R_i^p \overline{I}_{gi}^p + \overline{V}_{ti}^p - \overline{V}_{gi} \tag{3.2}$$

Let the AC voltage at the point of common coupling (PCC) of the *i*th converter station be expressed in the space-phasor form $\bar{V}_{gi} = V_{gi}e^{j\theta_{gi}}$. A PLL ensures that the X_i-axis of the rotating reference frame X_i–Y_i is locked with the voltage $\overline{V}_{gi}$ (Fig. 3.2) to ensure decoupled control of the active and reactive power, (2.78) and (2.79). Therefore, the AC system variables (say, $\overline{V}_{gi}$) need to be transformed to X_i–Y_i frame. The PLL is modeled algebraically to do this coordinate transformation, which leads to $V_{gXi} = V_{gi}$ and $V_{gYi} = 0$. Figure 3.3 also shows the feedback and measurable disturbance parameters (feed-forwards) used by the controller, which in turn generates the modulation signal m_{XYi}^p. The modulation signal m_{XYi}^p is then translated back to *abc* frame with the help of the reference angle θ_{gi} produced by the PLL to generate the firing pulse for the converters (refer to Fig. 2.28).

As described in Section 2.3.1, the converters are represented by their averaged model (Fig. 2.19). The positive pole converter model of the *i*th converter station (Fig. 3.4) uses the same reference frame as the controllers (i.e., X_i–Y_i frame). Note that the converter could also be modeled in any reference frame irrespective of its controls. The approach followed here is widely used because it reduces the number of times one needs to change the coordinate systems for interfacing variables in different frames.

After considering all above-mentioned modifications and transforming into the X_i–Y_i frame, the dynamics of the AC side of the converters are obtained from (2.67) and (2.68) as

$$L_i^p \frac{dI_{gXi}^p}{dt} = -R_i^p I_{gXi}^p + L_i^p \omega_{gi} I_{gYi}^p + V_{tXi}^p - V_{gi} \tag{3.3}$$

$$L_i^p \frac{dI_{gYi}^p}{dt} = -R_i^p I_{gYi}^p - L_i^p \omega_{gi} I_{gXi}^p + V_{tYi}^p \tag{3.4}$$

The DC bus dynamics at each converter end is modeled by the power balance equation between the AC and the DC sides of the converter station described in (2.95). Assuming the converters are lossless, the real power P_{ti}^p at the converter terminal is the same as that flowing into (or out of) the DC bus. Model of the DC bus is combined with the DC network model, which is elaborated in Section 3.2.4. It is a common practice to run dynamic simulations of power systems from the pre-disturbance steady-state.

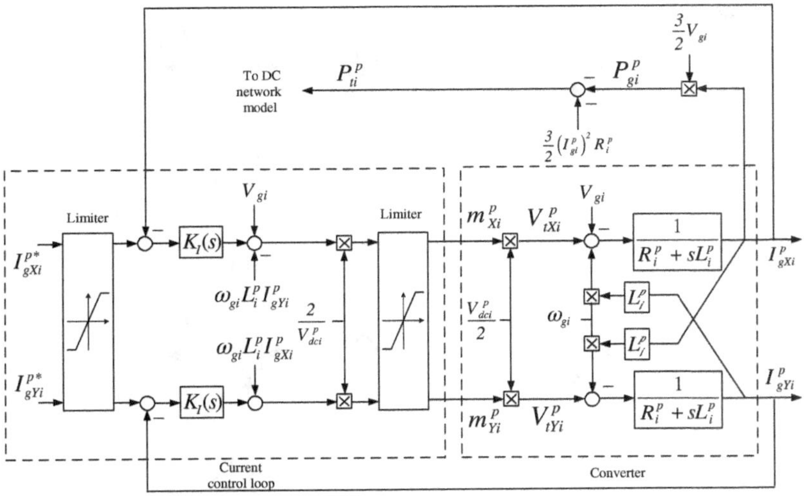

Figure 3.4 Model of the positive pole converter of the *i*th converter station with its inner current control loop. ©[36], 2011.

This necessitates appropriate initialization of all dynamic states of the system, as will be described in Section 3.5.1. This approach avoids the possibility of V_{dc} being zero in (2.95) and, therefore, one can drop ϵ from this equation for simulation studies.

3.2.3 Converter Controller Model

In an asymmetric bipole MTDC grid, there exists a wide-ranging option of control modes. A few options (e.g., P, Q, V_{dc}, V_{ac}, droop control, etc.) were highlighted in Sections 1.5 and 2.4. Not only different converter stations can be under different modes of control, but also the positive and negative poles within the same station can be operating under different modes. It was also mentioned in Section 2.4.1 that the current mode control strategy is invariably preferred over the voltage mode control strategy. The overall control loops in a converter station can be divided into (1) outer control loops (2) inner current control loop, which are described in detail in Chapter 2 and reviewed briefly in the next section.

3.2.3.1 *Outer Control Loops*

Mode I—*P–Q* Control In this control mode, both poles in a converter station control the individual real and the reactive power injected at the PCC. Control of real and reactive power was described in detail in Section 2.4.1. As shown in Fig. 3.5, the current references are derived using the following algebraic relationship obtained from (2.78) and (2.79) with the assumption $C = 0$:

$$I_{gXi}^{p*} = \frac{2P_{gi}^{p*}}{3V_{gi}}, I_{gYi}^{p*} = -\frac{2Q_{gi}^{p*}}{3V_{gi}} \tag{3.5}$$

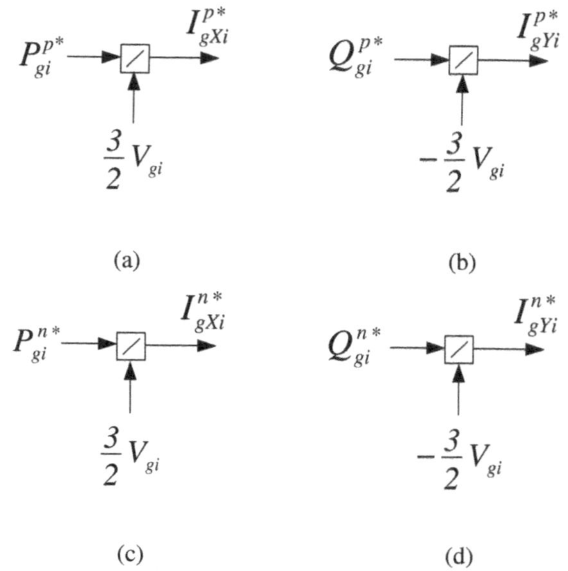

Figure 3.5 (a) and (b) Positive pole of the ith converter station in P–Q control mode, (c) and (d) Negative pole of the ith converter station in P–Q control mode.

$$I_{gXi}^{n*} = \frac{2P_{gi}^{n*}}{3V_{gi}}, I_{gYi}^{n*} = -\frac{2Q_{gi}^{n*}}{3V_{gi}} \tag{3.6}$$

Here, P_{gi}^{p*} and Q_{gi}^{p*} are the real and reactive power references for the positive pole, and P_{gi}^{n*} and Q_{gi}^{n*} are the corresponding reference quantities for the negative pole of the ith converter station.

Mode II—V_{dc}–Q Control In this mode of operation, both poles work in V_{dc}–Q control mode maintaining constant DC bus voltage and supply the desired reactive power at the PCC, as shown in Fig. 3.4(d). Section 2.4.3 described the DC voltage control in detail. V_{dci}^{p*} and V_{dci}^{n*} are the reference values for the positive and the negative pole of the ith converter station. The outer voltage control loop is slower than the inner current control loop and consists of the feed-forward terms to mitigate the measurable disturbances, as shown in Fig. 2.30. A compensator $-K_{Vdc}(s)$ drives the error in the DC link voltage to zero and generates the X-axis current reference for each pole, as shown in Fig. 3.6. The design methodology of such controllers was also described in Section 2.4.3.

Mode III A—P–V_g–Q Control Instead of controlling the reactive power injected by individual poles, the voltage at the PCC bus might be controlled. However, if both poles try to control the voltage, the controller will fight against each other. Therefore, a droop control can be employed to to avoid this issue, as described in

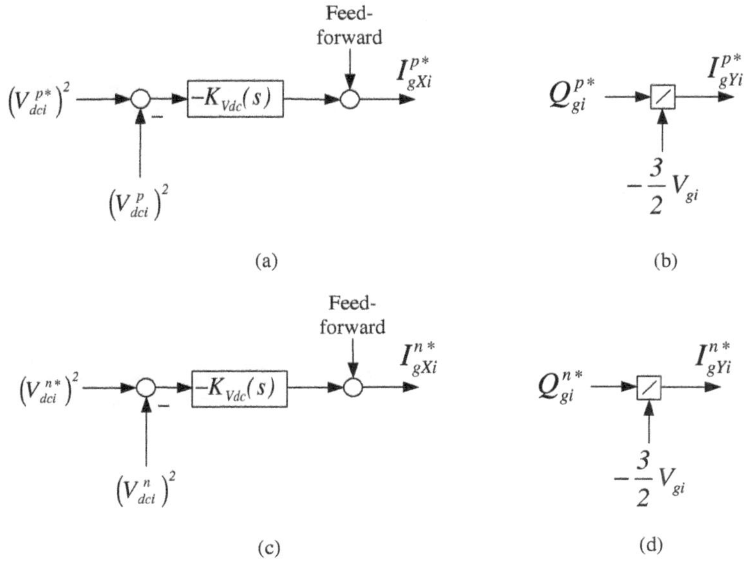

Figure 3.6 (a) and (b) Positive pole of the *i*th converter station in V_{dc}–Q control mode, (c) and (d) Negative pole of the *i*th converter station in V_{dc}–Q control mode.

Section 2.4.5. In case a flat PCC voltage is to be maintained, one pole can control the voltage and the other can control the reactive power distribution between the poles. As shown in Fig. 3.7(b), the positive pole converter is maintaining the voltage at the PCC constant at the reference V_{gi}^*. The outer loop compensator $-K_{Vac}(s)$ generates the Y component of the current reference for the inner current control loop. Section 2.4.4 elaborated the design of the AC voltage controller. If the objective is to equally distribute the reactive power amongst the poles, then the difference between reactive powers injected by the positive (Q_{gi}^p) and the negative (Q_{gi}^n) pole is driven toward zero by a compensator (Fig. 3.7(d)). The output of the compensator generates the reference reactive power Q_{gi}^{n*} for the negative pole converter.

Mode III B—V_{dc}–V_g–Q Control In this control mode, one converter pole controls the DC link voltage and the PCC voltage and the other pole controls the distribution of the reactive power injection in addition to the DC link voltage regulation. Figure 3.8 shows the outer loop control for this mode of operation.

Mode IV—V_g–ω_g Control The details of this control mode was described in Section 2.4.6.

For interconnecting offshore wind farms to the MTDC grids, control modes III B and IV can be used. Because of the remote location of the wind farms, it is not easy to phase-lock with respect to the PCC voltage. The converter station acts as

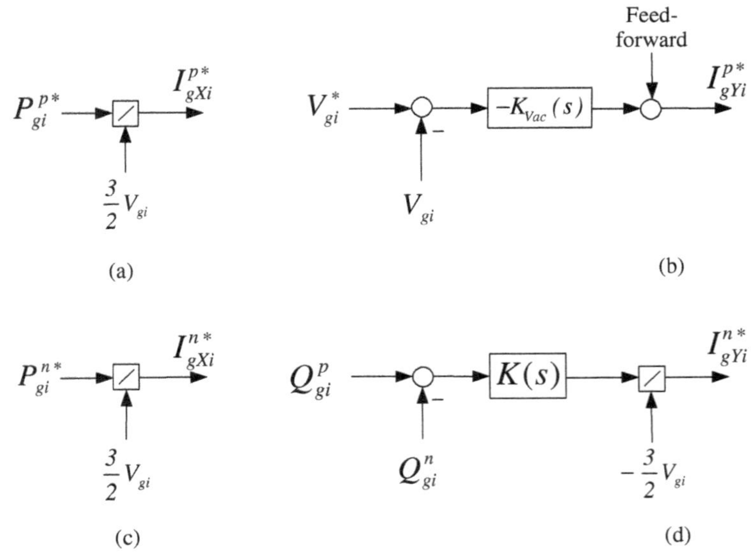

Figure 3.7 (a) and (b) Positive pole of the ith converter station in P–V_g control mode, (c) and (d) Negative pole of the ith converter station in P–Q control mode.

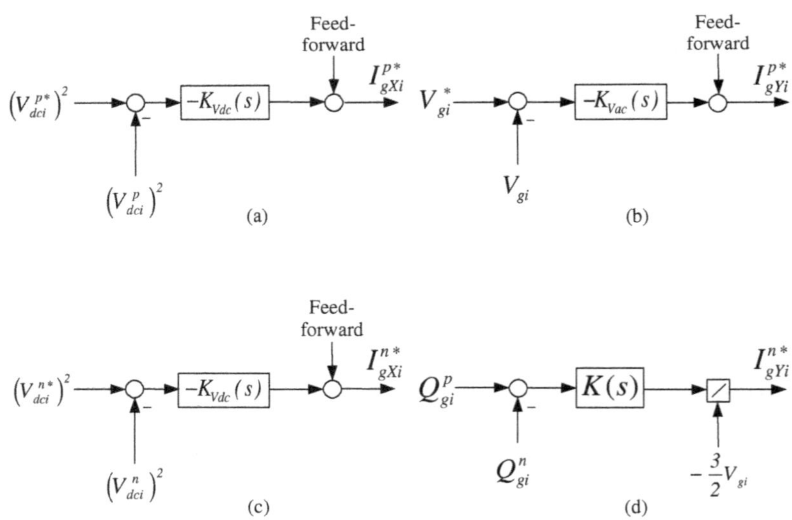

Figure 3.8 (a) and (b) Positive pole of the ith converter station in V_{dc}–V_g control mode, (c) and (d) Negative pole of the ith converter station in V_{dc}–Q control mode.

a slack bus absorbing the power generated by the wind farm. This control mode can also be used for feeding an islanded load center or used for "Black Start" and network restoration.

Other Control Modes It is possible to operate the converter poles in various other control modes. For example, the ground current can be regulated on the basis of the requirement. Different types of droop controls such as P–V_{dc} droop control, frequency droop control, etc. could be important depending on the system requirements. Section 2.4.5 mentioned this in brief, and Chapters 4 and 5 will further elaborate on modeling and analysis of the MTDC grid with droop controls.

3.2.3.2 Inner Current Control Loop All the converter stations operate with the decoupled current control strategy [37] described in Section 2.4.2.4. Figure 3.4 shows the current control loop, where I^p_{gXi}, I^p_{gYi}, V_{gi}, V^p_{dci} and ω_{gi} are measurable quantities, which can be used as feed-forward signals with appropriate signs to achieve decoupling between the X- and Y-axis current control loops. As described in Chapter 2, a moderate bandwidth (BW) is adequate in tracking I^{p*}_{gXi} and I^{p*}_{gYi} because they are DC in nature under steady-state. The reference values of the X- and Y-axis components of currents are determined by the outer loop controls, which in turn determine the mode of operation for each pole within a converter station. Section 2.4.2.4 introduced the limits on the reference values of the X and the Y components of current. Throughout the rest of the book, the limits on the current and modulation indices are imposed such that the magnitude of the phasor is limited while the phase angle is retained.

3.2.4 DC Network Model

The DC cables or overhead lines in the MTDC grid can be connected in several possible configurations. Therefore, it is challenging to develop a generic model of the DC network which is independent of the connectivity. The model of each line in the DC network is dependent on the following factors:

1. The length of the line
2. The range of the frequency of interest
3. Complexity vs accuracy tradeoff.

Fundamentally, parameters such as resistance, inductance, and capacitance are distributed over the length of the cable/line, which are expected to be of a few hundred kilometers in length. For example, the typical maximum subsea distances around the planned Pan-European subsea Supergrid or the offshore installations in the Atlantic would be more than 300 km. The most accurate model of such transmission lines takes into account the distributed parameters and are used in the EMTP type simulations. There are an infinite number of states associated with a distributed model of the cable and therefore it is not suitable for state-space representation and small-signal stability analysis. Depending upon the frequency range of interest, a cascaded pi-model of the cable network can be used. As the number of pi sections increases, so does the range of frequency of accurate representation. The maximum frequency up to which the model is accurate can be approximated by the following relation:

$$f_{\max} = \frac{Nv}{8l_e} \tag{3.7}$$

where
N: number of cascaded pi sections
v: propagation speed (km/s)
l_e: total length of the line (km).

A cascaded pi section approximation of a transmission line in the asymmetric bipolar configuration is shown in Fig. 3.1. As an example, Fig. 3.9 shows the comparison of the impedance spectra of the distributed model vis-a-vis the cascaded pi-section model for a 500-km-long cable with its parameters mentioned below:

- $r = 0.015$ Ω/km
- $l = 0.003$ H/km
- $c = 27.5$ μF/km
- $l_e = 500.0$ km.

Obviously, more pi sections will result in a better approximation at the cost of complexity and computational burden. As shown in Fig. 3.9, a close approximation upto the AC system frequency (60 Hz) can be obtained with four cascaded pi sections. A DC network model should therefore be made scalable in terms of the number of pi sections so that this can be changed according to the length of the cable.

Figure 3.1 shows the interconnection between the ith and the jth converter stations through the positive, negative, and the metallic return cable. Each converter station can be connected to multiple stations through the positive pole, negative pole, and metallic return cables or transmission lines. Note that in absence of either the positive/negative pole or one of the converters in a converter station, the metallic return network will carry a nonzero current. Such a scenario is termed henceforth as *unbalance (on the DC side)*.

A generic MTDC transmission network should be generic and modular enough to handle any

- configuration of bipolar MTDC grid converter stations
- grounding options along with metallic return network
- number of pi-section approximation of DC cable model
- type and location of fault on the DC side
- cable outage or converter outage leading to unbalanced (on DC side) operation.

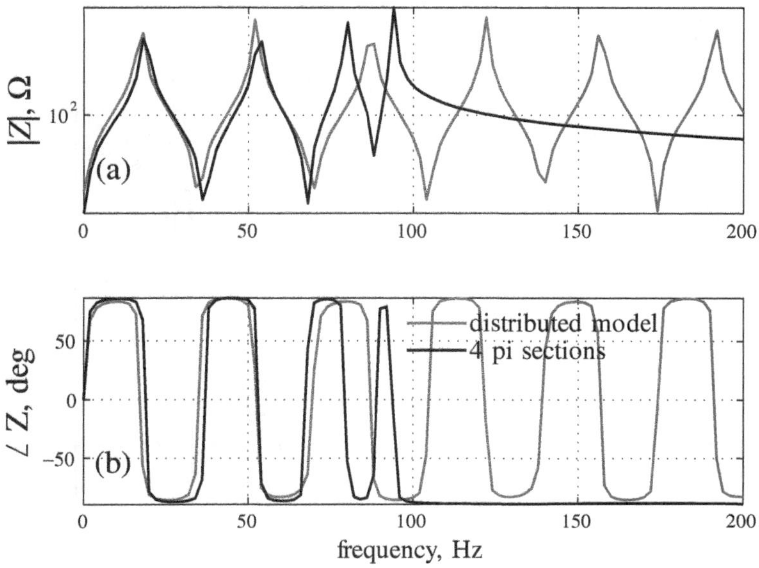

Figure 3.9 Comparison of impedance spectra of a 500-km-long cable with distributed model and cascaded pi-section model: (a) impedance magnitude (b) impedance phase angle. ©[36], 2011.

The station grounding options vary depending on the allowable ground current in a certain region. Especially for subsea converter stations, the grounding options could be limited because of maritime regulations.

The mathematical model of the DC network in Fig. 3.1 can be described by the following set of algebraic and differential equations which are shown in a block diagram in Fig. 3.10. The converter DC bus dynamics is also included in this model. All voltages are measured with respect to ground.

3.2.4.1 *Algebraic Equations* For the positive pole network shown in Fig. 3.1, the general form of current balance at any node is given by

$$I^p_{dc} - I^p_{line} + I^p_s + I^p_c + I_f = 0 \tag{3.8}$$

where $I^p_{dc} = \frac{P^p_t}{V^p_{dc}}$ is the current entering the DC side of the converter. Therefore, the DC side is modeled by a dependent current source, as described in Section 2.5. As shown in Fig. 3.4, the DC side power (P^p_t) of the converters is derived from the PCC real power after accounting for the losses. Depending on the connectivity and the location of the fault, some of the currents in (3.8) would be zeros (I_f is nonzero if the fault occurs at that node). For example, at the converter DC bus node all currents are nonzero. Any node other than that of a converter station would have zero I^p_{dc} and I^p_s.

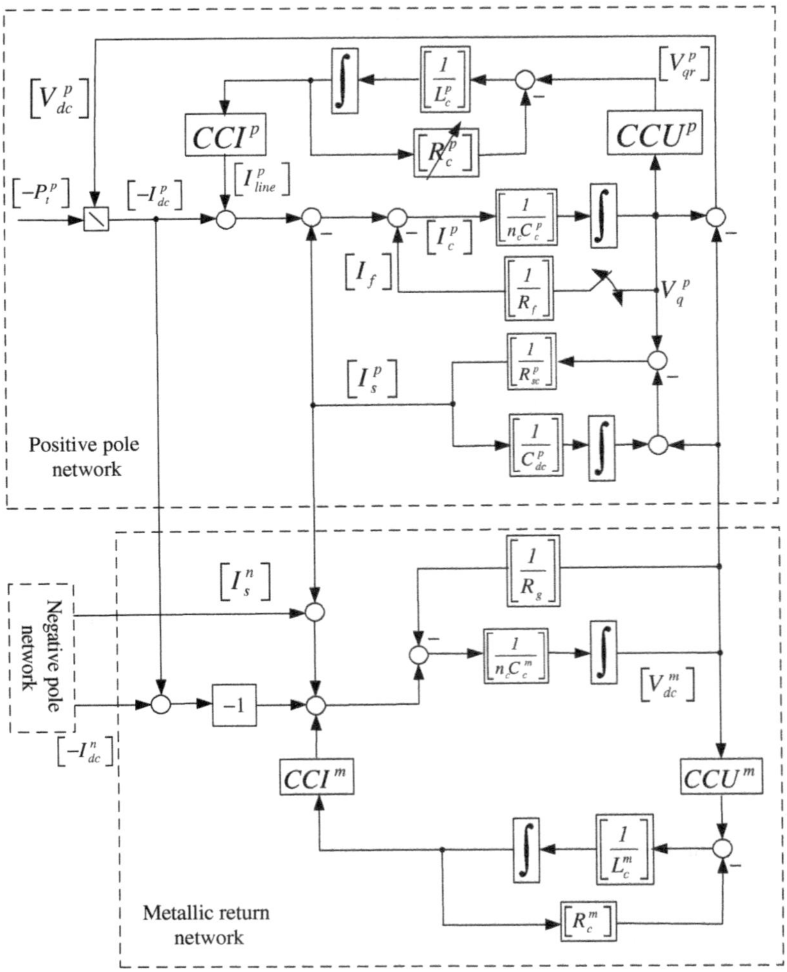

Figure 3.10 Block diagram of the DC network dynamic model including the converter DC bus. ©[36], 2011.

If the maximum possible number of nodes and the series *R–L* branches are N_{nodeP} and N_{brP}, respectively, then

$$\left[I^p_{line}\right] = CCI^p \times \left[I^p_{qr}\right] \tag{3.9}$$

where $\left[I^p_{qr}\right]$ is the vector of *R–L* branch currents flowing from any general node q^p to r^p, $\left[I^p_{line}\right]$ is the total node current injection vector, shown in Fig. 3.1, and CCI^p is the incidence matrix of dimension $[N_{nodeP} \times N_{brP}]$ representing on the connectivity of different branches. The entries of CCI^p are $CCI^p(i, j) = 1$ (-1), if the *i*th cable enters (leaves) the *j*th node and $CCI^p(i, j) = 0$, otherwise.

Similarly, the connectivity between the nodes is determined by the matrix CCU^p of dimension $[N_{brP} \times N_{nodeP}]$. It is assumed that the current in a branch will flow from the ith to the jth node if $i < j$. The elements of CCU^p are determined as $CCU^p(i,j) = 1$, if $i < j$, $CCU^p(i,j) = -1$, if $i > j$, all other entries are zeros where the ith and the jth node are not connected. Thus CCU^p relates the node voltages with the voltage drop across the series R–L branches by the equation

$$\left[V_{qr}^p\right] = CCU^p \times \left[V_q^p\right] \tag{3.10}$$

where $\left[V_q^p\right]$ is the node voltage vector and $\left[V_{qr}^p\right]$ is the vector of series branch voltage drops. Note that $CCI^p = -\left[CCU^p\right]^T$.

Similar set of equations are applicable for the negative pole network.

For the metallic return network, Fig. 3.1, equation (3.8) can be modified as

$$I_{dc}^p + I_{dc}^n + I_{line}^m + I_s^p + I_s^n - I_c^m - \frac{V_{dc}^m}{R_g} = 0 \tag{3.11}$$

All other above-mentioned equations are applicable for the metallic return network.

3.2.4.2 Differential Equations For the positive pole network, the converter station capacitor dynamics is given by

$$V_{dc}^p = \frac{1}{C_{dc}^p} \int I_s^p dt + I_s^p R_{sc}^p \tag{3.12}$$

where R_{sc}^p is the effective series resistance (ESR) (not shown explicitly in Fig. 3.1). For any cable node, q^p we can write

$$\frac{1}{n_c C_c^p} \int I_{cq}^p dt = V_q^p \tag{3.13}$$

Here, n_c depends on the node location, e.g., at any cable node other than the converter node, $n_c = 2$. At a converter node $n_c = 1$. Note that the parameter C_c^p is the capacitance at each end of a pi section in the cable model. Therefore, it is half the total capacitance in each pi section. The series R–L branch dynamics can be described by

$$L_c^p \frac{dI_{qr}^p}{dt} + R_c^p I_{qr}^p = V_q^p - V_r^p = V_{qr}^p \tag{3.14}$$

The negative pole and metallic return network dynamics can also be represented by similar equations.

3.2.5 State-Space Representation

It is convenient to express the dynamic behavior of a power system using a set of first-order nonlinear differential and algebraic equations (DAEs) in the following form:

$$\dot{x} = \varsigma(x, z, u) \tag{3.15}$$

$$0 = g(x, z, u) \tag{3.16}$$

$$y = h(x, z, u) \tag{3.17}$$

where ς and g are vectors of differential and algebraic equations. This state-space representation forms the basis of the small-signal stability analysis as described later in Section 3.6. The notations $x \in R^n, z \in R^m, u \in R^p$, and $y \in R^q$ denote the vectors of state variables, algebraic variables, inputs, and outputs, respectively. The algebraic variables are system voltage magnitudes and angles. The inputs are normally the reference values such as real or reactive power reference at a converter station or speed and voltage reference at a generating unit. The outputs can be DC bus voltage, generator power output, bus frequency, bus voltage, line power flow or current, etc.

In this book, the vectors for differential, algebraic, and output equations are expressed with ς, g, and h, respectively, for the positive pole converter, and all poles of the DC network. For the negative pole converter, they are denoted by ς', g', and h', respectively.

3.2.5.1 Dynamic Equations of Converters and Controllers The dynamic equations of the converters and the controllers are written in the form $\dot{x} = \varsigma(x, z, u)$ as described below:

The differential equations describing the AC side converter dynamics (3.3) can be rearranged as

$$\dot{I}^p_{gXi} = -\frac{R^p_i}{L^p_i} I^p_{gXi} + \omega_{gi} I_Y + \frac{V^p_{tXi}}{L^p_i} - \frac{V_{gi}}{L^p_i} \tag{3.18}$$

$$\dot{I}^p_{gYi} = -\frac{R^p_i}{L^p_i} I^p_{gYi} - \omega_{gi} I_X + \frac{V^p_{tYi}}{L^p_i} \tag{3.19}$$

For the inner current control loop, let us consider $K_I(s) = K_{pI} + \frac{K_{iI}}{s}$. Assuming that no limit violation has occurred, the following equations can be written for the decoupled current controller:

$$V^p_{tXi} = K_{pI}\left(I^{p*}_{gXi} - I^p_{gXi}\right) + K_{iI}\int\left(I^{p*}_{gXi} - I^p_{gXi}\right)dt + V_{gi} - \omega_{gi} L^p_i I^p_{gYi} \tag{3.20}$$

$$V^p_{tYi} = K_{pI}\left(I^{p*}_{gYi} - I^p_{gYi}\right) + K_{iI}\int\left(I^{p*}_{gYi} - I^p_{gYi}\right)dt + \omega_{gi} L^p_i I^p_{gXi} \tag{3.21}$$

Let us define the state variables for the current controller as

$$\dot{\kappa}_{IX} = I^{p*}_{gXi} - I^p_{gXi} \tag{3.22}$$

$$\dot{\kappa}_{IY} = I^{p*}_{gYi} - I^p_{gYi} \tag{3.23}$$

Replacing the values of V^p_{tXi} and V^p_{tYi} in (3.18) and (3.19) we get

$$\dot{I}^p_{gXi} = -\left(\frac{R^p_i + K_{pI}}{L^p_i}\right) I^p_{gXi} + \frac{K_{iI}}{L^p_i}\kappa_{IX} + \frac{K_{pI}}{L^p_i} I^{p*}_{gXi} \tag{3.24}$$

$$\dot{I}^p_{gYi} = -\left(\frac{R^p_i + K_{pI}}{L^p_i}\right) I^p_{gYi} + \frac{K_{iI}}{L^p_i}\kappa_{IY} + \frac{K_{pI}}{L^p_i} I^{p*}_{gYi} \tag{3.25}$$

Similar equations can be written for the negative pole DC network.

3.2.5.2 *Output Equations* The output equations of the converters and controllers are written in the form $y = h(x, z, u)$ as described below:
It was mentioned in Sections 2.5 and 3.2.2 that the real power balance between the converter terminal and the DC network forms the interface between these models. Therefore, the real power at the converter terminal is expressed in the form of an output equation:

$$P_{ti}^p = \frac{3}{2}\left(V_{tXi}^p I_{gXi}^p + V_{tYi}^p I_{gYi}^p\right) \tag{3.26}$$

Replacing the values of V_{tXi}^p and V_{tYi}^p in (3.26) from the converter controller equations, we get

$$\begin{aligned} P_{ti}^p = &\frac{3}{2}\left[K_{pI} I_{gXi}^{p*} I_{gXi}^p - K_{pI}\left(I_{gXi}^p\right)^2 + \kappa_{IX} K_{iI} I_{gXi}^p + V_{gi} I_{gXi}^p\right] \\ &+\frac{3}{2}\left[K_{pI} I_{gYi}^{p*} I_{gYi}^p - K_{pI}\left(I_{gYi}^p\right)^2 + \kappa_{IY} K_{iI} I_{gYi}^p\right] \end{aligned} \tag{3.27}$$

The current injected by the converter stations are also expressed in this form:

$$I_{gXi}^p = I_{gXi}^p, I_{gYi}^p = I_{gYi}^p \tag{3.28}$$

It can be seen that the state variables in the positive pole of the ith converter station and its controllers are $x = \left[I_{gXi}^p I_{gYi}^p \kappa_{IX} \kappa_{IY}\right]^T$. Depending on the control modes, the outer loop may introduce new state variables. The algebraic variable is $z = V_{gi}$ and the output variables are $y = \left[P_{ti}^p I_{gXi}^p I_{gYi}^p\right]^T$.

3.2.5.3 *Control Modes* The state-space representation for different control modes is summarized in this section.

Mode I—*P–Q* Control (Fig. 3.5) Equations (3.22–3.25) can be modified as

$$\varsigma_1 = \dot{I}_{gXi}^p = -\left(\frac{R_i^p + K_{pI}}{L_i^p}\right) I_{gXi}^p + \frac{K_{iI}}{L_i^p}\kappa_{IX} + \frac{2}{3}\frac{K_{pI}}{L_i^p}\frac{P_{gi}^{p*}}{V_{gi}} \tag{3.29}$$

$$\varsigma_2 = \dot{I}_{gYi}^p = -\left(\frac{R_i^p + K_{pI}}{L_i^p}\right) I_{gYi}^p + \frac{K_{iI}}{L_i^p}\kappa_{IY} - \frac{2}{3}\frac{K_{pI}}{L_i^p}\frac{Q_{gi}^{p*}}{V_{gi}} \tag{3.30}$$

$$\varsigma_3 = \dot{\kappa}_{IX} = \frac{2}{3}\frac{P_{gi}^{p*}}{V_{gi}} - I_{gXi}^p \tag{3.31}$$

$$\varsigma_4 = \dot{\kappa}_{IY} = -\frac{2}{3}\frac{Q_{gi}^{p*}}{V_{gi}} - I_{gYi}^p \tag{3.32}$$

The output equation (3.27) can be modified as

$$\begin{aligned} h_1 = P_{ti}^p = \frac{3}{2}\left[\frac{2}{3}K_{pI}I_{gXi}^p\frac{P_{gi}^{p*}}{V_{gi}} - K_{pI}\left(I_{gXi}^p\right)^2 + \kappa_{IX}K_{iI}I_{gXi}^p + V_{gi}I_{gXi}^p\right] \\ +\frac{3}{2}\left[-\frac{2}{3}K_{pI}I_{gYi}^p\frac{Q_{gi}^{p*}}{V_{gi}} - K_{pI}\left(I_{gYi}^p\right)^2 + \kappa_{IY}K_{iI}I_{gYi}^p\right] \end{aligned} \tag{3.33}$$

The other output equations are

$$h_2 = I_{gXi}^p = I_{gXi}^p, h_3 = I_{gYi}^p = I_{gYi}^p \tag{3.34}$$

These output equations are independent of the control modes.

Mode II—V_{dc}–Q Control (Fig. 3.6) Let us assume that the DC link voltage compensator $K_{Vdc}(s) = -K_{pdc} - \frac{K_{idc}}{s}$. Neglecting the feed-forward terms, the X component of reference current is

$$I_{gXi}^{p*} = -K_{pdc}\left\{\left(V_{dci}^{p*}\right)^2 - \left(V_{dci}^p\right)^2\right\} - K_{idc}\int\left\{\left(V_{dci}^{p*}\right)^2 - \left(V_{dci}^p\right)^2\right\}dt \tag{3.35}$$

Let us define the state variable κ_{dc} of the DC link voltage controller as

$$\varsigma_{10} = \dot{\kappa}_{dc} = \left(V_{dci}^{p*}\right)^2 - \left(V_{dci}^p\right)^2 \tag{3.36}$$

Equation (3.29) can be modified as

$$\begin{aligned} \varsigma_1 = \dot{I}_{gXi}^p = -\left(\frac{R_i^p + K_{pI}}{L_i^p}\right)I_{gXi}^p + \left(\frac{K_{iI}}{L_i^p}\right)\kappa_{IX} \\ -\left(\frac{K_{pI}K_{pdc}}{L_i^p}\right)\left\{\left(V_{dci}^{p*}\right)^2 - \left(V_{dci}^p\right)^2\right\} - \left(\frac{K_{pI}K_{idc}}{L_i^p}\right)\kappa_{dc} \end{aligned} \tag{3.37}$$

Equation (3.31) can be modified as

$$\varsigma_3 = \dot{\kappa}_{IX} = -K_{pdc}\left\{\left(V_{dci}^{p*}\right)^2 - \left(V_{dci}^p\right)^2\right\} - K_{idc}\kappa_{dc} - I_{Xi}^p \tag{3.38}$$

Equations for the Y-axis current remains the same as Mode I. The output equation (3.33) is modified by replacing I_{Xi}^{p*} from (3.35).

Mode III A—P–V_g–Q Control (Fig. 3.7) Let us assume that the PCC voltage compensator $-K_{Vac}(s) = K_{pvg} + \frac{K_{ivg}}{s}$. The positive pole Y component of reference current (neglecting the feed-forward terms) is

$$I_{gYi}^{p*} = K_{pvg}\left\{V_{gi}^* - V_{gi}\right\} + K_{ivg}\kappa_{vg} \tag{3.39}$$

where the controller state variable is defined such that

$$\varsigma_{10} = \dot{\kappa}_{vg} = V_{gi}^* - V_{gi} \tag{3.40}$$

Equation for ς_1 and ς_3 remains the same as in *P–Q* control mode, whereas ς_2 is defined as

$$\begin{aligned}\varsigma_2 = \dot{I}^p_{gYi} &= -\left(\frac{R^p_i+K_{pI}}{L^p_i}\right) I^p_{Yi} + \frac{K_{iI}\kappa_{IY}}{L^p_i} \\ &+\frac{K_{pI}}{L^p_i}\left[K_{pvg}\left\{V^*_{gi} - V_{gi}\right\} + K_{ivg}\kappa_{vg}\right]\end{aligned} \tag{3.41}$$

Expression for ς_4 is modified to

$$\varsigma_4 = \dot{\kappa}_{IY} = K_{pvg}\left\{V^*_{gi} - V_{gi}\right\} + K_{ivg}\kappa_{vg} - I^p_{gYi} \tag{3.42}$$

Let us assume that the reactive power distribution controller $K(s) = K_{pq} + \frac{K_{iq}}{s}$. The state equations for this control through the negative pole converter can be written as

$$\begin{aligned}\varsigma'_2 = \dot{I}^n_{gYi} &= -\left(\frac{R^n_i+K_{pI}}{L^n_i}\right) I^n_{Yi} + \frac{K_{iI}\kappa_{IY}}{L^n_i} \\ &-\frac{2K_{pI}}{3L^n_i}\left[K_{pq}\left\{-\frac{3}{2}I^p_{gYi} + \frac{3}{2}I^n_{gYi}\right\} + \frac{K_{iq}\kappa_q}{V_{gi}}\right]\end{aligned} \tag{3.43}$$

$$\varsigma'_4 = \dot{\kappa}'_{IY} = -\frac{2}{3}\left[K_{pq}\left\{-\frac{3}{2}I^p_{gYi} + \frac{3}{2}I^n_{gYi}\right\} + \frac{K_{iq}\kappa_q}{V_{gi}}\right] - I^n_{gYi} \tag{3.44}$$

$$\varsigma'_{10} = \dot{\kappa}_q = \frac{3}{2}V_{gi}\left(-I^p_{gYi} + I^n_{gYi}\right) \tag{3.45}$$

Expression for the output equation is obtained by replacing the values of I^{p*}_{gYi} and I^{n*}_{gYi} in (3.27).

Mode III B—V_{dc}–V_g–Q Control and Mode IV—V_g–ω_g Control The state-space model can be derived following a similar approach as used in Mode I, II, and III A.

The state-space model of the positive pole of the *i*th converter station is described here (except Mode III A, where part of the negative pole equations were also described). Similar equations can be written for all converter stations in the positive and negative poles and combined together to obtain the state-space model of all converters and the associated controllers.

3.2.5.4 Dynamic Equations of DC Network The dynamic equations of the DC network are written in the form $\dot{x} = \varsigma(x, z, u)$ as described below.

Let us analyze the state-space model at the *q*th node of the DC network assuming it is connected to the *i*th converter station. Assuming μ^p_i is the voltage drop across the DC bus capacitor of the positive pole converter station, we can express the DC bus dynamics in the following form:

$$\varsigma_5 = \dot{\mu}^p_i = \frac{V^p_q - V^m_q - \mu^p_i}{R^p_{sci}C^p_{dci}} \tag{3.46}$$

For the positive pole network, the dynamic equation of the series branch of a pi section can be rearranged in a vector-matrix form as

$$\varsigma_6 = \dot{I}^p_{qr} = \frac{V^p_q - V^p_r - I^p_{qr}R^p_c}{L^p_c} \tag{3.47}$$

The nodal equation of the positive pole network can be written as

$$\varsigma_7 = \dot{V}^p_q = \frac{1}{C^p_c}\left[-\frac{P^p_{ti}}{\left(V^p_q - V^m_q\right)} - \frac{V^p_q - V^m_q - \mu^p_i}{R^p_{sci}} + \left[CCI^p\right]\left[I^p_{qr}\right]\right] \tag{3.48}$$

Similar equations can be written for the negative pole network. The following branch and nodal equations can be written for the metallic return network:

$$\varsigma_8 = \dot{I}^m_{qr} = \frac{V^m_q - V^p_r - I^m_{qr}R^m_c}{L^m_c} \tag{3.49}$$

$$\begin{aligned}\varsigma_9 = \dot{V}^m_q = \frac{1}{C^m_c}\left[\frac{P^p_{ti}}{\left(V^p_q - V^m_q\right)} + \frac{P^n_{ti}}{\left(V^n_q - V^m_q\right)} + \frac{V^p_q - V^m_q - \mu^p_i}{R^p_{sci}}\right] \\ + \frac{1}{C^m_c}\left[\frac{V^n_q - V^m_q - \mu^n_i}{R^n_{sci}} - \frac{V^m_q}{R_{gi}} + \left[CCI^m\right]\left[I^m_{qr}\right]\right]\end{aligned} \tag{3.50}$$

Similar equations can be written for other nodes in the DC network.

3.2.5.5 *Output Equations of DC Network*

The DC bus voltage at the converter station is used as a feed-forward signal for the inner current control loop and a feedback signal in the outer DC voltage control loop as described in Section 2.4.3. Therefore, this voltage is expressed as output in the form $y = h(x, z, u)$. For the positive pole network

$$h_4 = V^p_{dci} = V^p_q - V^m_q \tag{3.51}$$

Similar equations can be written for the negative pole network. The state variables for the qth node of the DC network connected to the DC bus of the ith converter station, are: $x = \left[\mu^p_i \mu^n_i I^p_{qr} V^p_q I^m_{qr} V^m_q I^n_{qr} V^n_q\right]^T$. For the DC network connected to the ith converter station, the algebraic variables are $z = \left[P^p_{ti} P^n_{ti}\right]^T$ and the output variable is $y = \left[V^p_{dci} V^n_{dci}\right]^T$.

3.2.6 Phasor from Space Phasor

As discussed in Section 3.2.1, the AC network voltages and currents are expressed as phasors while the converters and their controllers are modeled using the space phasors. The magnitude of the voltage phasor in a balanced three-phase system is usually the phase-to-ground rms voltage, whereas the space phasor magnitude is the peak value of the instantaneous phase-to-ground voltage. Moreover, the phase angle of the phasor is determined in a synchronously rotating reference frame *Re–Im*. To interface the space-phasor-based converter model with the phasor-based AC system model, a two step process is considered:

1. Consider only the positive sequence fundamental frequency component of the dependent voltage source representing the AC side of the converter, as mentioned in Section 3.2.1, and scale the space phasors to account for the magnitude difference through the use of a p.u. system.

2. Express the angle of the space phasors with respect to the rotating reference frame *Re–Im*.

3.2.6.1 *Base Values and Per-unit Systems* A common MVA base S_B is used for the entire AC–MTDC grid.

Per-unit System for MTDC Grid The MTDC grid model can be expressed in per-unit (p.u.) quantities. Different voltage bases V_B are used for different sections of the DC grid. The voltage base in the AC side is defined as V_B = peak value of the rated phase-to-ground voltage (kV). The AC side base current (kA) is defined as $I_B = \frac{2S_B}{3V_B}$. The base frequency is defined as $\omega_B = \omega_s$. Similarly, the base impedance and inductance on the AC side are $Z_B = \frac{V_B}{I_B}, L_B = \frac{Z_B}{\omega_B}$.

The base voltage in the DC side is defined as V_{Bdc} = rated pole-to-ground DC voltage (kV). The base DC side current at the corresponding node is given by $I_{Bdc} = \frac{S_B}{2V_{Bdc}}$.

This results in the AC-side real and reactive power in p.u. as $P_{gpu} = V_{gpu} I_{Xpu}$ and $Q_{gpu} = -V_{gpu} I_{Ypu}$. Here, V_{gpu}, I_{Xpu}, and I_{Ypu} are the PCC voltage, X-axis component of current and Y-axis component of current in p.u., respectively. In the rest of this book, it is assumed that the MTDC grid parameters and variables are expressed in p.u. unless otherwise mentioned. For the sake of simplicity, the subscript "pu" is omitted from the notations.

Per-unit System for AC Grid The usual way of defining the p.u. system in the AC grid is followed, which can be obtained from any standard book on power systems analysis.

3.2.6.2 *Phase Angle of Space Phasors* As shown in Fig. 3.2, the phase angle of the space phasor $\overline{V}_{gi}$ with respect to the rotating frame *Re–Im* is denoted by Θ_{gi} as opposed to θ_{gi} with respect to a stationary frame (Fig. 2.18). For the rest of the book, the phasors are also denoted using an "over-bar." As an example, in the p.u. system mentioned above, the voltage space phasor $\bar{V}_{gi} = V_{gi} e^{j\theta_{gi}}$ is expressed in equivalent phasor notation $\bar{V}_{gi} = V_{gi} e^{j\Theta_{gi}}$.

The ith converter station with all quantities expressed in p.u. phasors are shown in Fig. 3.11.

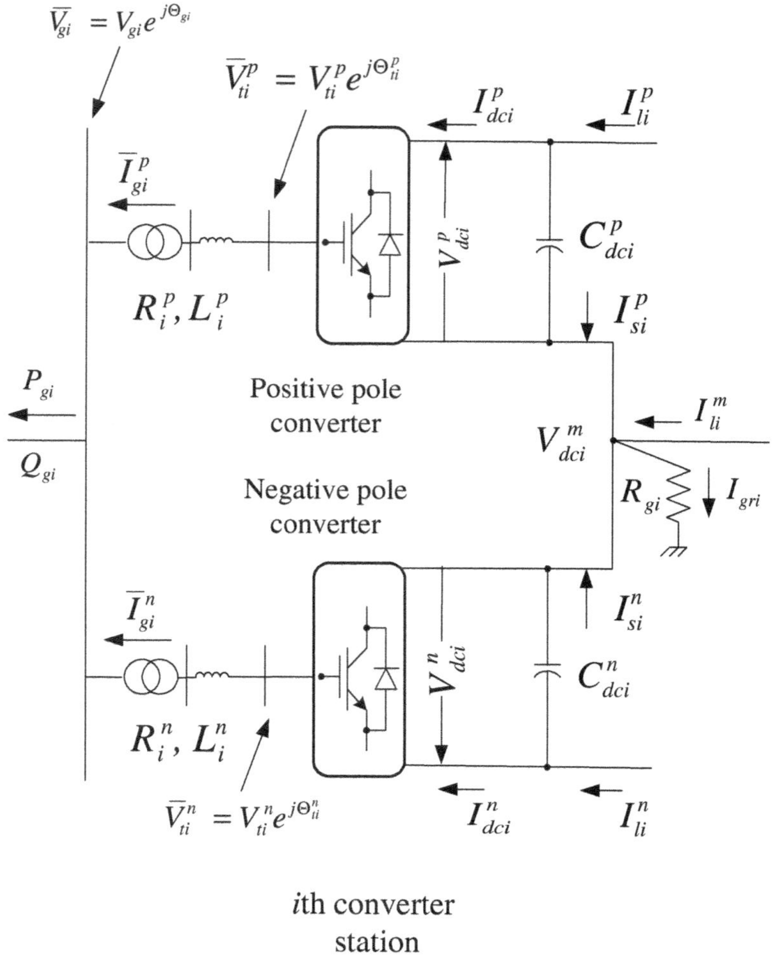

Figure 3.11 The *i*th converter station of the MTDC grid.

3.3 AC Grid Model

In the previous section, modeling of an MTDC grid was described considering voltage source representation of AC grids. In this section, modeling of an AC grid comprising generators, exciters, loads, etc. is presented. This would lead to a combined AC–MTDC model which is essential for planning and operation of future power systems. Although the dynamic modeling of an AC grid is well understood, we will very briefly review this topic for completeness. It should help the readers obtain a comprehensive picture of the integrated AC–MTDC system model.

In power systems, the primary sources of electrical energy are the synchronous generators. The problem of power system stability is primarily to keep the interconnected synchronous machines in synchronism [12]. Their stability is also dependent on several other components such as the speed governors, excitation systems of generators, loads, FACTS devices, etc. Therefore, accurate understanding of the characteristics and modeling of these components are of fundamental importance for stability studies and control design. The general approach to modeling of several power system components is quite standard. A quick overview of these models is given in this chapter. For further details, the readers are referred to [12, 38].

3.3.1 Generator Model

Accurate modeling of the synchronous generators and their excitation systems is of paramount importance for studying the dynamic behavior of AC power systems. Although today's power system is largely dominated by synchronous generators harnessing power from conventional sources, recently more and more power is being generated from renewable resources such as wind and solar. Most of the modern wind turbines are dominated by either the doubly fed induction generator (DFIG) or the full converter (FC) technology. These generators are controlled by power electronic converters. Solar thermal generation employs conventional generators, while the photovoltaic (PV) technology uses inverters to directly interconnect the energy resources to the AC grid.

In this chapter, modeling of synchronous generators, their excitation system, turbines, and governors is covered very briefly. The IEEE recommended practice regarding d–q axis orientation [39] of a synchronous generator with the d-axis leading the q-axis is used. This results in a negative d-axis component of stator current for an overexcited generator delivering power to the system. The equations of the generators are expressed in p.u. For a detailed treatment of the choice of the p.u. system for the generator model, the readers are referred to [12, 38].

3.3.1.1 *State-Space Representation of Synchronous Generator (SG) Model*

Synchronous generators (SGs) are usually represented by a sub-transient model [12, 38] with four equivalent coils on the rotor. Besides the field coil, there is one equivalent damper coil in the d-axis and two in the q-axis. The differential equations governing the sub-transient dynamic behavior of the ith SG are given by

$$\dot{\delta}_i = \omega_i - \omega_s \tag{3.52}$$

$$\dot{\omega}_i = \frac{\omega_s}{2H}[T_{mi} - D(\omega_i - \omega_s) - \frac{(X_{di}'' - X_{lsi})}{(X_{di}' - X_{lsi})}E_{qi}'I_{qi} - \frac{(X_{di}' - X_{di}'')}{(X_{di}' - X_{lsi})}\psi_{1di}I_{qi}$$
$$- \frac{(X_{qi}'' - X_{lsi})}{(X_{qi}' - X_{lsi})}E_{di}'I_{di} + \frac{(X_{qi}' - X_{qi}'')}{(X_{qi}' - X_{lsi})}\psi_{2qi}I_{di} + (X_{qi}'' - X_{di}'')I_{qi}I_{di}] \tag{3.53}$$

$$\dot{E}_{qi}' = \frac{1}{T_{doi}'}[-E_{qi}' - (X_{di} - X_{di}')\{-I_{di} - \frac{(X_{di}' - X_{di}'')}{(X_{di}' - X_{lsi})^2}$$

$$(\psi_{1di} - (X_{di}^{'} - X_{lsi})I_{di} - E_{qi}^{'})\} + E_{fdi}] \tag{3.54}$$

$$\dot{E}_{di}^{'} = -\frac{1}{T_{qoi}^{'}}[E_{di}^{'} + (X_{qi} - X_{qi}^{'})\{I_{qi} - \frac{(X_{qi}^{'} - X_{qi}^{''})}{(X_{qi}^{'} - X_{lsi})^2}$$

$$(-\psi_{2qi} + (X_{qi}^{'} - X_{lsi})I_{qi} - E_{di}^{'})\}] \tag{3.55}$$

$$\dot{\psi}_{1di} = \frac{1}{T_{doi}^{''}}[-\psi_{1di} + E_{qi}^{'} + (X_{di}^{'} - X_{lsi})I_{di}] \tag{3.56}$$

$$\dot{\psi}_{2qi} = -\frac{1}{T_{qo}^{''}}[\psi_{2qi} + E_{di}^{'} - (X_{qi}^{'} - X_{lsi})I_{qi}] \tag{3.57}$$

for $i = 1, 2, ..., m$, where
m : total number of SGs,
δ_i : SG rotor angle,
ω_i : rotor angular speed,
$E_{qi}^{'}$: transient emf due to field flux linkage,
$E_{di}^{'}$: transient emf due to flux linkage in q-axis damper coil,
ψ_{1di} : sub-transient emf due to flux linkage in d-axis damper,
ψ_{2qi} : sub-transient emf due to flux linkage in q-axis damper,
I_{di} : d-axis component of stator current,
I_{qi} : q-axis component of stator current,
$X_{di}, X_{di}^{'}, X_{di}^{''}$: synchronous, transient and sub-transient reactances, respectively, along d-axis,
$X_{qi}, X_{qi}^{'}, X_{qi}^{''}$: synchronous, transient and sub-transient reactances, respectively, along q-axis,
$T_{doi}^{'}, T_{doi}^{''}$: d-axis open-circuit transient and sub-transient time constants, respectively,
$T_{qoi}^{'}, T_{qoi}^{''}$: q-axis open-circuit transient and sub-transient time constants, respectively.

The stator electrical transients are generally much faster compared to the electromechanical (swing) dynamics. Hence, for stability studies, the stator quantities are assumed to be related to the terminal bus quantities through algebraic equations rather than state equations. The stator algebraic equations are given by

$$V_i \cos(\delta_i - \Theta_i) - \frac{(X_{di}^{''} - X_{lsi})}{(X_{di}^{'} - X_{lsi})}E_{qi}^{'} - \frac{(X_{di}^{'} - X_{di}^{''})}{(X_{di}^{'} - X_{lsi})}\psi_{1di} + R_{si}I_{qi} - X_{di}^{''}I_{di} = 0 \tag{3.58}$$

$$V_i \sin(\delta_i - \Theta_i) + \frac{(X_{qi}^{''} - X_{lsi})}{(X_{qi}^{'} - X_{lsi})}E_{di}^{'} - \frac{(X_{qi}^{'} - X_{qi}^{''})}{(X_{qi}^{'} - X_{lsi})}\psi_{2qi} - R_{si}I_{di} - X_{di}^{''}I_{qi} = 0 \tag{3.59}$$

where $E_{qi}^{''} = \frac{(X_{di}^{''} - X_{lsi})}{(X_{di}^{'} - X_{lsi})}E_{qi}^{'} + \frac{(X_{di}^{'} - X_{di}^{''})}{(X_{di}^{'} - X_{lsi})}\psi_{1di}$
and $E_{di}^{''} = \frac{(X_{qi}^{''} - X_{lsi})}{(X_{qi}^{'} - X_{lsi})}E_{di}^{'} - \frac{(X_{qi}^{'} - X_{qi}^{''})}{(X_{qi}^{'} - X_{lsi})}\psi_{2qi}$

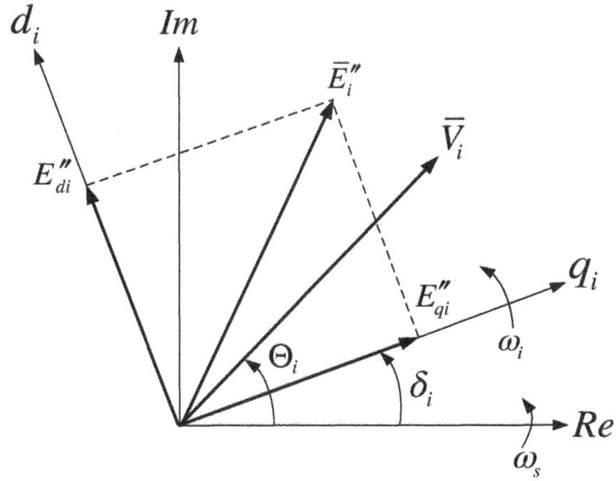

Figure 3.12 Reference frames for modeling the ith synchronous generator.

for $i = 1, 2, ..., m$, where
V_i : SG terminal voltage magnitude,
Θ_i : SG terminal voltage angle,
R_{si} : resistance of the armature,
X_{lsi} : armature leakage reactance,
The notation is standard as in [38].

As shown in Fig. 3.12, a reference frame denoted by d_i–q_i rotating at the rotor angular speed ω_i is used for modeling the ith SG. In this frame, the d-axis leads the q-axis. The generator terminal voltage V_i is expressed in the common network reference frame Re–Im rotating at the synchronous speed ω_s. The power angle δ_i is defined as the angular separation between the axes Re and q_i. The rotor current equations (3.58) and (3.59) are expressed with I_{di} and I_{qi} transformed in the d_i–q_i reference frame. In this context, refer to Fig. 3.2 for the reference frames used for MTDC converter modeling.

3.3.1.2 Inclusion of Generator in the Network At the bulk power transmission level, the high-voltage AC transmission networks are assumed to be balanced and symmetric. Hence for steady-state analysis, the network can be represented on a single-phase basis using phasor quantities (for slowly varying sinusoidal voltage and currents in the network). Thus the generators can also be represented on a single-phase basis.

Equations (3.58) and (3.59) can be expressed as a single equation in phasor quantities if sub-transient saliency is neglected, that is, $X_{di}'' = X_{qi}'' = X_i''$. In that case,

$$I_{qi} + jI_{di} = \frac{1}{(R_{si} + jX_i'')}[E_i - (V_{qi} + jV_{di})] \tag{3.60}$$

where

$$E_i = \left[\frac{(X_{di}'' - X_{lsi})}{(X_{di}' - X_{lsi})}E_{qi}' - \frac{(X_{di}' - X_{di}'')}{(X_{di}' - X_{lsi})}\psi_{1di}\right] + j\left[\frac{(X_{qi}'' - X_{lsi})}{(X_{qi}' - X_{lsi})}E_{di}' - \frac{(X_{qi}' - X_{qi}'')}{(X_{qi}' - X_{lsi})}\psi_{2qi}\right] \quad (3.61)$$

The generator bus current I_i can be represented on a synchronously rotating common network reference frame *Re–Im* as

$$\overline{I}_i = I_{Re(i)} + jI_{Im(i)} = (I_{qi} + jI_{di})e^{j\delta_i} = \frac{1}{(R_{si} + jX_i'')}[E_i - (V_{qi} + jV_{di})]e^{j\delta_i} \quad (3.62)$$

Equation (3.62) represents an equivalent circuit shown in Fig. 3.13 where

$$\overline{Y}_{gi} = \frac{1}{(R_{si} + jX_i'')} \quad (3.63)$$

$$\overline{I}_{gi} = E_i\overline{Y}_{gi} = E_i\frac{1}{(R_{si} + jX_i'')}e^{j\delta_i} \quad (3.64)$$

It is to be noted that $\overline{I}_{gi}$ is a function of state variables only. Hence it does not change suddenly whenever there is a network switching. The equivalent circuit shown in Fig. 3.13 can readily be merged with the AC network (external to the generator).

3.3.1.3 *Treatment of Sub-transient Saliency* If sub-transient saliency is to be considered, the stator cannot be represented by the single-phase equivalent circuit shown in Fig. 3.13. Hence, there is a need for special techniques to handle sub-transient saliency. One way is to use the concept of dummy rotor coil [39].

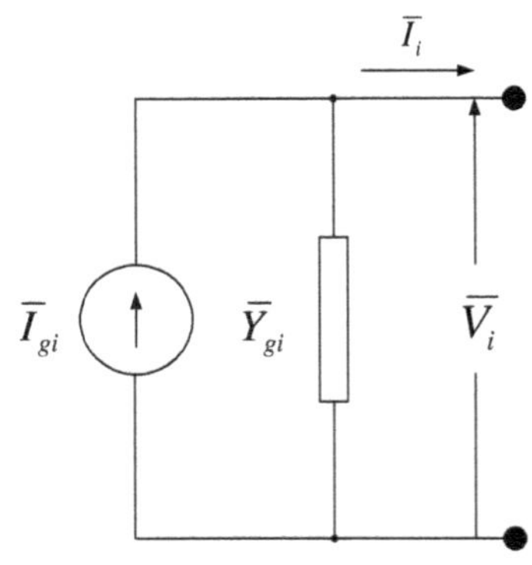

Figure 3.13 Generator equivalent circuit.

The stator equation can be rewritten as

$$V_{qi} = \frac{(X_{di}'' - X_{lsi})}{(X_{di}' - X_{lsi})} E_{qi}' + \frac{(X_{di}' - X_{di}'')}{(X_{di}' - X_{lsi})} \psi_{1di} - R_{si} I_{qi} + X_{di}'' I_{di} \tag{3.65}$$

$$V_{di} = \frac{(X_{qi}'' - X_{lsi})}{(X_{qi}' - X_{lsi})} E_{di}' - \frac{(X_{qi}' - X_{qi}'')}{(X_{qi}' - X_{lsi})} \psi_{2qi} + E_{dci} - R_{si} I_{di} - X_{di}'' I_{qi} \tag{3.66}$$

where

$$E_{dci} = -(X_{qi}'' - X_{di}'') I_{qi} \tag{3.67}$$

Then (3.60) can be rewritten as

$$I_{qi} + jI_{di} = \frac{1}{(R_{si} + jX_{di}'')} [\overline{E}_{eqi} - (V_{qi} + jV_{di})] \tag{3.68}$$

where

$$\overline{E}_{eqi} = [\frac{(X_{di}'' - X_{lsi})}{(X_{di}' - X_{lsi})} E_{qi}' + \frac{(X_{di}' - X_{di}'')}{(X_{di}' - X_{lsi})} \psi_{1di}] + j[\frac{(X_{qi}'' - X_{lsi})}{(X_{qi}' - X_{lsi})} E_{di}' - \frac{(X_{qi}' - X_{qi}'')}{(X_{qi}' - X_{lsi})} \psi_{2qi} + E_{dci}] \tag{3.69}$$

The equivalent circuit of Fig. 3.13 applies if

$$\overline{I}_i = I_{Re(i)} + jI_{Im(i)} = (I_{qi} + jI_{di}) e^{j\delta_i} = \frac{1}{(R_{si} + jX_i'')} [\overline{E}_{eqi} - (V_{qi} + jV_{di})] e^{j\delta_i} \tag{3.70}$$

The motivation behind this approach is that if E_{dci} is a state variable (proportional to the flux linkage of a dummy rotor coil in the q-axis), then the problem of subtransient saliency is properly addressed. This is an approximate treatment of subtransient saliency, but the degree of approximation can be directly chosen by proper selection of T_{dci} to achieve acceptable accuracy.

Considering a rotor dummy coil in q-axis, which is linked only with the q-axis coil in the armature but has no coupling with other coils, E_{dci} can be treated as a voltage source proportional to the flux linkage of the dummy rotor coil. The differential equation for E_{dci} can be expressed as

$$\dot{E}_{dci} = \frac{1}{T_{dci}} [-E_{dci} - (X_{qi}'' - X_{di}'') I_{qi}] \tag{3.71}$$

where T_{dci} is the open circuit constant of the dummy coil, which can be arbitrarily selected. In computing (3.71) with (3.67), it is seen that the latter is a steady-state solution of the former. As T_{dci} tends to zero, the solution of (3.71) is given by (3.67). From a numerical standpoint, it was observed that T_{dci} need to be smaller than 0.01 s for acceptable accuracy. This is of similar order as the time constant of a high-resistance damper winding.

3.3.1.4 *State-Space Model of Excitation Systems for SGs* The excitation system of the SG controls the terminal voltage by changing the field flux of the generator through a closed-loop control. Depending upon the source of power, the excitation system can be of three broad types: DC excitation systems, AC excitation systems, and static excitation systems. The IEEE has proposed standard models to represent different types of excitation systems [40], from which two models are described here.

The differential equations governing the behavior of an IEEE-DC1A type excitation system of the ith SG are given by

$$\dot{V}_{tri} = \frac{1}{T_{ri}}[-V_{tri} + V_i] \tag{3.72}$$

$$\dot{E}_{fdi} = -\frac{1}{T_{Ei}}[K_{Ei}E_{fdi} + E_{fdi}A_{ex}e^{B_{ex}E_{fdi}} - V_{ri}] \tag{3.73}$$

$$\dot{V}_{ri} = \frac{1}{T_{Ai}}[\frac{K_{Ai}K_{Fi}}{T_{Fi}}R_{Fi} + K_{Ai}(V_{refi} - V_{tri}) - \frac{K_{Ai}K_{Fi}}{T_{Fi}}E_{fdi} - V_{ri}] \tag{3.74}$$

$$\dot{R}_{Fi} = \frac{1}{T_{Fi}}[-R_{Fi} + E_{fdi}] \tag{3.75}$$

where
E_{fdi} : field voltage,
V_{tri} : measured voltage state variable after sensor lag block,
and the rest of the notations carry their standard meaning [38].

The governing equations for the IEEE-ST1A type excitation system are given by

$$\dot{V}_{tri} = \frac{1}{T_{ri}}[-V_{tri} + V_i] \tag{3.76}$$

$$E_{fdi} = K_{ai}(V_{refi} - V_{tri}) \tag{3.77}$$

The notations carry their standard meaning as in [38].

Figure 3.14 shows the block diagram of the model of the ith synchronous generator and the corresponding exciter. The current injected by this generator is $\overline{I}_i = I_{Re(i)} + jI_{Im(i)}$. This current will be used to solve the AC network voltage as elaborated later in Section 3.3.3, and the generator terminal voltage V_i is measured and fed back to the exciter.

3.3.1.5 *State-Space Model of Turbine and Governor* The turbine acts as a prime-mover which produces the mechanical power to rotate the generator shaft. The speed governors changes the mechanical power reference in response to the change in speed of the generator. In its simplest form, the model of the turbine is given by

$$\Delta\dot{P}_{Ti} = \frac{1}{T_{Ti}}\left(-\Delta P_{Ti} + \Delta P_i\right) \tag{3.78}$$

$$\Delta P_{Ti} = \omega_i\left(T_{mi} - T^*_{mi}\right) \tag{3.79}$$

where
T_{Ti}: is the time-constant of the ith turbine.

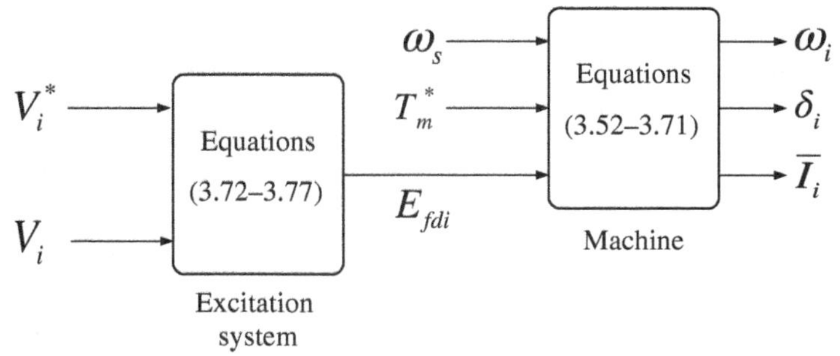

Figure 3.14 Block diagram of synchronous generator and exciter model.

The governor model can be written as

$$\Delta\dot{P}_i = \frac{1}{T_{Gi}}\left[-\Delta P_i + \frac{1}{R_{goi}}\left(\frac{\omega_s - \omega_i}{\omega_s}\right)\right] \tag{3.80}$$

$$\Delta P_i = P_i - P_i^0 \tag{3.81}$$

where
T_{Gi}: the time-constant of the ith governor,
R_{goi}: the governor droop constant,
P_i^0, P_i: mechanical power input to the generator under nominal and off-nominal condition.
Under steady-state, (3.80) describes the power–frequency droop characteristics of governor, which is shown in Fig. 4.2 and will be discussed in more detail in Chapters 4 and 5 in the context of autonomous power sharing in MTDC grids and frequency support of the surrounding AC systems, respectively. The governor droop constant R_{goi} determines the slope of the droop characteristics. It is a common practice to chose the droop coefficient according to the rating of the generator.

3.3.2 Load Model

Accurate modeling of loads is critical because it significantly impacts the power system stability analysis. Unlike generators, accurate load modeling has proven to be much more challenging. This is primarily due to the fact that the load composition is less known, it changes over time, and it has seasonal variation.

The details of load modeling is outside the scope of this chapter. A widely used static load model called the constant impedance–constant current–constant power (ZIP) model and its integration into the AC–MTDC system studies is presented here very briefly. The equations describing the model of the loads are expressed in p.u. The ZIP load model at the ith bus represents the voltage dependency of the loads in

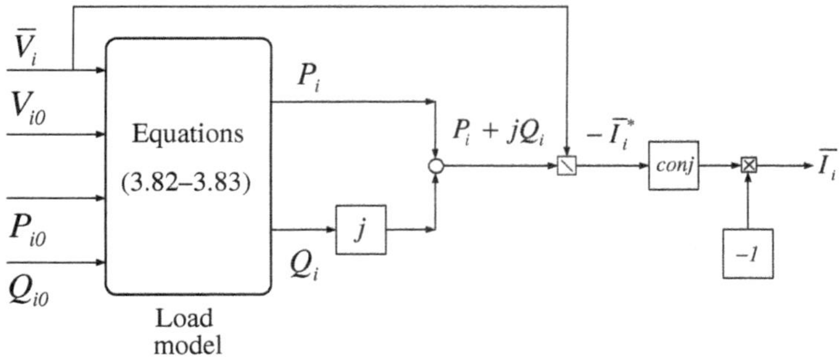

Figure 3.15 Block diagram of the static load model.

the form of a polynomial:

$$P_i = P_{i0}\left[p_{i1}\left(\frac{V_i}{V_{i0}}\right)^2 + p_{i2}\left(\frac{V_i}{V_{i0}}\right) + p_{i3}\right] \tag{3.82}$$

$$Q_i = Q_{i0}\left[q_{i1}\left(\frac{V_i}{V_{i0}}\right)^2 + q_{i2}\left(\frac{V_i}{V_{i0}}\right) + q_{i3}\right] \tag{3.83}$$

Here, P_i, Q_i, and V_i are the load real power, reactive power, and voltage at the ith bus, respectively. The variables with the subscript 0 denote the values under nominal condition. The coefficients with subscripts 1, 2, and 3 correspond to the fraction of constant impedance (Z) constant current (I) and constant power (P) component of the load, respectively. Therefore, $p_{i1} + p_{i2} + p_{i3} = 1$ and $q_{i1} + q_{i2} + q_{i3} = 1$.

Figure 3.15 shows the block diagram of the polynomial load model. The input to the model are the nominal values of load voltage V_{i0}, load real power P_{i0}, reactive power Q_{i0}, and also the actual load voltage V_i. The load current drawn from the system is given by $-\overline{I}_i^* = \frac{P_i+jQ_i}{\overline{V}_i}$. The injected current $\overline{I}_i$ is derived and used to solve the AC network equation, which in turn gives $\overline{V}_i$.

Dynamic loads such as induction/synchronous motors are modeled in the same manner as the synchronous generators. The readers are referred to [12, 38] for further details.

3.3.3 AC Network Model

For most stability studies, the dynamics of the AC transmission network is considered to be much faster compared to other components. Therefore, these networks are represented by algebraic equations. Each transmission line and transformer is represented by an equivalent pi model [41]. As before, balanced operation of power systems is assumed, and the positive sequence parameters on a per-phase basis are considered. The relationship between the bus/node voltages and currents can be described either by loop equations or node equations [41]. The node equations are usually preferred because the number of independent node equations are less than the number of independent loop equations. The network equations can be expressed as

$$\begin{bmatrix} \overline{I}_1 \\ \overline{I}_2 \\ \dots \\ \overline{I}_i \\ \dots \\ \overline{I}_{n_{ac}} \end{bmatrix} = \begin{bmatrix} \overline{Y}_{11} & \overline{Y}_{12} & \dots & \overline{Y}_{1i} & \dots & \overline{Y}_{1n_{ac}} \\ \overline{Y}_{21} & \overline{Y}_{22} & \dots & \overline{Y}_{2i} & \dots & \overline{Y}_{2n_{ac}} \\ \dots & \dots & \dots & \dots & \dots & \dots \\ \overline{Y}_{i1} & \overline{Y}_{i2} & \dots & \overline{Y}_{ii} & \dots & \overline{Y}_{in_{ac}} \\ \dots & \dots & \dots & \dots & \dots & \dots \\ \overline{Y}_{n_{ac}1} & \overline{Y}_{n_{ac}2} & \dots & \overline{Y}_{n_{ac}i} & \dots & \overline{Y}_{n_{ac}n_{ac}} \end{bmatrix} \begin{bmatrix} \overline{V}_1 \\ \overline{V}_2 \\ \dots \\ \overline{V}_i \\ \dots \\ \overline{V}_{n_{ac}} \end{bmatrix} \tag{3.84}$$

where n_{ac} is the total number of buses in the AC network,
$\overline{Y}_{ii} = G_{ii} + jB_{ii}$ is the self admittance at node i
= sum of all admittance terminating at node i,
$\overline{Y}_{ij} = G_{ij} + jB_{ij}$ is the mutual admittance between nodes i and j
= sum of all admittance between nodes i and j,
$\overline{V}_i = V_i e^{j\Theta_i}$ is the voltage at the ith node
expressed in the common network reference frame *Re–Im*,
$\overline{I}_i$ is the current injected at the ith node in the same reference frame.

The AC network equations are expressed in p.u. with respect to appropriate base values.

Figure 3.16 shows the model of the AC network and the interface of different components to this network. Components such as generators, loads, and MTDC stations

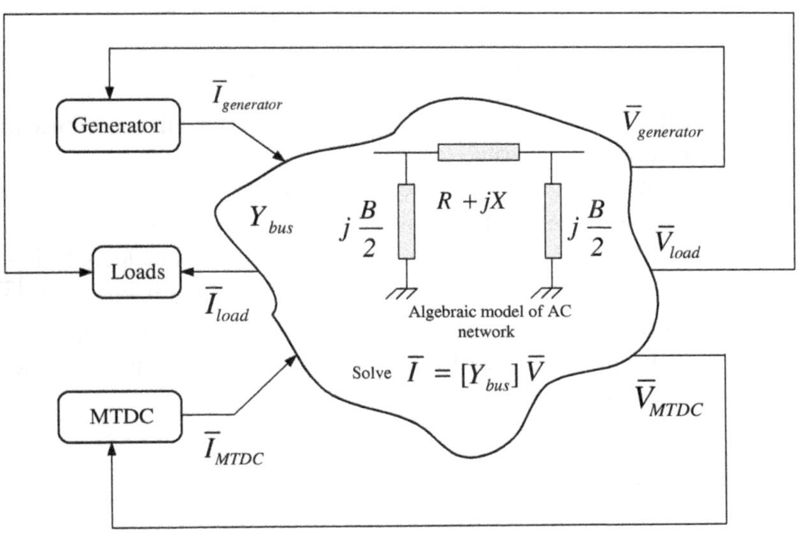

Figure 3.16 AC network model with different components interfaced with the network.

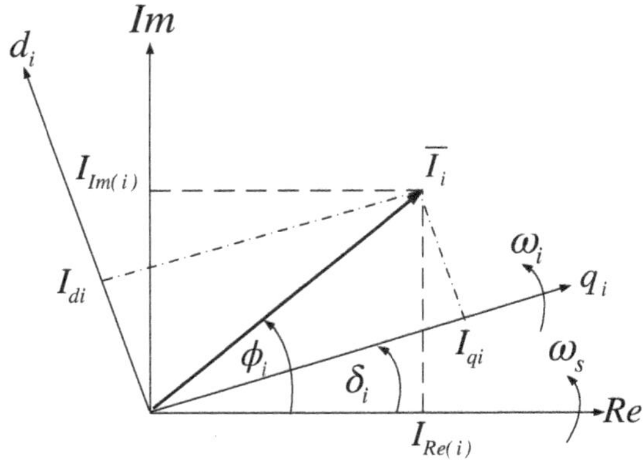

Figure 3.17 Coordinate transformation of the stator current of the *i*th generator from the d_i–q_i frame into the common network frame of reference *Re*–*Im*.

are modeled with differential and algebraic equations, and the current injections from them are derived as described in Sections 3.3.1.2, 3.3.2, and 3.2, respectively. These injected currents are expressed in their respective reference frames. For example, generator currents I_{di}, I_{qi} are expressed in their individual d_i–q_i frames, HVDC converter current injections I_{gXi}, I_{gYi} in X_i–Y_i axis, and so on. Therefore, a coordinate transformation is performed on the injected current vector $\left[\overline{I}\right]$ to express it in the common network reference frame *Re*–*Im*. As an example, Fig. 3.17 shows the components of the current injected by the *i*th generator in the d_i–q_i reference frame and the *Re*–*Im* reference frame. The coordinate transformation is done using the following equation:

$$\begin{bmatrix} I_{Re(i)} \\ I_{\mathrm{Im}(i)} \end{bmatrix} = \begin{bmatrix} \cos\delta_i & -\sin\delta_i \\ \sin\delta_i & \cos\delta_i \end{bmatrix} \begin{bmatrix} I_{qi} \\ I_{di} \end{bmatrix} \tag{3.85}$$

This transformation was used in (3.63) and (3.70). Currents in other reference frames can also be transformed in the same manner.

The node voltages are computed to satisfy the equation $\left[\overline{I}\right] - \left[\overline{Y}_{bus}\right]\left[\overline{V}\right] = 0$. For generators, the measured terminal voltage is used as a feedback signal for the excitation system. For the ZIP loads, the node voltage forms an algebraic loop. It should be noted that constant impedance loads can be usually included in the network $\left[\overline{Y}_{bus}\right]$. The MTDC grid model uses the AC node voltage at the PCC as feedback for the PLL action and also uses it as feed-forward measurement for disturbance rejection, as described in Sections 3.2 and 3.2.3.

3.4 AC–MTDC Load flow Analysis

Load flow study is an essential part of the planning and operation of a power system. It is also the first step before doing the dynamic simulation. The load flow analysis of combined AC–MTDC network is presented in the following sections. All AC side voltages and currents are expressed with respect to the common network reference frame *Re–Im* and common base values.

3.4.1 AC Grid Load flow Model

The objective of the AC grid load flow is to determine the voltage magnitude and angle at each bus of the AC system. If the current injection at each bus of the AC system was known *a priori*, then (3.84) could be solved for the voltage phasors using the relation $\left[\overline{V}\right] = \left[\overline{Y_{bus}}\right]^{-1}\left[\overline{I}\right]$, which is linear in nature. Unfortunately, the current injection is not known beforehand and it is related to the the injected power and bus voltage at the ith bus through the following load flow equation:

$$\overline{I_i} = \frac{P_i - jQ_i}{\overline{V_i}^*} \tag{3.86}$$

Depending on the constraints imposed by the nodes of the AC system, the problem becomes nonlinear in nature. Such constraints divide the nodes in the AC grid into four types:

Slack bus: buses where the voltage magnitude V_i and angle Θ_i are specified.
PV bus: buses where the real power injected to the network P_i and the voltage magnitude V_i are known.
PQ bus: buses where the real power injected to the network P_i and the reactive power injected to the network Q_i are known.
MTDC PCC bus: buses where special constraints are imposed based on the control philosophy of the converter stations–refer to Section 3.2.3.1 for different control modes.

From (3.84), the injected current at the ith bus can be expressed as

$$\overline{I_i} = \sum_{k=1}^{n_{ac}} \overline{Y_{ik} V_k} \tag{3.87}$$

From (3.86) and (3.87), separating the real and the imaginary part we get:

$$f_{1i} = P_i = \sum_{k=1}^{n_{ac}} V_i V_k [G_{ik} \cos(\Theta_i - \Theta_k) + B_{ik} \sin(\Theta_i - \Theta_k)] \tag{3.88}$$

$$f_{2i} = Q_i = \sum_{k=1}^{n_{ac}} V_i V_k [G_{ik} \sin(\Theta_i - \Theta_k) - B_{ik} \cos(\Theta_i - \Theta_k)] \tag{3.89}$$

for $i = 1, ..., n_{ac}$
where n_{ac} is the total number of buses in the AC system and $\overline{Y}_{ik} = G_{ik} + jB_{ik}$.

The AC system buses are inclusive of the PCC buses of the MTDC grid. However, the AC side of the converter stations connecting the PCC bus to the converter terminal bus is not considered within the AC network. Instead, they are included within the MTDC grid. For the ith bus connected to a generator, the bus could either be considered a slack or a PV bus. For the ith bus connected to a load, it is considered to be a PQ bus where $P_i = P_{i0}$ and $Q_i = Q_{i0}$, (3.82) and (3.83). For the ith bus connected to the PCC of the MTDC converter station, the load flow equations are derived in the next section.

3.4.2 MTDC Grid Load flow Model

The objective of the MTDC grid load flow is to determine the DC voltage magnitudes at each converter station and the power flows within the DC grid. Let us assume there are n converter stations in the MTDC grid. The load flow equations for the ith converter station of the MTDC grid will be derived considering the following:

1. Interface of the converter with the AC system
2. AC side of the converter stations
3. Interface of the converter station AC and DC sides
4. DC side of the MTDC grid
5. Converter control methods.

The ith converter station of the MTDC grid and the corresponding interfaces are shown in Fig. 3.11.

3.4.2.1 *MTDC Interface with AC System* As shown in Fig. 3.11, the load flow equation at the PCC bus of the ith converter station interfacing with the AC system can be written as

$$f_{3i} = \sum_{k=1}^{n_{ac}} V_{gi}V_k[G_{ik}\cos(\Theta_{gi} - \Theta_k) + B_{ik}\sin(\Theta_{gi} - \Theta_k)] - P_{gi} = 0 \tag{3.90}$$

$$f_{4i} = \sum_{k=1}^{n_{ac}} V_{gi}V_k[G_{ik}\sin(\Theta_{gi} - \Theta_k) - B_{ik}\cos(\Theta_{gi} - \Theta_k)] - Q_{gi} = 0 \tag{3.91}$$

where $\overline{V}_{gi} = V_{gi}e^{j\Theta_{gi}}$ is the bus voltage at the ith PCC bus. For a sequential method of AC–MTDC load flow solution, $\overline{V}_k, k \neq i$ obtained from the AC load flow are used while solving the MTDC grid load flow equations.

3.4.2.2 *MTDC AC Side Load flow Model* For an asymmetric bipolar MTDC grid, the real and reactive power balance equation at the PCC of the ith converter station is

$$f_{5i} = P_{gi} - P^p_{gi} - P^n_{gi} = 0 \tag{3.92}$$

$$f_{6i} = Q_{gi} - Q^p_{gi} - Q^n_{gi} = 0 \tag{3.93}$$

For the positive and the negative poles of the ith converter station, the apparent power injection at the PCC by the positive and negative poles are

$$S^p_{gi} = \overline{V}_{gi}\overline{I}^{p*}_{gi} = P^p_{gi} + jQ^p_{gi} \tag{3.94}$$

$$S^n_{gi} = \overline{V}_{gi}\overline{I}^{n*}_{gi} = P^n_{gi} + jQ^n_{gi} \tag{3.95}$$

Applying KCL at the PCC of the ith converter station we get

$$\overline{I}_{gi} = \overline{I}^p_{gi} + \overline{I}^n_{gi} \tag{3.96}$$

Using KVL between the PCC bus and the converter terminal bus, we get

$$\overline{I}^p_{gi} = \frac{\overline{V}^p_{ti} - \overline{V}_{gi}}{R^p_i + jX^p_i} = \left(\overline{V}^p_{ti} - \overline{V}_{gi}\right)\left(G^p_i + jB^p_i\right) \tag{3.97}$$

$$\overline{I}^n_{gi} = \frac{\overline{V}^n_{ti} - \overline{V}_{gi}}{R^n_i + jX^n_i} = \left(\overline{V}^n_{ti} - \overline{V}_{gi}\right)\left(G^n_i + jB^n_i\right) \tag{3.98}$$

where $\overline{V}^p_{ti} = V^p_{ti}e^{j\Theta^p_{ti}}$, and $\overline{V}^n_{ti} = V^n_{ti}e^{j\Theta^n_{ti}}$ are the positive and negative pole converter terminal voltages, respectively. Substituting the values of $\overline{I}^p_{gi}$ and $\overline{I}^n_{gi}$ into (3.94) and separating the real and the imaginary parts, we get

$$\begin{aligned} f_{7i} = P^p_{gi} - V_{gi}[-V_{gi}G^p_i + V^p_{ti}\{G^p_i\cos(\Theta_{gi} - \Theta^p_{ti}) \\ +B^p_i\sin(\Theta_{gi} - \Theta^p_{ti})\}] = 0 \end{aligned} \tag{3.99}$$

$$\begin{aligned} f_{8i} = Q^p_{gi} - V_{gi}[V_{gi}B^p_i + V^p_{ti}\{G^p_i\sin(\Theta_{gi} - \Theta^p_{ti}) \\ -B^p_i\cos(\Theta_{gi} - \Theta^p_{ti})\}] = 0 \end{aligned} \tag{3.100}$$

$$\begin{aligned} f_{9i} = P^n_{gi} - V_{gi}[-V_{gi}G^n_i + V^n_{ti}\{G^n_i\cos(\Theta_{gi} - \Theta^n_{ti}) \\ +B^n_i\sin(\Theta_{gi} - \Theta^n_{ti})\}] = 0 \end{aligned} \tag{3.101}$$

$$\begin{aligned} f_{10i} = Q^n_{gi} - V_{gi}[V_{gi}B^n_i + V^n_{ti}\{G^n_i\sin(\Theta_{gi} - \Theta^n_{ti}) \\ -B^n_i\cos(\Theta_{gi} - \Theta^n_{ti})\}] = 0 \end{aligned} \tag{3.102}$$

3.4.2.3 *Interface of MTDC AC and DC Sides* Neglecting converter switching losses, the real power balance between converter AC and DC sides can be expressed as

$$f_{11i} = P^p_{ti} - V^p_{dci}I^p_{dci} = 0 \tag{3.103}$$

$$f_{12i} = P^n_{ti} - V^n_{dci}I^n_{dci} = 0 \tag{3.104}$$

Here, the real power output at the AC side terminal of the positive and the negative pole of the converter stations are

$$P^p_{ti} - V^p_{ti}[V^p_{ti}G^p_i - V_{gi}\{G^p_i\cos(\Theta^p_{ti} - \Theta_{gi}) + B^p_i\sin(\Theta^p_{ti} - \Theta_{gi})\}] = 0 \tag{3.105}$$

$$P^n_{ti} - V^n_{ti}[V^n_{ti}G^n_i - V_{gi}\{G^n_i\cos(\Theta^n_{ti} - \Theta_{gi}) + B^n_i\sin(\Theta^n_{ti} - \Theta_{gi})\}] = 0 \tag{3.106}$$

The DC bus voltage of the positive and the negative poles of the ith converter station are given by V^p_{dci} and V^n_{dci}, respectively.

3.4.2.4 MTDC DC Side Load flow Model Referring to Fig. 3.11, it can be seen that under steady-state, $I^p_{dci} = I^p_{li}$ and $I^n_{dci} = I^n_{li}$ since current flowing through the DC bus capacitor is zero. The current injection is determined by the conductance matrix of the DC grid. The DC current injection at the positive and the negative poles of the DC bus is given by

$$\begin{bmatrix} -I^p_{dc1} \\ \cdots \\ -I^p_{dci} \\ \cdots \\ -I^p_{dcN} \\ -I^n_{dc1} \\ \cdots \\ -I^n_{dci} \\ \cdots \\ -I^n_{dcN} \end{bmatrix} = \begin{bmatrix} G_{pp11} & \cdots & G_{pp1i} & \cdots & G_{pp1N} & G_{pn11} & \cdots & G_{pn1i} & \cdots & G_{pn1N} \\ \cdots & \cdots & \cdots & \cdots & \cdots & \cdots & \cdots & \cdots & \cdots & \cdots \\ G_{ppi1} & \cdots & G_{ppii} & \cdots & G_{ppiN} & G_{pni1} & \cdots & G_{pnii} & \cdots & G_{pniN} \\ \cdots & \cdots & \cdots & \cdots & \cdots & \cdots & \cdots & \cdots & \cdots & \cdots \\ G_{ppN1} & \cdots & G_{ppNi} & \cdots & G_{ppNN} & G_{pnN1} & \cdots & G_{pnNi} & \cdots & G_{pnNN} \\ G_{np11} & \cdots & G_{np1i} & \cdots & G_{np1N} & G_{nn11} & \cdots & G_{nn1i} & \cdots & G_{nn1N} \\ \cdots & \cdots & \cdots & \cdots & \cdots & \cdots & \cdots & \cdots & \cdots & \cdots \\ G_{npi1} & \cdots & G_{npii} & \cdots & G_{npiN} & G_{nni1} & \cdots & G_{nnii} & \cdots & G_{nniN} \\ \cdots & \cdots & \cdots & \cdots & \cdots & \cdots & \cdots & \cdots & \cdots & \cdots \\ G_{npN1} & \cdots & G_{npNi} & \cdots & G_{npNN} & G_{nnN1} & \cdots & G_{nnNi} & \cdots & G_{nnNN} \end{bmatrix} \begin{bmatrix} V^p_{dc1} \\ \cdots \\ V^p_{dci} \\ \cdots \\ V^p_{dcN} \\ V^n_{dc1} \\ \cdots \\ V^n_{dci} \\ \cdots \\ V^n_{dcN} \end{bmatrix} \tag{3.107}$$

Therefore, current injection at the ith bus in the DC side is given by

$$f_{13i} = I^p_{dci} + \sum_{k=1}^{N} (G_{ppki} V^p_{dck} + G_{pnki} V^n_{dck}) = 0 \tag{3.108}$$

$$f_{14i} = I^n_{dci} + \sum_{k=1}^{N} (G_{npki} V^p_{dck} + G_{nnki} V^n_{dck}) = 0 \tag{3.109}$$

G_{ppij}, G_{nnij}, G_{pnij}, and G_{npij} are the elements of the conductance matrix of the DC transmission network.

3.4.2.5 MTDC Converter Control Modes The other steady-state equations come in the form of equality constraints as obtained from the converter control modes described in Section 3.2.3.1. Converter stations in the MTDC grid can operate under different control modes. The most basic mode of converter control in an MTDC grid could be one converter station controlling the DC link voltage at the DC bus (thus acting as a slack bus) while the remaining ones controlling real power flow at the PCC—as mentioned in Section 1.5. Additionally, the converter stations can control the reactive power or voltage at the PCC.

Also, there are certain inequality constraints originating from the limits imposed by the voltage and the current ratings. These inequality constraints are shown below:

$$V^{\min}_{p(i)} \leq V^m_{dci} + V^p_{dci} \leq V^{\max}_{p(i)} \tag{3.110}$$

$$V^{\min}_{n(i)} \leq V^m_{dci} + V^n_{dci} \leq V^{\max}_{n(i)} \tag{3.111}$$

$$I^p_{gi} \leq I^{p\max}_{gi} \tag{3.112}$$

$$I^n_{gi} \leq I^{n\max}_{gi} \tag{3.113}$$

Let us assume that the limits of the control loops are not violated.

- Mode I—*P–Q* control: Under normal condition, the desired real and reactive powers P_{gi}^* and Q_{gi}^* at the PCC are equally divided between the positive and the negative pole converters. Therefore,

$$f_{15i} = \frac{P^*_{gi}}{2} - P^p_{gi} = 0 \tag{3.114}$$

$$f_{16i} = \frac{P^*_{gi}}{2} - P^n_{gi} = 0 \tag{3.115}$$

$$f_{17i} = \frac{Q^*_{gi}}{2} - Q^p_{gi} = 0 \tag{3.116}$$

$$f_{18i} = \frac{Q^*_{gi}}{2} - Q^n_{gi} = 0 \tag{3.117}$$

- Mode II—V_{dc}–*Q* control: Similar to the *P–Q* control mode, the DC link voltage reference for the individual pole of a converter station can be determined based on the desired pole-to-pole DC link voltage V^*_{dci} as shown below:

$$f_{15i} = \left(\frac{V^*_{dci}}{2}\right)^2 - \left(V^p_{dci}\right)^2 = 0 \tag{3.118}$$

$$f_{16i} = \left(\frac{V^*_{dci}}{2}\right)^2 - \left(V^n_{dci}\right)^2 = 0 \tag{3.119}$$

$$f_{17i} = \frac{Q^*_{gi}}{2} - Q^p_{gi} = 0 \tag{3.120}$$

$$f_{18i} = \frac{Q^*_{gi}}{2} - Q^n_{gi} = 0 \tag{3.121}$$

- Mode III A and B—*P*-V_t-*Q* or V_{dc}–V_t–*Q* control: In addition to the real power or DC link voltage control, the following constraints are imposed:

$$f_{17i} = V^*_{gi} - V_{gi} = 0 \tag{3.122}$$

$$f_{18i} = Q^p_{gi} - Q^n_{gi} = 0 \tag{3.123}$$

  This ensures that the distribution of Q^p_{gi} and Q^n_{gi} is maintained in addition to the terminal voltage.

- Special Case—Converter pole outage: If the positive pole outage is considered, the following constraints are imposed:

$$f_{17i} = |\overline{I}^p_{gi}| = 0 \tag{3.124}$$

$$f_{18i} = I^p_{dci} = 0 \tag{3.125}$$

The control for the other pole of the station remains unaffected.

There are many other control modes possible, e.g., power droop control, ground current control, frequency droop control, etc., which will impose different constraints. Details of the power droop control and frequency droop control will be elaborated in Chapter 4 and 5, respectively.

From the above equations, we can see that there are 16 unknowns for each MTDC converter station: P_{gi}, P^p_{gi}, P^n_{gi}, Q_{gi}, Q^p_{gi}, Q^n_{gi}, V_{gi}, V^p_{ti}, V^n_{ti}, V^p_{dci}, V^n_{dci}, I^p_{dci}, I^n_{dci}, Θ_{gi}, Θ^p_{ti}, Θ^n_{ti}. There are 16 equations (3.90–3.93), (3.99–3.104), (3.108), (3.109), (3.114–3.117) (or 3.118–3.121, and so on) to solve for 16 variables. The technique used for solving these equations is described in the next section.

3.4.3 AC–MTDC Grid Load flow Solution

The AC–DC load flow equations can be solved sequentially using the Newton–Raphson (NR) method [41], which is described here briefly. Let us assume that there are n non-linear equations with n unknowns.

$$
\begin{aligned}
&f_1(x_1, x_2, ..., x_i, ...x_n) - c_1 = 0\\
&f_2(x_1, x_2, ..., x_i, ...x_n) - c_2 = 0\\
&\dots\\
&f_i(x_1, x_2, ..., x_i, ...x_n) - c_i = 0\\
&\dots\\
&f_n(x_1, x_2, ..., x_i, ...x_n) - c_n = 0
\end{aligned}
\tag{3.126}
$$

Let us assume that an initial guess of the n unknowns is $x_1^0, x_2^0, ..., x_i^0, ..., x_n^0$. If the actual solution of the equations is given by $x_1^0+\Delta x_1, x_2^0+\Delta x_2, ..., x_i^0+\Delta x_i, ..., x_n^0+\Delta x_n$, then we can write

$$
\begin{aligned}
&f_1(x_1^0 + \Delta x_1, x_2^0 + \Delta x_2, ..., x_i^0 + \Delta x_i, ..., x_n^0 + \Delta x_n) - c_1 = 0\\
&f_2(x_1^0 + \Delta x_1, x_2^0 + \Delta x_2, ..., x_i^0 + \Delta x_i, ..., x_n^0 + \Delta x_n) - c_2 = 0\\
&\dots\\
&f_i(x_1^0 + \Delta x_1, x_2^0 + \Delta x_2, ..., x_i^0 + \Delta x_i, ..., x_n^0 + \Delta x_n) - c_i = 0\\
&\dots\\
&f_n(x_1^0 + \Delta x_1, x_2^0 + \Delta x_2, ..., x_i^0 + \Delta x_i, ..., x_n^0 + \Delta x_n) - c_n = 0
\end{aligned}
\tag{3.127}
$$

Expanding the ith equation into Taylor's Series, we get

$$
\begin{aligned}
&f_i(x_1^0 + \Delta x_1, x_2^0 + \Delta x_2, ..., x_i^0 + \Delta x_i, ..., x_n^0 + \Delta x_n) - c_i\\
&= f_i(x_1^0, x_2^0, ..., x_i^0, ..., x_n^0) + \left(\frac{\partial f_i}{\partial x_1}\right)\bigg|_0 \Delta x_1 + \left(\frac{\partial f_i}{\partial x_2}\right)\bigg|_0 \Delta x_2 + ...\\
&+ \left(\frac{\partial f_i}{\partial x_i}\right)\bigg|_0 \Delta x_i + ... + \left(\frac{\partial f_i}{\partial x_n}\right)\bigg|_0 \Delta x_n + higherorderterms - c_i\\
&= 0
\end{aligned}
\tag{3.128}
$$

Rearranging the above equation and neglecting higher order terms, the following set of linear equations can be derived:

$$\begin{bmatrix} c_1 - f_1(x_1^0 + \Delta x_1, x_2^0 + \Delta x_2, ..., x_i^0 + \Delta x_i, ..., x_n^0 + \Delta x_n) \\ c_2 - f_2(x_1^0 + \Delta x_1, x_2^0 + \Delta x_2, ..., x_i^0 + \Delta x_i, ..., x_n^0 + \Delta x_n) \\ \cdots \\ c_i - f_i(x_1^0 + \Delta x_1, x_2^0 + \Delta x_2, ..., x_i^0 + \Delta x_i, ..., x_n^0 + \Delta x_n) \\ \cdots \\ c_n - f_n(x_1^0 + \Delta x_1, x_2^0 + \Delta x_2, ..., x_i^0 + \Delta x_i, ..., x_n^0 + \Delta x_n) \end{bmatrix}$$

$$= \begin{bmatrix} \left(\frac{\partial f_1}{\partial x_1}\right)\Big|_0 & \left(\frac{\partial f_1}{\partial x_2}\right)\Big|_0 & \cdots & \left(\frac{\partial f_1}{\partial x_i}\right)\Big|_0 & \cdots & \left(\frac{\partial f_1}{\partial x_n}\right)\Big|_0 \\ \left(\frac{\partial f_2}{\partial x_1}\right)\Big|_0 & \left(\frac{\partial f_2}{\partial x_2}\right)\Big|_0 & \cdots & \left(\frac{\partial f_2}{\partial x_i}\right)\Big|_0 & \cdots & \left(\frac{\partial f_2}{\partial x_n}\right)\Big|_0 \\ \cdots & \cdots & \cdots & \cdots & \cdots & \cdots \\ \left(\frac{\partial f_i}{\partial x_1}\right)\Big|_0 & \left(\frac{\partial f_i}{\partial x_2}\right)\Big|_0 & \cdots & \left(\frac{\partial f_i}{\partial x_i}\right)\Big|_0 & \cdots & \left(\frac{\partial f_i}{\partial x_n}\right)\Big|_0 \\ \cdots & \cdots & \cdots & \cdots & \cdots & \cdots \\ \left(\frac{\partial f_n}{\partial x_1}\right)\Big|_0 & \left(\frac{\partial f_n}{\partial x_2}\right)\Big|_0 & \cdots & \left(\frac{\partial f_n}{\partial x_i}\right)\Big|_0 & \cdots & \left(\frac{\partial f_n}{\partial x_n}\right)\Big|_0 \end{bmatrix} \begin{bmatrix} \Delta x_1 \\ \Delta x_2 \\ \cdots \\ \Delta x_i \\ \cdots \\ \Delta x_n \end{bmatrix}$$

This can be expressed in a symbolic form as

$$\Delta f = J\Delta x \tag{3.129}$$

J contains the partial derivative terms and is known as the *Jacobian*. The higher order terms are neglected with the assumption that the initial guess is close to the actual solution. Equation (3.129) can be solved for Δx. The ith variable x_i^0 is then updated using the relation $x_i^1 = x_i^0 + \Delta x_i$. The process of such iterations continue till the mismatch Δf falls below a predefined threshold. The Jacobian has to be recomputed during each iteration.

In the sequential method of AC–MTDC load flow, the steps mentioned below are followed:

Step I: AC Load flow First, the load flow equations for the AC grid are solved. From the load flow equations of the AC system described in Section 3.4.1, we can see that the real and reactive power at the ith bus can be expressed as a function of voltage magnitudes and angles at all other buses. Since these equations are nonlinear in voltage magnitudes and angles, they can be solved by using numerical techniques like the Newton–Raphson method.

As described in Section 3.4.1, there are four types of buses in the system. For the slack buses, the voltage magnitudes and angles are known and are usually assumed to be equal to $1\angle 0$. In an AC–MTDC grid, there could be multiple AC systems interconnected with each other through the DC grid. Those AC systems could be asynchronous with respect to each other. Therefore, there could be more than one slack bus in the AC system.

Assuming that the limits on the control loops are not hit, the MTDC PCC buses are treated as *P–Q* buses or *P–V* buses depending on different control laws described above. If the *i*th converter station is in *P–Q* control mode (Mode I), then $P_{gi} = P^*_{gi}$ and $Q_{gi} = Q^*_{gi}$ is assumed. For the converter stations with V_{dc}–Q control (Mode II), the real power is unknown at the corresponding PCC bus. Therefore, an initial guess of the real power $P^{(0)}_{gi}$ is made. It can be based on the fact that the sum of total real power injected into the MTDC grid is zero when losses are neglected. The reactive power is again assumed to be equal to Q^*_{gi}. For Mode III A, the PCC bus can be considered as a PV bus with $P_{gi} = P^*_{gi}$ and $V_{gi} = V^*_{gi}$. For Mode III B and Mode IV, an initial guess is made for the real power, while the bus voltage is specified. To consider a PCC bus with a converter pole outage, the control mode of the healthy converter will impose constraints according their control laws.

Let there be n_s slack buses and n_{pv} PV buses. Therefore, there are $r = n_{ac} - n_s - n_{pv}$ unknown voltage magnitudes and $s = n_{ac} - n_s$ unknown voltage angles. To determine the bus voltages, an initial guess of the unknown voltage magnitudes and angles (e.g., $1\angle 0$) is made. As described above, the following equation can be written:

$$\begin{bmatrix} [\Delta f_1]_{r\times 1} \\ --- \\ [\Delta f_2]_{s\times 1} \end{bmatrix}_{l\times 1} = \begin{bmatrix} \left[\frac{\partial f_1}{\partial V}\right]_{r\times r} & | & \left[\frac{\partial f_1}{\partial \Theta}\right]_{r\times s} \\ ---- & -- & ---- \\ \left[\frac{\partial f_2}{\partial V}\right]_{s\times r} & | & \left[\frac{\partial f_2}{\partial \Theta}\right]_{s\times s} \end{bmatrix}_{l\times l} \begin{bmatrix} [\Delta V]_{r\times 1} \\ --- \\ [\Delta \Theta]_{s\times 1} \end{bmatrix}_{l\times 1} \tag{3.130}$$

where $l = r + s$.
The element of the *i*th row and the *k*th column of the submatrix $\left[\frac{\partial f_1}{\partial V}\right]_{r\times r}$ is given by $\frac{\partial f_{1i}}{\partial V_k}$. The corresponding elements of the submatrices $\left[\frac{\partial f_1}{\partial \Theta}\right]_{r\times s}$, $\left[\frac{\partial f_2}{\partial V}\right]_{s\times r}$ and $\left[\frac{\partial f_2}{\partial \Theta}\right]_{s\times s}$ are given by $\frac{\partial f_{1i}}{\partial \Theta_k}$, $\frac{\partial f_{2i}}{\partial V_k}$, and $\frac{\partial f_{2i}}{\partial \Theta_k}$ respectively. The elements of the Jacobian are
for $k = i$,

$$\frac{\partial f_{1i}}{\partial V_k} = 2V_i G_{ii} + \sum_{k\neq i}^{n_{ac}} V_k \left[G_{ik} \cos\left(\Theta_i - \Theta_k\right) + B_{ik} \sin\left(\Theta_i - \Theta_k\right)\right] \tag{3.131}$$

$$\frac{\partial f_{1i}}{\partial \Theta_k} = \sum_{k=1}^{n_{ac}} V_i V_k \left[-G_{ik} \sin(\Theta_i - \Theta_k) + B_{ik} \cos(\Theta_i - \Theta_k)\right] \tag{3.132}$$

$$\frac{\partial f_{2i}}{\partial V_k} = -2V_i B_{ii} + \sum_{k\neq i}^{n_{ac}} V_k \left[G_{ik} \sin\left(\Theta_i - \Theta_k\right) - B_{ik} \cos\left(\Theta_i - \Theta_k\right)\right] \tag{3.133}$$

$$\frac{\partial f_{2i}}{\partial \Theta_k} = \sum_{k=1}^{n_{ac}} V_i V_k \left[G_{ik}\cos(\Theta_i - \Theta_k) + B_{ik}\sin(\Theta_i - \Theta_k)\right] \tag{3.134}$$

for $k \neq i$,

$$\frac{\partial f_{1i}}{\partial V_k} = V_i\left[G_{ik}\cos(\Theta_i - \Theta_k) + B_{ik}\sin(\Theta_i - \Theta_k)\right] \tag{3.135}$$

$$\frac{\partial f_{1i}}{\partial \Theta_k} = V_i V_k \left[G_{ik}\sin(\Theta_i - \Theta_k) - B_{ik}\cos(\Theta_i - \Theta_k)\right] \tag{3.136}$$

$$\frac{\partial f_{2i}}{\partial V_k} = V_i\left[G_{ik}\sin(\Theta_i - \Theta_k) - B_{ik}\cos(\Theta_i - \Theta_k)\right] \tag{3.137}$$

$$\frac{\partial f_{2i}}{\partial \Theta_k} = -V_i V_k \left[G_{ik}\cos(\Theta_i - \Theta_k) + B_{ik}\sin(\Theta_i - \Theta_k)\right] \tag{3.138}$$

The unknown bus voltage magnitudes and angles are determined iteratively as described previously. Readers are referred to the standard literature [41] on AC load flow for further details. After convergence, the voltage magnitudes and angles of all AC buses including the PCC buses are obtained.

Step II: MTDC Load flow Following Step I, $\overline{V}_k, k \neq i$ (i is a PCC bus for an MTDC converter station), obtained from the solution of the AC load flow is used for (3.90) and (3.87). $\overline{V}_{gi}$ is treated as a variable in the set of MTDC load flow equations and the value obtained from Step I can only be used as an initial guess. In line with (3.129), we can write

$$[\Delta f] = \left[[\Delta f_3]^T_{1\times N}\ [\Delta f_4]^T_{1\times N} \cdots [\Delta f_{18}]^T_{1\times N}\right]^T_{16N\times 1} \tag{3.139}$$

$$[J] = \begin{bmatrix} \left[\frac{\partial f_3}{\partial X}\right]_{N\times 16N} \\ \left[\frac{\partial f_4}{\partial X}\right]_{N\times 16N} \\ \vdots \\ \left[\frac{\partial f_{18}}{\partial X}\right]_{N\times 16N} \end{bmatrix}_{16N\times 16N} \tag{3.140}$$

where the unknown variables are

$$[\Delta X] = \begin{bmatrix} [\Delta V_g]^T_{1\times N}\ [\Delta \Theta_g]^T_{1\times N}\ [\Delta V_t^p]^T_{1\times N}\ [\Delta \Theta_t^p]^T_{1\times N} \cdots \\ [\Delta V_t^n]^T_{1\times N}\ [\Delta \Theta_t^n]^T_{1\times N}\ [\Delta P_g]^T_{1\times N}\ [\Delta Q_g]^T_{1\times N} \cdots \\ [\Delta P_g^p]^T_{1\times N}\ [\Delta Q_g^p]^T_{1\times N}\ \left[\Delta P_g^n\right]^T_{1\times N}\ \left[\Delta Q_g^n\right]^T_{1\times N} \cdots \\ [\Delta V_{dc}^p]^T_{1\times N}\ [\Delta V_{dc}^n]^T_{1\times N}\ [\Delta I_{dc}^p]^T_{1\times N}\ [\Delta I_{dc}^n]^T_{1\times N} \end{bmatrix}^T_{16N\times 1} \tag{3.141}$$

Expressions for the nonzero elements of the Jacobian are shown below.

For the interface of the ith converter station with the AC system

$$\frac{\partial f_{3i}}{\partial P_{gi}} = -1 \tag{3.142}$$

$$\frac{\partial f_{3i}}{\partial V_{gi}} = 2V_{gi}G_{ii} + \sum_{k\neq i}^{n_{ac}} V_k \left[G_{ik}\cos\left(\Theta_{gi} - \Theta_k\right) + B_{ik}\sin\left(\Theta_{gi} - \Theta_k\right)\right] \tag{3.143}$$

$$\frac{\partial f_{3i}}{\partial \Theta_{gi}} = \sum_{k=1}^{n_{ac}} V_{gi}V_k \left[-G_{ik}\sin(\Theta_{gi} - \Theta_k) + B_{gik}\cos(\Theta_{gi} - \Theta_k)\right] \tag{3.144}$$

$$\frac{\partial f_{4i}}{\partial Q_{gi}} = -1 \tag{3.145}$$

$$\frac{\partial f_{4i}}{\partial V_{gi}} = -2V_{gi}B_{ii} + \sum_{k\neq i}^{n_{ac}} V_k \left[G_{ik}\sin\left(\Theta_{gi} - \Theta_k\right) - B_{ik}\cos\left(\Theta_{gi} - \Theta_k\right)\right] \tag{3.146}$$

$$\frac{\partial f_{4i}}{\partial \Theta_{gi}} = \sum_{k=1}^{n_{ac}} V_{gi}V_k \left[G_{gik}\cos(\Theta_{gi} - \Theta_k) + B_{gik}\sin(\Theta_{gi} - \Theta_k)\right] \tag{3.147}$$

$$\frac{\partial f_{5i}}{\partial P_{gi}} = 1,\ \frac{\partial f_{5i}}{\partial P^p_{gi}} = -1,\ \frac{\partial f_{5i}}{\partial P^n_{gi}} = -1 \tag{3.148}$$

$$\frac{\partial f_{6i}}{\partial Q_{gi}} = 1,\ \frac{\partial f_{6i}}{\partial Q^p_{gi}} = -1,\ \frac{\partial f_{6i}}{\partial Q^n_{gi}} = -1 \tag{3.149}$$

$$\frac{\partial f_{7i}}{\partial P^p_{gi}} = 1 \tag{3.150}$$

$$\frac{\partial f_{7i}}{\partial V_{gi}} = 2V_{gi}G^p_i - V^p_{ti}\left[G^p_i\cos(\Theta_{gi} - \Theta^p_{ti}) + B^p_i\sin(\Theta_{gi} - \Theta^p_{ti})\right] \tag{3.151}$$

$$\frac{\partial f_{7i}}{\partial V^p_{ti}} = -V_{gi}\left[G^p_i\cos(\Theta_{gi} - \Theta^p_{ti}) + B^p_i\sin(\Theta_{gi} - \Theta^p_{ti})\right] \tag{3.152}$$

$$\frac{\partial f_{7i}}{\partial \Theta_{gi}} = V_{gi}V^p_{ti}\left[G^p_i\sin(\Theta_{gi} - \Theta^p_{ti}) - B^p_i\cos(\Theta_{gi} - \Theta^p_{ti})\right] \tag{3.153}$$

$$\frac{\partial f_{7i}}{\partial \Theta^p_{ti}} = -V_{gi}V^p_{ti}\left[G^p_i\sin(\Theta_{gi} - \Theta^p_{ti}) - B^p_i\cos(\Theta_{gi} - \Theta^p_{ti})\right] \tag{3.154}$$

$$\frac{\partial f_{8i}}{\partial Q_{gi}} = 1 \tag{3.155}$$

$$\frac{\partial f_{8i}}{\partial V_{gi}} = -2V_{gi}B^p_i - V^p_{ti}\left[G^p_i\sin(\Theta_{gi} - \Theta^p_{ti}) - B^p_i\cos(\Theta_{gi} - \Theta^p_{ti})\right] \tag{3.156}$$

$$\frac{\partial f_{8i}}{\partial V^p_{ti}} = -V_{gi}\left[G^p_i\sin(\Theta_{gi} - \Theta^p_{ti}) - B^p_i\cos(\Theta_{gi} - \Theta^p_{ti})\right] \tag{3.157}$$

$$\frac{\partial f_{8i}}{\partial \Theta_{gi}} = -V_{gi}V^p_{ti}\left[G^p_i\cos(\Theta_{gi} - \Theta^p_{ti}) + B^p_i\sin(\Theta_{gi} - \Theta^p_{ti})\right] \tag{3.158}$$

$$\frac{\partial f_{8i}}{\partial \Theta^p_{ti}} = V_{gi}V^p_{ti}\left[G^p_i \cos(\Theta_{gi} - \Theta^p_{ti}) + B^p_i \sin(\Theta_{gi} - \Theta^p_{ti})\right] \quad (3.159)$$

$$\frac{\partial f_{11i}}{\partial V^p_{dci}} = -I^p_{dci} \quad (3.160)$$

$$\frac{\partial f_{11i}}{\partial I^p_{dci}} = -V^p_{dci} \quad (3.161)$$

$$\frac{\partial f_{11i}}{\partial V_{gi}} = V^p_{ti}\left[G^p_i \cos(\Theta^p_{ti} - \Theta_{gi}) + B^p_i \sin(\Theta^p_{ti} - \Theta_{gi})\right] \quad (3.162)$$

$$\frac{\partial f_{11i}}{\partial V^p_{ti}} = -2V^p_{ti}G^p_i + V_{gi}\left[G^p_i \cos(\Theta^p_{ti} - \Theta_{gi}) + B^p_i \sin(\Theta^p_{ti} - \Theta_{gi})\right] \quad (3.163)$$

$$\frac{\partial f_{11i}}{\partial \Theta_{gi}} = V^p_{ti}V_{gi}\left[G^p_i \sin(\Theta^p_{ti} - \Theta_{gi}) - B^p_i \cos(\Theta^p_{ti} - \Theta_{gi})\right] \quad (3.164)$$

$$\frac{\partial f_{11i}}{\partial \Theta^p_{ti}} = V^p_{ti}V_{gi}\left[-G^p_i \sin(\Theta^p_{ti} - \Theta_{gi}) + B^p_i \cos(\Theta^p_{ti} - \Theta_{gi})\right] \quad (3.165)$$

$$\frac{\partial f_{13i}}{\partial I^p_{dci}} = 1, \frac{\partial f_{13i}}{\partial V^p_{dck}} = G_{ppki}, \frac{\partial f_{13i}}{\partial V^n_{dck}} = G_{pnki} \quad (3.166)$$

For Mode I:

$$\frac{\partial f_{15i}}{\partial P^p_{gi}} = -1, \frac{\partial f_{16i}}{\partial P^n_{gi}} = -1, \frac{\partial f_{17i}}{\partial Q^p_{gi}} = -1, \frac{\partial f_{18i}}{\partial Q^n_{gi}} = -1 \quad (3.167)$$

For Mode II:

$$\frac{\partial f_{15i}}{\partial V^p_{dci}} = -2V^p_{dci}, \frac{\partial f_{16i}}{\partial V^n_{dci}} = -2V^n_{dci}, \frac{\partial f_{17i}}{\partial Q^p_{gi}} = -1, \frac{\partial f_{18i}}{\partial Q^n_{gi}} = -1 \quad (3.168)$$

Similar expressions can be derived for Modes III and IV. Expressions for the elements of the Jacobian related to $f_{9i}, f_{10i}, f_{12i}, f_{14i}$, can be obtained by using the superscript n in place of p in the equations relating f_{7i}, f_{8i}, f_{11i}, and f_{13i}, respectively. Depending on the control mode certain elements of $[\Delta X]$ are zeros. For example, $\Delta V_g = 0$ for control Modes III A and III B. After the convergence of the iterative solution process, the boundary conditions of the PCC buses, i.e., voltage magnitudes, real power, and reactive power, are changed by the updated values obtained from the MTDC load flow.

Step III: Repetition of Steps I and II The AC load flow (i.e., Step I) is solved using the updated values of the boundary conditions at the PCC buses. The MTDC load flow (Step II) is also repeated. This process is continued till the kth iteration when the maximum value of $\left|\Delta P^k_{gi}\right| = \left|P^k_{gi} - P^{(k-1)}_{gi}\right|, \forall i = 1(1)n$ falls below a predefined threshold.

3.5 AC–MTDC Grid Model for Nonlinear Dynamic Simulation

In Section 3.2, the MTDC grid model was described assuming the AC system represented by voltage sources. Section 3.3 described the model of an AC grid. In this section, these models will be interfaced with each other to develop an integrated framework for nonlinear dynamic simulation of the AC–MTDC grid. The generic structure of the dynamic models of the individual components and the interfacing variables of the combined AC–MTDC grid is shown in Fig. 3.18.

For the multi-machine AC system shown in Fig. 3.18, the generators, excitation systems, loads, and other dynamic components are modeled by their respective

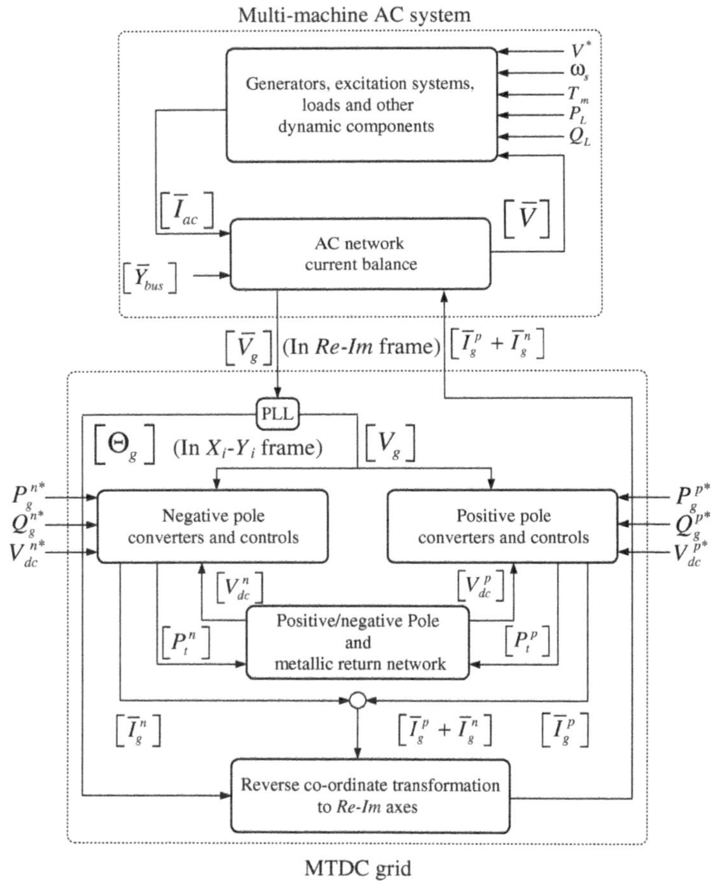

Figure 3.18 Block diagram of AC–MTDC grid model for nonlinear dynamic simulation. ©[36], 2011.

differential and algebraic equations (DAEs) as described in Section 3.3. In the interconnected AC–MTDC system, the dynamic equations associated with the ith generator, corresponding exciter, governor, and prime mover are expressed on the reference frame denoted by d_i–q_i rotating at the rotor angular speed ω_i. Since the stator dynamics are neglected, the stator voltages and currents are expressed in the common network reference frame *Re–Im* which rotates at the synchronous speed ω_s. Figure 3.12 shows the orientation of these frames of reference. The rotor angle of all the generators are measured with respect to the common reference frame. It should be noted that the speed of the d_i–q_i reference frame for each individual generator varies depending upon the solution of the swing dynamics (3.53).

For the MTDC grid model shown in Fig. 3.18, the voltages at the PCC of the ith converter station of the MTDC grid ($\overline{V}_{gi}$) is locked with the X-axis of the X_i–Y_i reference frame rotating at speed ω_{gi} (Fig. 3.2). As described in Section 3.2.2, the PLL dynamics (Section 2.4.2.5) is neglected. Therefore, the function of PLL here is to do coordinate transformation of $\overline{V}_{gi}$ from *Re–Im* to X_i–Y_i frame. The AC currents injected by the positive and the negative pole converters $\overline{I}_{gi}^{p/n} = I_{gXi}^{p/n} + jI_{gYi}^{p/n}$ are calculated using the converter and controller model shown in Fig. 3.4. The DC side power ($P_{dci}^{p/n}$) of the converters is derived from the PCC real power after accounting for the losses. ($P_{dci}^{p/n}$) is further used as input to the DC network model (Fig. 3.10) to obtain the DC link voltages.

The AC transmission network is modeled with algebraic equations, and the voltage and current phasors of this network are expressed in the common network reference frame *Re–Im*. As described in Section 3.3.3, the AC network is represented algebraically by the bus admittance matrix $\left[\overline{Y}_{bus}\right]$. A set of current balance equations including the current injection from the MTDC grid are solved to find the AC network bus voltages, as shown in Fig. 3.18. Since the converter AC side current injection is expressed in the X_i–Y_i frame, it is transformed into the common network reference frame *Re–Im* before using it in the AC current balance equation (3.84).

3.5.1 Initialization of Dynamic Models

It was mentioned in Section 3.2.2 that it is a common practice to start the dynamic simulation studies from the pre-disturbance steady-state condition. This is done to avoid lengthy simulation time needed for a large system before it reaches steady-state. To start the dynamic simulation from a particular steady-state operating condition (i.e., to obtain a "flat start"), the values of the state variables x, algebraic variables z, and control variables u, Section 3.2.5, need to be calculated. The input variables are directly obtained from their respective reference setpoints. The algebraic variables are determined from the load flow solution described in Section 3.4.3.

The state variables are initialized on the basis of the steady-state values of the algebraic and the input variables. Initialization of different components in the dynamic model of the AC–MTDC grid is described below:

3.5.1.1 MTDC Grid

Converter There are two state variables in the model of each converter pole of the ith converter station $I_{gXi}^{p/n}, I_{gYi}^{p/n}$. Equations (3.97) and (3.98) are used to obtain $I_{gXi}^{p/n}, I_{gYi}^{p/n}$ in the *Re–Im* frame. Coordinate transformation is used to obtain the X_i- and Y_i-axis components.

Controller A converged load flow solution ensures that all the reference and actual values of the variables are equal, thereby making sure the input errors to the compensators are zeros. For example, the current controller states κ_{IX} and κ_{IY} are zeros in the initial condition.

DC Network The AC–DC load flow solution gives V_{dci} and I_{dci} for each pole and metallic return path of the DC transmission line connected to the ith converter station. Equations (3.8) and (3.11) gives the net DC current going into each of the positive, negative, and metallic return networks. From (3.9), the current injected at each node can be determined. Equation (3.10) is used to find voltage at each node of the cable from the line section drops, which accounts for only the resistive drop during the dynamic condition.

3.5.1.2 AC Grid

Generator For the ith generator, six state variables δ_i, ω_i, E'_{qi}, E'_{di}, ψ_{1di}, and ψ_{2qi} need to be initialized (Section 3.3.1.1). If sub-transient saliency is taken into account, it will introduce another state variable E_{dci} (Section 3.3.1.3).

First, generator terminal voltages and current injections are obtained from AC–DC load flow. Equation (3.62) (or (3.70) with sub-transient saliency) is solved to obtain E_i (or E_{eqi}) linking the current injection with the state variables. The state variables at steady-state are calculated by solving the set of nonlinear algebraic equations obtained by setting the derivative terms in (3.52–3.57) (and (3.71) when sub-transient saliency is considered).

Excitation System For the IEEE-DC1A excitation system of the ith generator, the steady-state parameters of the four state variables V_{tri}, E_{fdi}, V_{ri}, and R_{Fi} are obtained by equating the derivative terms of (3.72–3.75) to zeros. Since the resulting equations are nonlinear, an iterative method like the N–R method described in Section 3.4.3 can be used to solve them.

The IEEE-ST1 model has only one state, which is solved by equating the derivative term in (3.76) to zero.

Turbine and Governor The state variables are initialized by equating (3.78) and (3.80) to zero.

3.6 Small-signal Stability Analysis of AC–MTDC Grid

The modeling framework for the nonlinear dynamic simulation of the AC–MTDC grid was described in Section 3.5. However, the interaction among the state variables is often not clear from such nonlinear time-domain simulations. This insight is important in root-cause analysis of the system dynamic behavior and also for control design. Small-signal analysis in frequency-domain is very useful in this regard. A linearized model around an operating condition is required for such small-signal analysis, which is described in this section.

The set of nonlinear DAEs in (3.15–3.17) can be linearized around the equilibrium point $\{x_0, z_0, u_0\}$. The steps for obtaining the equilibrium point was described in 3.5.1. The linearization is done by perturbing x, z, and u by a small amount Δx, Δz, and Δu; expanding the equations into a Taylor series as described in Section 3.4.3; and neglecting the second and higher order terms:

$$\Delta \dot{x} = \frac{\partial \varsigma}{\partial x}\Delta x + \frac{\partial \varsigma}{\partial z}\Delta z + \frac{\partial \varsigma}{\partial u}\Delta u \tag{3.169}$$

$$0 = \frac{\partial g}{\partial x}\Delta x + \frac{\partial g}{\partial z}\Delta z + \frac{\partial g}{\partial u}\Delta u \tag{3.170}$$

$$\Delta y = \frac{\partial h}{\partial x}\Delta x + \frac{\partial h}{\partial z}\Delta z + \frac{\partial h}{\partial u}\Delta u \tag{3.171}$$

The architecture of the model used for small-signal stability analysis is the same as that of the nonlinear dynamic model shown in Fig. 3.18. However, the components are represented by their linearized models and the variables are replaced by small perturbations as shown in Fig. 3.19. The linear model of SGs, excitation systems, and other components are well documented in literature. They can be derived from their respective state-space models as mentioned briefly in this chapter. Interested readers are referred to [12, 38]. The linear model of the MTDC grid is derived next.

3.6.1 Linear Model of Converters and Controllers

Following the per-unit system mentioned in Section 3.2.6.1, the factor of $\frac{3}{2}$ (or $\frac{2}{3}$) associated with the real and reactive power relations is omitted from the state-space equations in Section 3.2.5. These equations are linearized around a nominal condition $\{x_0, z_0, u_0\}$ and the linear model is expressed in terms of small perturbations in the state (Δx), algebraic (Δz), and output (Δy) variables.

Mode I—*P*–*Q* Control The linear model for this control mode is described below:

- The state variables of the linear model are $\Delta x = \left[\Delta I^p_{gXi} \Delta I^p_{gYi} \Delta \kappa_{IX} \Delta \kappa_{IY}\right]^T$, the control variables are $\Delta u = \left[\Delta P^{p*}_{gi}\, \Delta Q^{p*}_{gi}\right]^T$, the algebraic variable is $\Delta z = \left[\Delta V_{gi}\right]$, and the output variables are $\Delta y = \left[\Delta P^p_{ti} \Delta I^p_{gXi} \Delta I^p_{gYi}\right]^T$.
- The parameters of the linearized state matrix $\left[\frac{\partial \varsigma}{\partial x}\right]$ are given by the elements of Jacobian at the nominal condition. Only nonzero elements are mentioned

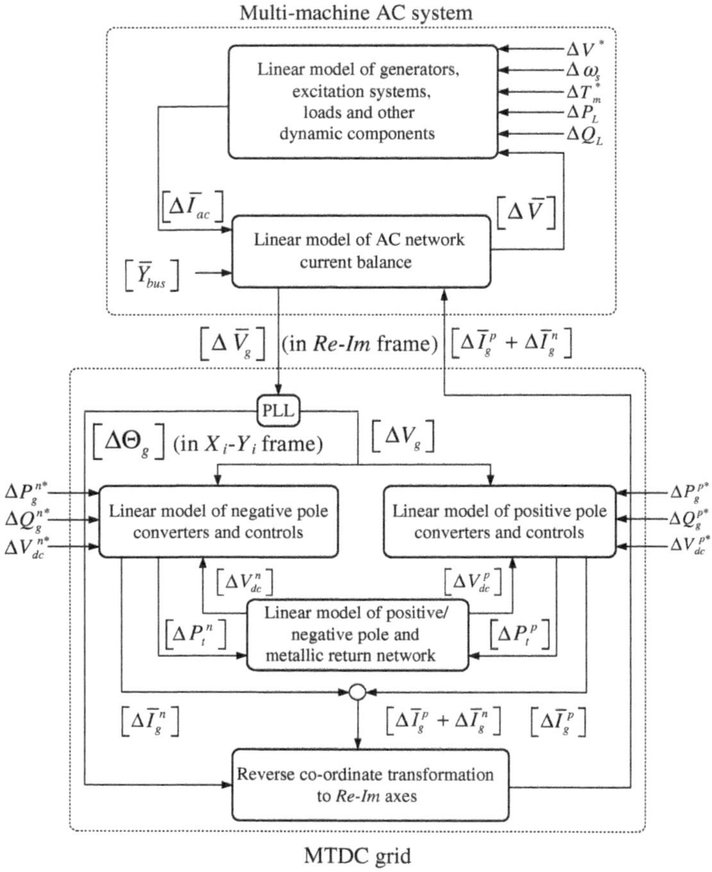

Figure 3.19 Block diagram of the linear model of AC–MTDC grid.

here.

$$\frac{\partial \varsigma_1}{\partial I_{gXi}^p} = -\left(\frac{R_i^p + K_{pI}}{L_i^p}\right), \frac{\partial \varsigma_1}{\partial \kappa_{IX}} = \frac{K_{iI}}{L_i^p} \tag{3.172}$$

$$\frac{\partial \varsigma_2}{\partial I_{gYi}^p} = -\left(\frac{R_i^p + K_{pI}}{L_i^p}\right), \frac{\partial \varsigma_2}{\partial \kappa_{IY}} = \frac{K_{iI}}{L_i^p} \tag{3.173}$$

$$\frac{\partial \varsigma_3}{\partial I_{gXi}^p} = -1, \frac{\partial \varsigma_4}{\partial I_{gYi}^p} = -1 \tag{3.174}$$

(3.175)

- The nonzero elements of the matrix $\left[\frac{\partial \varsigma}{\partial z}\right]$ are given by

$$\frac{\partial \varsigma_1}{\partial V_{gi}} = -\frac{K_{pI} P^{p*}_{gi}}{L^p_i V^2_{gi}}, \frac{\partial \varsigma_2}{\partial V_{gi}} = \frac{K_{pI} Q^{p*}_{gi}}{L^p_i V^2_{gi}} \tag{3.176}$$

$$\frac{\partial \varsigma_3}{\partial V_{gi}} = -\frac{P^{p*}_{gi}}{V^2_{gi}}, \frac{\partial \varsigma_4}{\partial V_{gi}} = \frac{Q^{p*}_{gi}}{V^2_{gi}} \tag{3.177}$$

- The nonzero elements of the matrix $\left[\frac{\partial \varsigma}{\partial u}\right]$ are given by

$$\frac{\partial \varsigma_1}{\partial P^{p*}_{gi}} = \frac{K_{pI}}{L^p_i V_{gi}}, \frac{\partial \varsigma_2}{\partial Q^{p*}_{gi}} = -\frac{K_{pI}}{L^p_i V_{gi}} \tag{3.178}$$

$$\frac{\partial \varsigma_3}{\partial P^{p*}_{gi}} = \frac{1}{V_{gi}}, \frac{\partial \varsigma_4}{\partial Q^{p*}_{gi}} = -\frac{1}{V_{gi}} \tag{3.179}$$

- The nonzero elements of the matrix $\left[\frac{\partial h}{\partial x}\right]$ are given by

$$\frac{\partial h_1}{\partial I^p_{gXi}} = K_{pI} \frac{P^{p*}_{gi}}{V_{gi}} - 2K_{pI} I^p_{gXi} + \kappa_{IX} K_{iI} + V_{gi} \tag{3.180}$$

$$\frac{\partial h_1}{\partial I^p_{gYi}} = -K_{pI} \frac{Q^{p*}_{gi}}{V_{gi}} - 2K_{pI} I^p_{gYi} + \kappa_{IX} K_{iI} \tag{3.181}$$

$$\frac{\partial h_1}{\partial \kappa_{IX}} = K_{iI} I^p_{gXi}, \frac{\partial h_{1i}}{\partial \kappa_{IY}} = K_{iI} I^p_{gYi} \tag{3.182}$$

$$\frac{\partial h_2}{\partial I^p_{gXi}} = 1, \frac{\partial h_3}{\partial I^p_{gYi}} = 1 \tag{3.183}$$

- The nonzero elements of the matrix $\left[\frac{\partial h}{\partial u}\right]$ are given by

$$\frac{\partial h_1}{\partial P^{p*}_{gi}} = K_{pI} \frac{I^p_{gXi}}{V_{gi}}, \frac{\partial h_1}{\partial Q^{p*}_{gi}} = -K_{pI} \frac{I^p_{gYi}}{V_{gi}} \tag{3.184}$$

- The nonzero elements of the matrix $\left[\frac{\partial h}{\partial z}\right]$ are given by

$$\frac{\partial h_1}{\partial V_{gi}} = -K_{pI} \frac{P^{p*}_{gi}}{\left(V_{gi}\right)^2} I^p_{gXi} + I^p_{gXi} + K_{pI} \frac{Q^{p*}_{gi}}{\left(V_{gi}\right)^2} I^p_{gYi} \tag{3.185}$$

Mode II—V_{dc}–Q Control The linear model for this control mode is described below.

- The state variables are $\Delta x = \left[\Delta I^p_{gXi} \Delta I^p_{gYi} \Delta\kappa_{IX} \Delta\kappa_{IY} \Delta\kappa_{dc}\right]^T$, the control variables are $\Delta u = \left[\Delta V^{p*}_{dci} \Delta Q^{p*}_{gi}\right]^T$, and the algebraic variables are $\Delta z = \left[\Delta V_{gi} \Delta V^p_{dci}\right]^T$.
- The nonzero elements of $\left[\frac{\partial\varsigma}{\partial x}\right]$ are given by

$$\frac{\partial\varsigma_1}{\partial\kappa_{dc}} = -\frac{K_{pI}K_{idc}}{L^p_i}, \frac{\partial\varsigma_3}{\partial\kappa_{dc}} = -K_{idc} \tag{3.186}$$

The values of $\frac{\partial\varsigma_1}{\partial I^p_{gXi}}$, $\frac{\partial\varsigma_1}{\partial\kappa_{IX}}$, $\frac{\partial\varsigma_2}{\partial I^p_{gYi}}$, $\frac{\partial\varsigma_2}{\partial\kappa_{IY}}$, $\frac{\partial\varsigma_3}{\partial I^p_{gXi}}$, and $\frac{\partial\varsigma_4}{\partial I^p_{gYi}}$ are the same as those for *P–Q* control.

- The nonzero elements of $\left[\frac{\partial\varsigma}{\partial u}\right]$ are given by

$$\frac{\partial\varsigma_1}{\partial V^{p*}_{dci}} = -2V^{p*}_{dci}\frac{K_{pI}K_{pdc}}{L^p_i}, \frac{\partial\varsigma_3}{\partial V^{p*}_{dci}} = -2K_{pdc}V^{p*}_{dci}, \frac{\partial\varsigma_{10}}{\partial V^{p*}_{dci}} = 2V^{p*}_{dci} \tag{3.187}$$

Other related elements of $\frac{\partial\varsigma_2}{\partial Q^{p*}_{gi}}$, $\frac{\partial\varsigma_4}{\partial Q^{p*}_{gi}}$ remain the same as in *P–Q* control.

- The nonzero elements of $\left[\frac{\partial\varsigma}{\partial z}\right]$ are given by

$$\frac{\partial\varsigma_1}{\partial V^p_{dci}} = 2V^p_{dci}\frac{K_{pI}K_{pdc}}{L^p_i}, \frac{\partial\varsigma_3}{\partial V^p_{dci}} = 2K_{pdc}V^p_{dci}, \frac{\partial\varsigma_{10}}{\partial V^p_{dci}} = -2V^p_{dci} \tag{3.188}$$

Elements of $\frac{\partial\varsigma_2}{\partial V_{gi}}$ remain the same as in *P–Q* control.

- The nonzero elements of $\left[\frac{\partial h_1}{\partial x}\right]$ are given by

$$\frac{\partial h_1}{\partial I^p_{Xi}} = -K_{pI}\left[K_{pdc}\left\{\left(V^{p*}_{dci}\right)^2 - \left(V^p_{dci}\right)^2\right\} - K_{idc}\kappa_{dc}\right] -2K_{pI}I^p_{Xi} + \kappa_{IX}K_{iI} + V_{gi} \tag{3.189}$$

$$\frac{\partial h_1}{\partial\kappa_{dc}} = -K_{pI}K_{idc}I^p_{Xi} \tag{3.190}$$

The values of $\frac{\partial h_1}{\partial I^p_{Yi}}$, $\frac{\partial h_1}{\partial\kappa_{IX}}$, $\frac{\partial h_1}{\partial\kappa_{IY}}$ are the same as those in the *P–Q* control mode.

- The nonzero elements of $\left[\frac{\partial h_1}{\partial u}\right]$ are given by

$$\frac{\partial h_1}{\partial V^{p*}_{dci}} = -2K_{pI}K_{pdc}V^{p*}_{dci} \tag{3.191}$$

Value of $\frac{\partial h_1}{\partial Q^{p*}_{gi}}$ is the same as in *P–Q* control.

- The nonzero elements of $\left[\frac{\partial h_1}{\partial z}\right]$ are given by

$$\frac{\partial h_1}{\partial V_{gi}} = I^p_{Xi} + \frac{K_{pI} Q^{p*}_{gi}}{V^2_{gi}} I^p_{Xi}, \frac{\partial h_1}{\partial V^p_{dci}} = 2K_{pI} K_{pdc} V^p_{dci} I^p_{Xi} \tag{3.192}$$

Mode III A—*P*–V_g–*Q* Control The linear model for this control mode is described below.

- The state variables, control variables, and algebraic variables are $\Delta x = \left[\Delta I^p_{gXi} \Delta I^p_{gYi} \Delta \kappa_{IX} \Delta \kappa_{IY} \Delta \kappa_{vg}\right]^T$, $\Delta u = \left[\Delta P^{p*}_{gi} \Delta V^*_{gi}\right]^T$, and $\Delta z = \left[\Delta V_{gi}\right]^T$, respectively.
  Partial derivatives of ς_1 and ς_3 are the same as in *P–Q* control mode.
- The nonzero elements of $\left[\frac{\partial \varsigma}{\partial x}\right]$ are given by

$$\frac{\partial \varsigma_2}{\partial \kappa_{vg}} = \frac{K_{pI} K_{ivg}}{L^p_i} \tag{3.193}$$

$$\frac{\partial \varsigma_4}{\partial I^p_{gYi}} = -1, \frac{\partial \varsigma_4}{\partial \kappa_{vg}} = K_{ivg} \tag{3.194}$$

  Expressions of $\frac{\partial \varsigma_2}{\partial I^p_{gYi}}$ and $\frac{\partial \varsigma_2}{\partial \kappa_{IY}}$ remain the same as in *P–Q* mode.
- The nonzero elements of $\left[\frac{\partial \varsigma}{\partial u}\right]$ are given by

$$\frac{\partial \varsigma_2}{\partial V^*_{gi}} = \frac{K_{pI} K_{pvg}}{L^p_i}, \frac{\partial \varsigma_{10}}{\partial V^*_{gi}} = 1 \tag{3.195}$$

$$\frac{\partial \varsigma_4}{\partial V^*_{gi}} = K_{pvg} \tag{3.196}$$

- The nonzero elements of $\left[\frac{\partial \varsigma}{\partial z}\right]$ are given by

$$\frac{\partial \varsigma_2}{\partial V_{gi}} = -\frac{K_{pI} K_{pvg}}{L^p_i}, \frac{\partial \varsigma_4}{\partial V_{gi}} = -K_{pvg}, \frac{\partial \varsigma_{10}}{\partial V_{gi}} = -1 \tag{3.197}$$

- The nonzero elements of $\left[\frac{\partial \varsigma'}{\partial x}\right]$ are given by

$$\frac{\partial \varsigma'_2}{\partial I^p_{gYi}} = \frac{K_{pI} K_{pq}}{L^n_i}, \frac{\partial \varsigma'_2}{\partial I^n_{gYi}} = -\frac{R^n_i + K_{pI} + K_{pI} K_{pq}}{L^n_i}, \frac{\partial \varsigma'_2}{\partial \kappa_{IY}} = \frac{K_{iI}}{L^n_i} \tag{3.198}$$

$$\frac{\partial \varsigma_2'}{\partial \kappa_{IY}} = \frac{K_{iI}}{L_i^n}, \frac{\partial \varsigma_2'}{\partial \kappa_q} = -\frac{K_{pI}K_{iq}}{L_i^n V_{gi}} \tag{3.199}$$

$$\frac{\partial \varsigma_4'}{\partial I_{gYi}^p} = K_{pq}, \frac{\partial \varsigma_4'}{\partial I_{gYi}^n} = -K_{pq} - 1 \tag{3.200}$$

$$\frac{\partial \varsigma_{10}'}{\partial I_{gYi}^p} = -V_{gi}, \frac{\partial \varsigma_{10}'}{\partial I_{gYi}^n} = V_{gi} \tag{3.201}$$

$$\frac{\partial \varsigma_4'}{\partial \kappa_q} = \frac{K_{iq}}{V_{gi}} \tag{3.202}$$

There exists no nonzero elements in $\left[\frac{\partial \varsigma}{\partial u}\right]$.

- The nonzero elements of $\left[\frac{\partial \varsigma'}{\partial z}\right]$ are given by

$$\frac{\partial \varsigma_2'}{\partial V_{gi}} = \frac{K_{pI}K_{iq}\kappa_q}{L_i^n \left(V_{gi}\right)^2}, \frac{\partial \varsigma_4'}{\partial V_{gi}} = \frac{K_{iq}\kappa_q}{\left(V_{gi}\right)^2}, \frac{\partial \varsigma_{10}'}{\partial V_{gi}} = -I_{gYi}^p + I_{gYi}^n \tag{3.203}$$

Expressions for the partial derivatives of h_1 and h_1' can be derived in a similar manner and are not included here.

Mode III B—V_{dc}–V_g–Q Control and Mode IV—V_g–ω_g Control The linear model can be derived using expressions from Modes I, II, and III A.

The linear model of the positive pole of the ith converter station is described here. Similar equations can be written for all converter stations in the positive and negative poles and combined together to obtain the linearized state-space model of all converters and controllers.

3.6.2 Linear Model of DC Network

- The linear model of the DC network is represented in terms of the following state, algebraic, and output variables as described earlier in Sections 3.2.5.4 and 3.2.5.5. The state variables are $\Delta x = \left[\Delta\mu_i^p \Delta\mu_i^n \Delta I_{qr}^p \Delta V_q^p \Delta I_{qr}^m \Delta V_q^m \Delta I_{qr}^n \Delta V_q^n\right]^T$, the algebraic variable is $\Delta z = \left[\Delta P_{ti}^p \Delta P_{ti}^n\right]^T$, and the output variable is $\Delta y = \left[\Delta V_{dci}^p \Delta V_{dci}^n\right]^T$.

- The parameters of the linearized state matrix $\left[\frac{\partial \varsigma}{\partial x}\right]$ are given by the elements of Jacobian at the nominal condition. Only nonzero elements are mentioned here.

$$\frac{\partial \varsigma_5}{\partial \mu_i^p} = -\frac{1}{R_{sci}^p C_{dci}^p}, \frac{\partial \varsigma_5}{\partial V_q^p} = \frac{1}{R_{sci}^p C_{dci}^p}, \frac{\partial \varsigma_5}{\partial V_q^m} = -\frac{1}{R_{sci}^p C_{dci}^p} \tag{3.204}$$

$$\frac{\partial \varsigma_6}{\partial I_{qr}^p} = -\frac{R_c^p}{L_c^p}, \frac{\partial \varsigma_6}{\partial V_q^p} = \frac{1}{L_c^p} \tag{3.205}$$

$$\frac{\partial \varsigma_7}{\partial \mu_i^p} = \frac{1}{R_{sci}^p C_c^p}, \frac{\partial \varsigma_7}{\partial I_{qr}^p} = \frac{1}{C_c^p} \tag{3.206}$$

$$\frac{\partial \varsigma_7}{\partial V_q^p} = \frac{1}{C_c^p}\left[\frac{P_{ti}^p}{\left(V_q^p - V_q^m\right)^2} - \frac{1}{R_{sci}^p}\right] \tag{3.207}$$

$$\frac{\partial \varsigma_7}{\partial V_q^m} = \frac{1}{C_c^p}\left[-\frac{P_{ti}^p}{\left(V_q^p - V_q^m\right)^2} + \frac{1}{R_{sci}^p}\right] \tag{3.208}$$

$$\frac{\partial \varsigma_8}{\partial I_{qr}^m} = -\frac{R_c^m}{L_c^m}, \frac{\partial \varsigma_8}{\partial V_q^m} = \frac{1}{L_c^m} \tag{3.209}$$

$$\frac{\partial \varsigma_9}{\partial \mu_i^p} = -\frac{1}{R_{sci}^p C_c^m}, \frac{\partial \varsigma_9}{\partial \mu_i^n} = -\frac{1}{R_{sci}^n C_c^m}, \frac{\partial \varsigma_9}{\partial I_{qr}^m} = \frac{1}{C_c^m} \tag{3.210}$$

$$\frac{\partial \varsigma_9}{\partial V_q^p} = \frac{1}{C_c^m}\left[-\frac{P_{ti}^p}{\left(V_q^p - V_q^m\right)^2} + \frac{1}{R_{sci}^p}\right] \tag{3.211}$$

$$\frac{\partial \varsigma_9}{\partial V_q^n} = \frac{1}{C_c^m}\left[-\frac{P_{ti}^n}{\left(V_q^n - V_q^m\right)^2} + \frac{1}{R_{sci}^n}\right] \tag{3.212}$$

$$\frac{\partial \varsigma_9}{\partial V_q^m} = \frac{1}{C_c^m}\left[\frac{P_{ti}^p}{\left(V_q^p - V_q^m\right)^2} + \frac{P_{ti}^n}{\left(V_q^n - V_q^m\right)^2} - \frac{1}{R_{sci}^p} - \frac{1}{R_{sci}^n} - \frac{1}{R_{gi}}\right] \tag{3.213}$$

- The nonzero elements of the matrix $\left[\frac{\partial \varsigma}{\partial z}\right]$ are given by

$$\frac{\partial \varsigma_7}{\partial P_{ti}^p} = -\frac{1}{C_c^p\left(V_q^p - V_q^m\right)} \tag{3.214}$$

$$\frac{\partial \varsigma_9}{\partial P_{ti}^p} = \frac{1}{C_c^m\left(V_q^p - V_q^m\right)}, \frac{\partial \varsigma_9}{\partial P_{ti}^n} = \frac{1}{C_c^m\left(V_q^n - V_q^m\right)} \tag{3.215}$$

- The nonzero elements of the matrix $\left[\frac{\partial h}{\partial z}\right]$ are given by

$$\frac{\partial h_4}{\partial V_q^p} = 1, \frac{\partial h_4}{\partial V_q^m} = -1 \tag{3.216}$$

3.6.3 Eigenvalue, Eigenvector, and Participation Factor

The linear model of the AC–MTDC grid can be used to obtain the eigenvalues and participation factors. Eigenvalue analysis provides valuable insight into the stability of the system, and eigenvectors, and participation factors are important for the root-cause analysis of time-domain behavior [12]. The overall state matrix (A) for the AC–MTDC system is required for eigenvalue analysis, which is obtained as follows: A linearized state-space model is derived by elimination of the vector of algebraic variables Δz from (3.169) to (3.171), which yields the following:

$$\begin{aligned} \Delta \dot{x} &= A\Delta x + B\Delta u \\ \Delta y &= C\Delta x + D\Delta u \end{aligned} \tag{3.217}$$

where A, B, C, D can be derived from the partial derivatives in (3.169–3.171) evaluated at equilibrium $\{x_0, z_0, u_0\}$ as follows:

$$A = \left[\frac{\partial \varsigma}{\partial x} - \frac{\partial \varsigma}{\partial z}\left(\frac{\partial g}{\partial z}\right)^{-1}\frac{\partial g}{\partial x}\right],\ B = \left[\frac{\partial \varsigma}{\partial u} - \frac{\partial \varsigma}{\partial z}\left(\frac{\partial g}{\partial z}\right)^{-1}\frac{\partial g}{\partial u}\right] \tag{3.218}$$

$$C = \left[\frac{\partial h}{\partial x} - \frac{\partial h}{\partial z}\left(\frac{\partial g}{\partial z}\right)^{-1}\frac{\partial g}{\partial x}\right],\ D = \left[\frac{\partial h}{\partial u} - \frac{\partial h}{\partial z}\left(\frac{\partial g}{\partial z}\right)^{-1}\frac{\partial g}{\partial u}\right] \tag{3.219}$$

The characteristic equation of the system can be obtained from this linearized state-space model as

$$|sI - A| = 0 \tag{3.220}$$

The roots of the characteristic equation are the eigenvalues of the system. The right (ϕ_i) and left (ψ_i) eigen vectors corresponding to the eigenvalues λ_i , $i = 1, 2, ..., n$ satisfy

$$A\phi_i = \lambda_i\phi_i,\ \psi_i A = \lambda_i\psi_i \tag{3.221}$$

The kth element of the right eigenvector ϕ_i measures the activity of the state variable x_k in the ith mode, while that of the left eigenvector ψ_i weighs the contribution of this activity to the ith mode [12].

To get around the scaling problem, the above measures are often combined to form the participation matrix P whose elements $p_{ki} = \phi_{ki}\psi_{ik}$ are termed as the participation factor. It is a measure of the relative participation of the kth state variable in the ith mode, and vice versa [12].

3.7 Transient Stability Analysis of AC–MTDC Grid

Small-signal stability analysis does not capture the system behavior accurately for large disturbances where the effect of nonlinearity could be pronounced. Hence, transient stability analysis based on nonlinear simulation is essential. The transient stability analysis involves nonlinear time-domain simulation of the system where a set of nonlinear ordinary differential equations (ODEs) are solved step by step using numerical integration techniques. Large disturbances such as faults and line or converter outages, etc., are analyzed.

The initial condition of the system denoted by (x_0, z_0, u_0) is obtained as described in Section 3.5.1. There are different approaches that can be used to do the transient stability analysis depending on

1. the integration method used (implicit or explicit, fixed-step or variable-step) to solve the differential equations (3.15);

2. the method used to solve the algebraic equations (3.16) and (3.17);

3. the sequential vs the simultaneous solution of the differential and algebraic equations.

Readers are referred to [12] for a detailed description.

In this context, the stiffness of the system equations is defined as the ratio of the smallest to the largest time constant of the system model: the smaller the ratio, the greater the stiffness. For the AC–MTDC system model, the problem could be very stiff. For such systems, implicit integration methods are more suitable than the explicit integration methods. They provide a better stability compared to explicit methods, but are computationally expensive. In the implicit integration technique, the Jacobian matrix is first derived from the above-mentioned equations and then they are solved using Newton-like methods in each time step. These equations are very sparse and therefore sparsity-oriented techniques are used to exploit this property [42].

Unlike a fixed-step solver, a variable-step solver is more suitable to speed up the simulation while maintaining a desired degree of accuracy. It works with lower integration step size when faster transients are present in the response. As the faster modes die out, the step size is increased dynamically to meet a specific error tolerance.

3.7.1 Large Disturbance Simulation

Large-scale disturbances which could cause transient stability problems in a AC–MTDC grid can be classified under the following three categories. Ways of simulating these disturbances are described next.

AC Fault A three-phase AC fault near a bus is simulated by appropriately modifying the self-conductance of the $\left[\overline{Y}_{bus}\right]$ at the faulted bus. Outage of an AC line following the fault is also simulated by using the post-fault $\left[\overline{Y}_{bus}\right]$ with appropriate modifications.

DC Fault The DC fault and DC cable outage are simulated by reducing the shunt resistance at the fault point. Unlike the AC network, the DC network is represented by a dynamic model. As shown in Fig. 3.10, the fault resistance R_f is inserted selectively corresponding to the faulted node. Depending on the number of cascaded pi sections used in the cable model (Fig. 3.1), the location of fault can be varied across the length of the cable. During fault, the value of $\frac{1}{R_f}$ increases from zero to a high value.

The faulted cable is isolated by opening the DC breaker at either end, which is emulated by increasing the series resistance of the pi sections at both ends to $10^6\ \Omega$ [26]. Although no DC breaker is commercially available at high power levels, a recent technological development has shown successful testing of a hybrid DC breaker which relies on power electronic switches [43]. The breaker can open in less than 5 ms. Details of the hybrid DC breaker is presented in Chapter 6. In this chapter, a fault clearing time of 5 ms is assumed while simulating the DC breaker.

Converter Outage Converter outage is simulated by opening both the DC side and the AC side breakers at the respective converter pole. However, the DC cable connections to the switchyard of the affected converter are retained. The opening of AC side breaker is simulated by instantaneously reducing the current injected by the corresponding pole to zero.

3.7.2 Representation of Rotor and Phase Angles

One important consideration is representation of generator rotor angles and also phase angles of the bus voltages in absolute or relative terms. During the step-by-step numerical integration process, the solution of the generator swing equation determines the speed of rotation ω_i of the d_i–q_i axis (Fig. 3.12). Under post-contingency steady-state condition, ω_i might not be equal to ω_s, resulting in monotonous increase in the average value of the rotor angle δ_i. Therefore, it is a common practice to express the rotor angles with respect to a reference generator.

As opposed to ω_i, the speed of rotation of the X_i–Y_i frame, ω_{gi}, is not a state variable. The speed of rotation is determined by the equation $\omega_{gi} = \left(\frac{d\Theta_{gi}}{dt} + \omega_s\right)$. The value of angle Θ_{gi} is obtained by solving the algebraic equations at each time step. Under the post-contingency steady-state condition, the value of ω_{gi} may be differ from ω_s, leading to similar variations as that of the rotor angle. Many commercial stability analysis programs express the bus voltage angles with respect to the slack bus voltage angle to avoid this problem.

3.8 Case Studies

Case studies on three test systems are presented in the following sections to demonstrate the usefulness of the proposed framework in analyzing different aspects of AC–MTDC systems. These case studies make use of the modeling framework presented in Sections 3.2–3.5 and the methodologies for small-signal and transient stability analysis described in Sections 3.6–3.7. First, a simplified version of the North-Sea benchmark system proposed in [44, 45] is studied. More complexity is considered in the second case study where a cascaded pi-section model of the DC transmission system is introduced [36]. Finally, the third case study considers integration of dynamic models of an interconnected AC system with the MTDC grid [36].

3.9 Case Study 1: The North Sea Benchmark System

3.9.1 Study Network

A simplified version of the envisioned offshore MTDC grid in the North Sea, shown in Fig. 3.20 [44], is considered here for the case study.

The study network comprises four bipole converter stations which are connected through three trans-national subsea DC cables as shown in Fig. 3.20. The system data

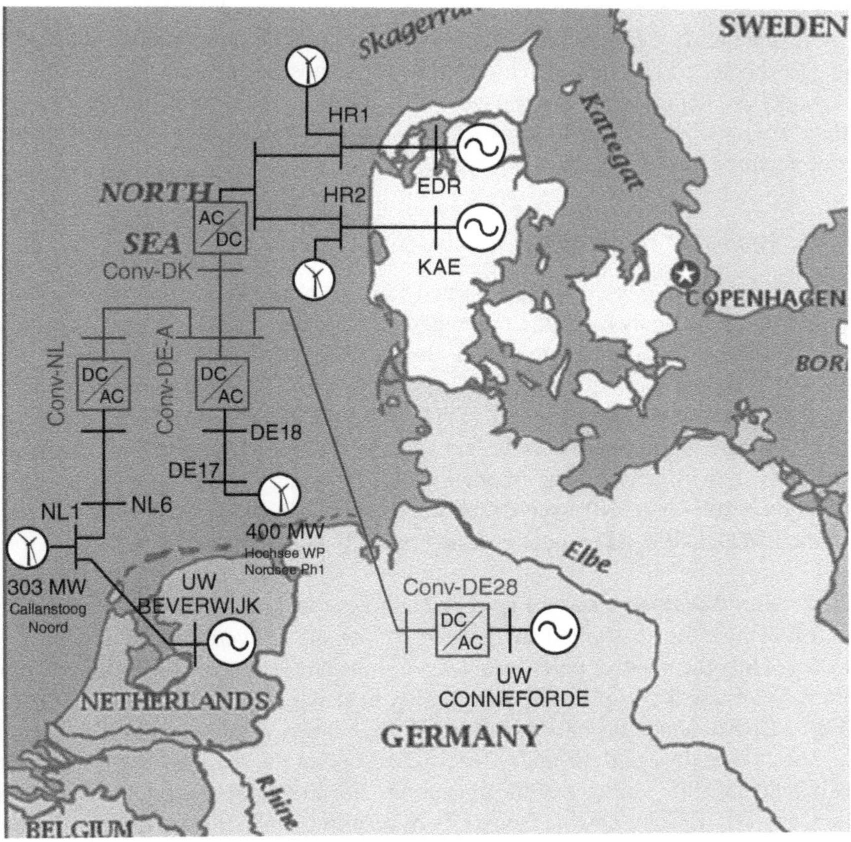

Figure 3.20 Case Study 1: Simplified benchmark test system for the envisioned offshore MTDC grid in the North Sea, corresponding to the year 2011. ©[45], 2011.

is taken from [44] and is not repeated here except for the converter parameters shown in Table 3.1. As shown in Fig. 3.20, the converter stations Conv-DK and Conv-NL are associated with Denmark and Netherlands, respectively, while Conv-DE28 and Conv-DE-A are part of Germany.

The converters are represented by their averaged model, Section 3.2.2, in MATLAB®/SIMULINK®. Conv-DK is considered to be operating as an inverter under Mode II–V_{dc}–Q control, Section 3.2.3.1, transporting about 800 MW under the nominal condition. Conv-DE28 and Conv-DE-A carry 400 MW each, operating as an inverter and rectifier, respectively. Conv-NL is in rectifying mode carrying 800 MW power.

The metallic return network of Conv-DK is grounded with 0.5 Ω resistance, while the rest are left ungrounded. A single pi section is used for modeling the DC transmission system, Section 3.2.3. Conv-DK keeps the DC link voltage constant at ±150 kV while the remaining stations operate in Mode I–P–Q control mode, Section 3.2.3.1, to maintain unity power factor at their respective PCC. Real power injected into the AC system is considered to be positive for the rest of the book.

Note that the AC network representation in this model is not complete because the actual European grid is interconnected. However, in absence of such data we consider this model as a starting point for analysis. The AC sources are modeled as ideal voltage sources behind impedances. The wind farms are represented as constant-power negative loads with unity power factor.

3.9.2 Nonlinear Simulation

Transient stability analysis was performed through a nonlinear time-domain simulation. Load flow analysis of the combined AC–MTDC grid was done to initialize the dynamic states of the system model, Section 3.5.1. To assess the accuracy of the proposed averaged model, a switched model, Section 2.5, of the converters and the rest of the system was built in the EMTDC/PSCAD [26] platform. Two-level converter topology is considered, and a sinusoidal PWM switching technique is used, Section 2.3.1, with a switching frequency of 1980 Hz. Validation of the nonlinear simulation results obtained from the averaged model in MATLAB®/SIMULINK® with the EMTDC/PSCAD model is done for small and large disturbances.

3.9.2.1 Small Disturbances Figure 3.21 shows the dynamic response of the system following a pulse applied to P_g^{p*} and P_g^{n*} of the converters in Conv-DE28 at $t = 0.1$ s. Only the positive pole variables are shown here because of symmetry. Note that the DC link voltage of Conv-DK is tightly held at ±150 kV (check the tight scale of Fig. 3.21(a)), hence visually appearing to have larger ripple content than the others. The power-flow pattern under the studied operating condition determined the steady-state DC link voltages of the converters. For example, around 800 MW flows from Conv-NL to Conv-DK via Conv-DE-A, resulting in the highest DC link voltage (±154 kV) at Conv-NL followed by Conv-DE-A (±152 kV).

Overall, the responses exhibit very good match between the switched model in EMTDC/PSCAD and the averaged model in MATLAB®/SIMULINK®. The former

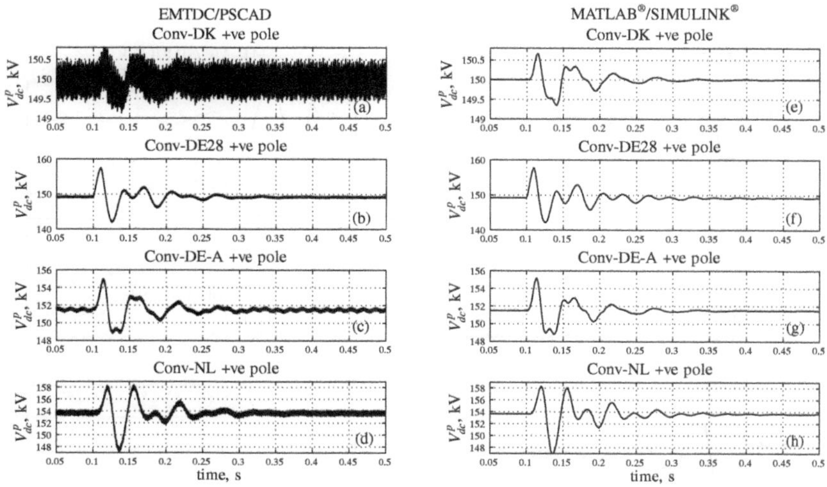

Figure 3.21 (a-h) Case Study 1: Validation of the averaged model in MATLAB®/SIMULINK® against the switched model in EMTDC/PSCAD. A pulse change to the real power reference of both the converters in Conv-DE28 is applied. ©[45], 2011.

was run with a fixed step size of 5.0 μs, while the latter was run with a variable step solver based on implicit integration technique, Section 3.7. The dynamic variations observed from the EMTDC/PSCAD model are clearly captured in the MATLAB®/SIMULINK® model except for the high-frequency switching ripples.

3.9.2.2 ***Converter Outage*** As described in Section 3.7.1, the converter outage was simulated by opening both the AC and the DC side breakers. Figures 3.22 and 3.23 show the dynamic behavior of the MTDC grid following the outage of the positive pole converter of Conv-DE-A.

Although the positive pole converter in Conv-DE-A is absent, the switchyard in the DC side remains connected to the DC cable network. As shown in Fig. 3.22(b), the switching ripple in this switchyard voltage is noticeably reduced after the outage at $t = 0.1$ s, compare with Fig. 3.22(c) showing the DC link voltage of the negative pole at the same converter station.

Figures 3.23(a), (d) and (b), (e) shows the real power at the PCC (P_g) of the converter stations from the averaged model and the switched model. Note that P_g at the positive pole of Conv-DE-A reduces to zero at a slower rate in the EMTDC/PSCAD simulation because of the operation of the AC side breaker at respective current zero of each phase—which is not the case for the phasor model. Since the slack converter station (Conv-DK) adjusts for the change in power, the dip in the corresponding P_g is higher in the MATLAB®/SIMULINK® model. In absence of one converter pole, an unbalance in the DC side occurs and the metallic return network connecting Conv-DK and Conv-DE-A carries current as high as 1.5 kA (Fig. 3.23(c) and (f)).

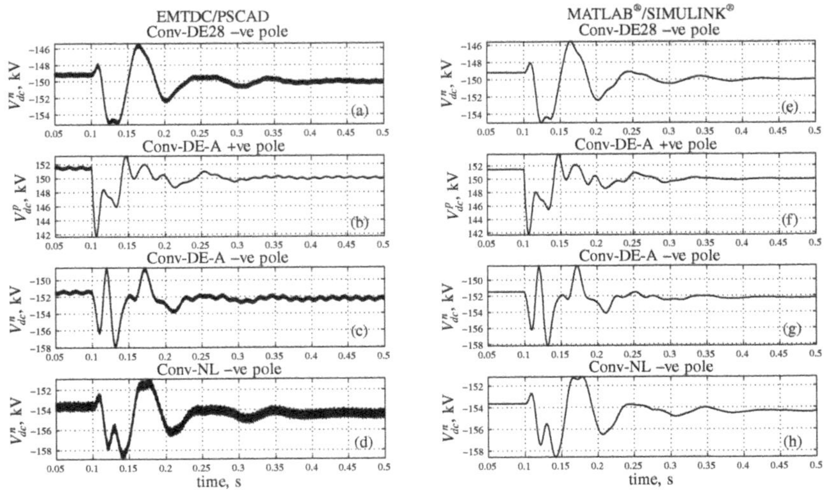

Figure 3.22 (a-h) Case Study 1: Comparison of the DC link voltage dynamics of the averaged model in MATLAB®/SIMULINK® against the switched model in EMTDC/PSCAD following the outage of the positive pole converter in Conv-DE-A. ©[45], 2011.

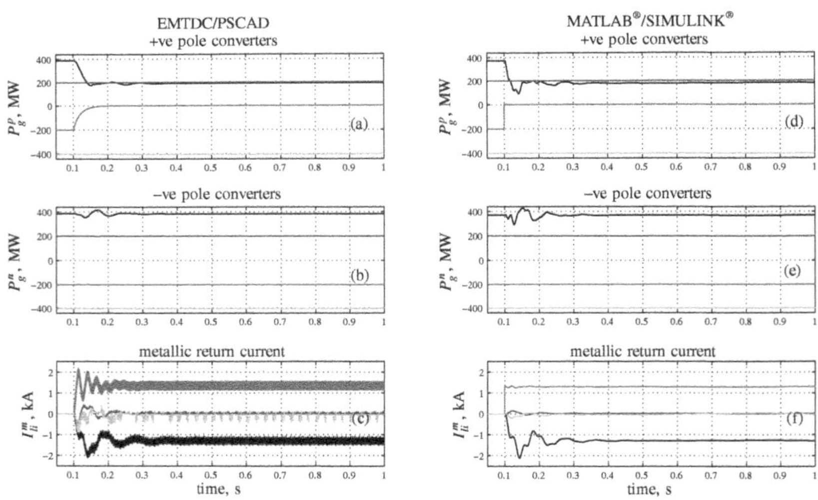

Figure 3.23 (a-f) Case Study 1: Comparison of the dynamic behavior of the averaged model in MATLAB®/SIMULINK® against the switched model in EMTDC/PSCAD following the outage of the positive pole converter in Conv-DE-A. ©[45], 2011.

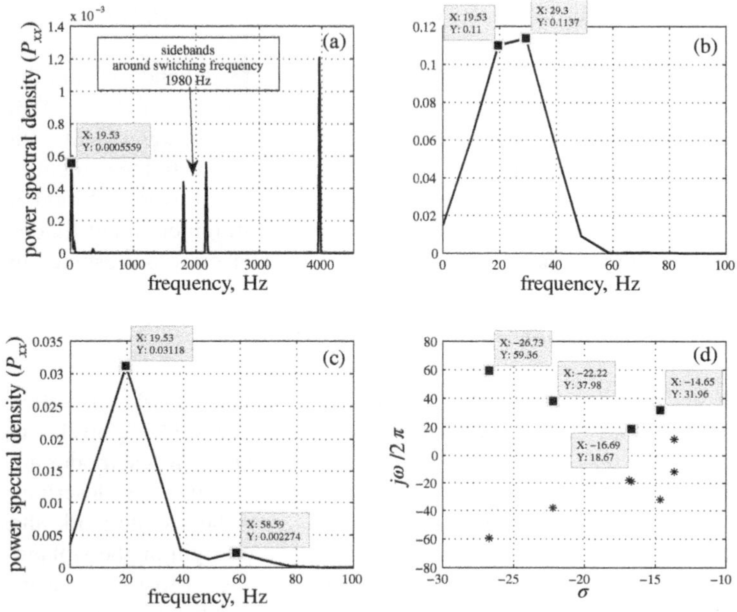

Figure 3.24 Case Study 1: Subplots (a), (b), (c): Power spectral density (P_{XX}) calculated from Welch Periodogram for determining the dominant modes in the DC link voltages of Conv-DK, Conv-DE28, and Conv-DE-A obtained from EMTDC/PSCAD as shown in Fig, 3.21. Subplot (d): modes obtained from eigenvalue analysis of the averaged model. ©[45], 2011.

As only Conv-DK is in V_{dc}–Q control mode, it shares the full burden of unbalance in current. Autonomous power sharing can ensure division of this burden amongst all the converters based on their ratings. This issue will be highlighted further in Case Study 3. A detailed treatment of autonomous power sharing is covered in Chapter 4.

Overall, it can be seen that there has been a close match between the dynamic responses observed in the EMTDC/PSCAD model and the MATLAB®/SIMULINK® model. This confirms the validity of the proposed averaged model for modal analysis and stability studies.

3.9.3 Small-signal Stability Analysis

3.9.3.1 Eigenvalue Analysis The nonlinear model was linearized as described in Section 3.6. Eigenvalue analysis was performed on the linear model, Section 3.6.3. As shown in Fig. 3.24(d), the eigenvalue plot obtained from the linear model highlights the presence of three modes with frequencies 18.7, 31.96, and 59.36 Hz, respectively. The accuracy of the results obtained from the linear model was validated by computing the power spectral density (P_{XX}) of the waveforms obtained from the EMTDC/PSCAD model using the Welch Periodogram technique to ascertain the principal modal content of the waveforms. The results for the DC link voltages of Conv-DK, Conv-DE28, and Conv-DE-A are shown in Fig. 3.24(a), (b), and (c), respectively.

As expected, the Conv-DK DC link voltage has prominent sidebands around the switching frequency of 1980 Hz. Presence of lower frequency modes of 19.53, 29.3, and 58.6 Hz are also apparent from Fig. 3.24(a-c), which is in close agreement with the modal frequencies obtained from linear analysis. Sidebands due to the switching ripples were not captured in the eigenvalues because of the use of averaged model. Participation factor analysis, Section 3.6.3, shows that the aforementioned modes have high participation from the states in the DC side of the converter as presented in the next section.

3.9.3.2 *Participation Factor Analysis*

Nominal Condition Modal participation factor analysis was carried out to ascertain the nature and the root cause of the dynamic response of the MTDC grid. It can be seen from the pole map of Fig. 3.25 (upper subplot) that the fastest modes are associated with the AC network (modeled by tie reactor dynamics as described in Sections 2.2, and 2.3) states, and the participation of the DC network and the converter controller states increases as we move toward right on the s-plane. The lower subplot shows a zoomed view of the critical poles. Most of these are associated with the DC network except one pole pair which has high participation from the current controllers of the converters. Refer to Section 3.2.5 for the

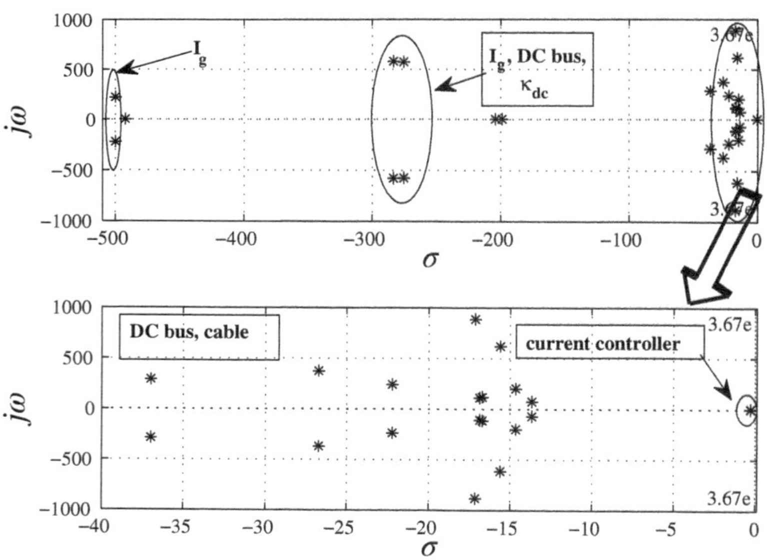

Figure 3.25 Case Study 1: Clustering of system poles based on participation factors in the nominal condition. ©[45], 2011.

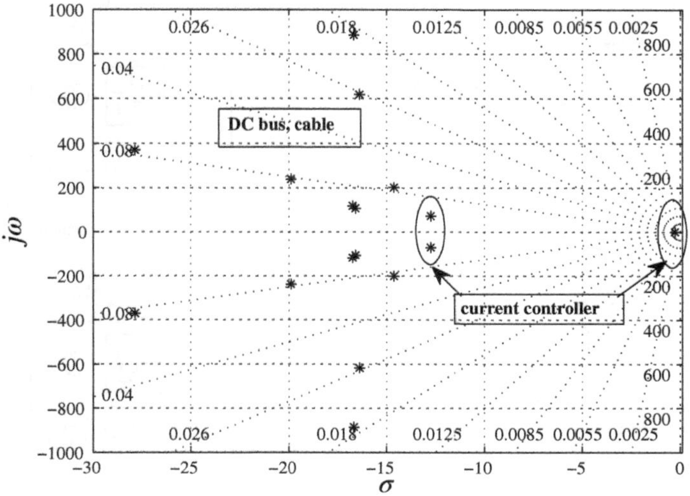

Figure 3.26 Case Study 1: Clustering of system poles based on participation factors with the outage of the positive pole converter in Conv-DE-A. ©[45], 2011.

dynamic states of the nonlinear model, and Section 3.6 for the corresponding state variables of the linear model.

Converter Outage Linear analysis was also done considering the outage of the positive pole converter in Conv-DE-A. It can be seen from Fig. 3.26 that the eigenvalue plot shows minimal change from the nominal condition except for a new pole pair $-12.76 \pm j71.45$ having high participation from the converter current controller states.

3.10 Case Study 2: MTDC Grid Connected to Equivalent AC Systems

3.10.1 Study Network

The study network consisting four bipole converter stations connected with 500 km of cables is shown in Fig. 3.27. Two converter stations (#2 and #3) are considered to be acting as rectifiers and the others (#1 and #4) as inverters under nominal condition. Three converter stations are in Mode I—*P*–*Q* control mode, Section 3.2.3.1 – each maintaining 900 MW real power at unity power factor (upf) at the PCC buses #2, #3, and #4 as shown in Fig. 3.27. Real power injected into the AC system is considered to be positive, and vice versa, for the rest of the paper. Converter station #1 operates in Mode II—V_{dc}–Q control mode, Section 3.2.3.1, and keeps the DC link voltage at ±350 kV with upf at the PCC. The metallic return is grounded with a 0.5-Ω resistance at converter station #1 while the other stations are left ungrounded.

The parameters of the above study network are shown in Table 3.1.

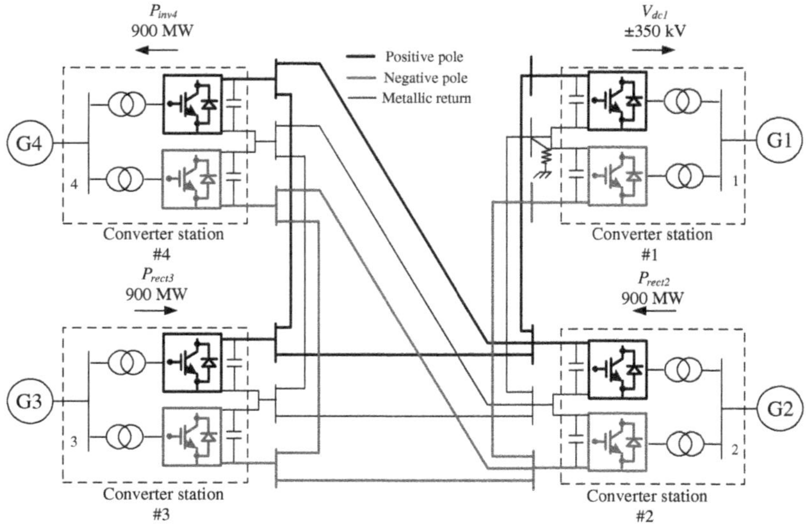

Figure 3.27 Case Study 2: Bipole MTDC grid with four converter stations. ©[36], 2011.

Table 3.1 MTDC grid Parameters from Figs 3.1 and 3.4

Parameters	Values	Parameters	Values
R^p/R^n	0.07 Ω	C^p_{dc}/C^n_{dc}	0.4 mF
L^p/L^n	40.0 mH	R^p_{sc}/R^n_{sc}	1.0 μΩ
R_g	0.5 Ω	$R^p_c/R^n_c/R^m_c$	1.5 Ω
L^m_c	0.30 H	L^p_c/L^n_c	0.30 H
C^m_c	40.0 μF	C^p_c/C^n_c	14.00 μF
R_f	0.01 Ω	$K_I(s)$	$300(\frac{R_c + sL_c}{s})$

The surrounding AC system was modeled by an ideal 400-kV voltage source behind a series impedance ($0.3 + j4.0\ \Omega$). The steady-state operating condition was determined by solving the AC–DC load flow equations described in Section 3.4.3.

3.10.2 Nonlinear Simulation

As discussed in Section 3.2.4, the approximation of the distributed cable model by four pi-sections can be considered adequate for the 500-km cable to reasonably capture frequency up to 60 Hz (Fig. 3.9). Therefore, four pi sections were used to model each DC cable in EMTDC/PSCAD. The following sections describe the comparison of response obtained from the nonlinear averaged model of the MTDC grid developed in MATLAB®/SIMULINK® with a switched model in PSCAD.

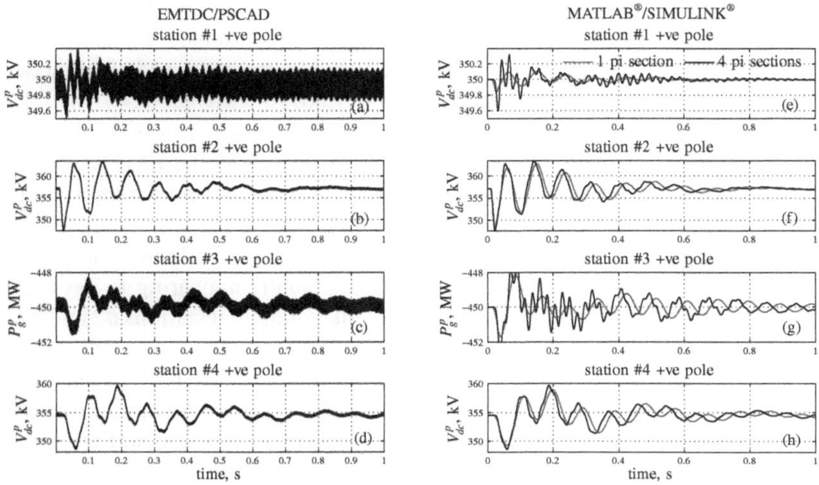

Figure 3.28 (a-h) Case Study 2: Validation of the averaged model in MATLAB®/SIMULINK® against switched model in EMTDC/PSCAD. A pulse change to the real power reference of both the converters in converter station #2 is applied. ©[36], 2011.

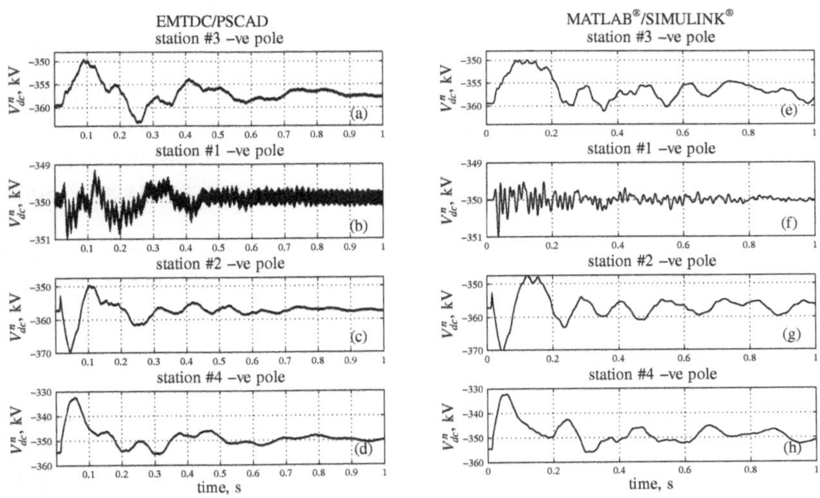

Figure 3.29 (a-h) Case Study 2: Comparison of the dynamic behavior of the averaged model in MATLAB®/SIMULINK® against switched model in EMTDC/PSCAD. A cable fault at the mid-point of the negative pole connecting converter stations #2 and #4 followed by its outage is simulated. ©[36], 2011.

3.10.2.1 Small Disturbances Figure 3.28 shows the dynamic behavior of the system in response to a pulse change applied to P_g^* of both the converters in converter station #2. Due to symmetry, only the positive pole variables are shown here. A close look at the responses shows very good match between the switched model in EMTDC/PSCAD and the averaged model in MATLAB®/SIMULINK®. The former was run with a fixed step size of 0.5 μs, while the latter was run with a variable-step solver, Section 3.7. The dynamic response observable in the EMTDC/PSCAD is clearly captured in the MATLAB®/SIMULINK® model except for the switching ripples.

Note that the DC link voltage of converter station #1 is tightly held at ±350 kV (check the tight scale of top left subplot) – hence visually appearing to have larger ripple content than the others. The resulting DC link voltages at stations #2 and #3 are slightly higher (±355 kV) as a result of the direction of power flow in the DC network. The subplots on the right column of Fig. 3.28 also compares the performance of the averaged model in MATLAB®/SIMULINK® when the cables are approximated by only one pi section instead of four sections. A single pi section could not capture certain high-frequency modes and showed substantial phase-shift in the response of the low-frequency modes.

3.10.2.2 Large Disturbances Section 3.7.1 described the method of simulating fault in DC cable and the outage of the cable by opening the DC breakers. The dynamic response of the system following a line-to-ground fault at the mid-point of the negative pole cable connecting converter stations #2 and #4 followed by the outage of the faulted cable is shown in Fig. 3.29. The DC bus voltages of the converter stations show large peak deviations from the pre-fault steady-state values except for converter #1.

It can be seen that the response of the averaged model is reasonably similar to that of the switched model—although not as close as that in the previous case study with a pulse disturbance. This is possibly due to the occurrence of transients in presence of cable capacitance and to other probable factors including different types of solvers and simulation step sizes used in the two platforms.

Successful validation of the averaged model against the switched model gives the confidence for adopting it in small-signal stability analysis of AC–MTDC grids.

3.10.3 Small-signal Stability Analysis

The importance of small-signal stability analysis for the root-cause analysis and control design was highlighted in Section 3.6. To that end, the nonlinear dynamic model was linearized, Section 3.6.3, around the nominal condition. Modal participation factor analysis was done to figure out the nature and the root cause of the dynamic response of the MTDC grid. It can be seen from the pole map of Fig. 3.30 (upper subplot) that the fastest modes are associated with the AC network (modeled by tie reactor impedance as mentioned in Sections 2.2 and 2.3) states (equations (3.24) and (3.25)), and the participation of the DC network and the converter controller states (Sections 3.2.5.4, 3.2.5.1) increases as we move toward right on the s-plane. The

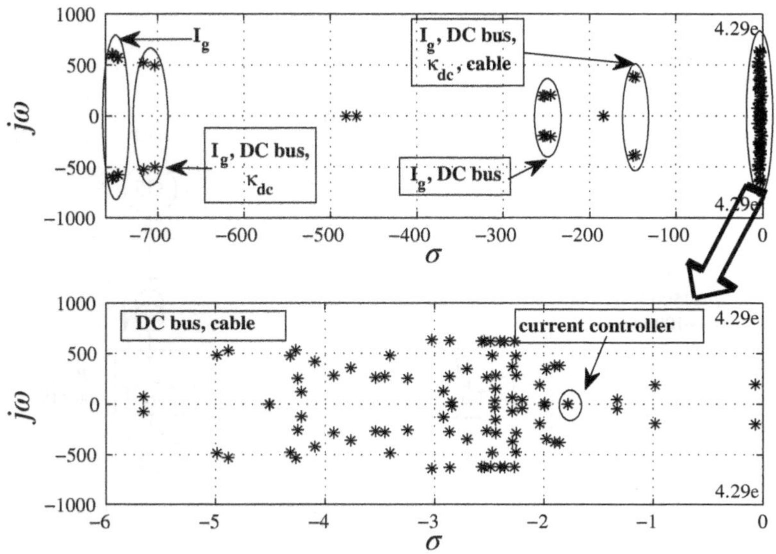

Figure 3.30 Case Study 2: Clustering of system poles based on participation factors. ©[36], 2011.

lower subplot shows a zoomed view of the critical poles. Most of these are associated with the DC network except one pole pair which has high participation from the current controllers of the converters (equations (3.22) and (3.33)).

3.11 Case Study 3: MTDC Grid Connected to Multi-machine AC System

3.11.1 Study Network

In Case Studies 1 and 2, the AC system was represented by isolated ideal voltage sources behind impedance. In reality, power system planners and operators need to understand the dynamic behavior of the MTDC grid surrounded by interconnected AC systems which include the dynamic model of AC system components. Figure 3.31 shows the test system used for steady-state analysis, transient stability study, and small-signal stability study of such a network. The same MTDC grid analyzed in Case Study 2 is used in this study. Since the averaged model of the MTDC grid was benchmarked with a switched model in EMTDC/PSCAD, Section 3.10.2, it was directly integrated with the surrounding AC system model following the modeling framework described in Section 3.5. The details of the connectivity of this grid will be described below.

The AC system consists of three asynchronous grids. The first grid has two generating areas each having two generators $G1, G2$ and $G3, G4$, respectively. The two areas are connected by a 230-kV AC tie-line which is 220 km long. Two asymmetric

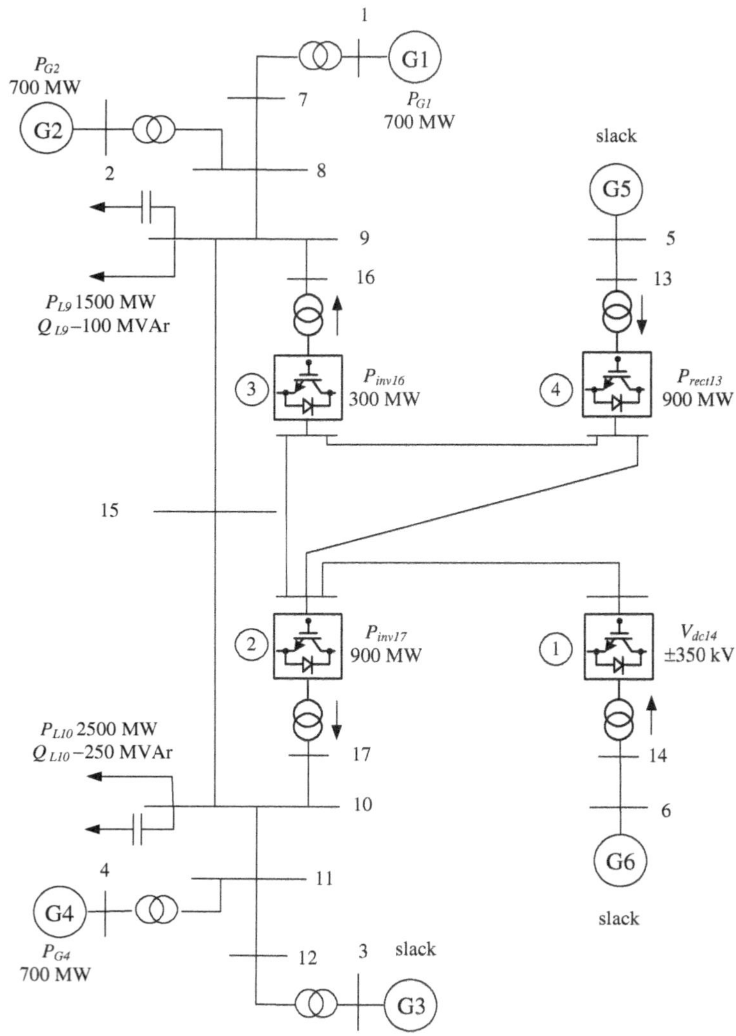

Figure 3.31 Case Study 3: Asymmetric bipole MTDC grid (single line diagram) connected to a multi-machine AC system. ©[36], 2011.

bipole HVDC converter stations (#2 and #3) connect each area to an MTDC grid as shown in Fig. 3.31 through a 500-km-long cable. The converter station bus 16 and 17 are connected to bus 9 and 10, respectively, through two feeders. Each generator ($G1$–$G6$) was represented by a sub-transient model described in Section 3.3.1. The generators are equipped with IEEE DC1A exciters described in Section 3.3.1.4. No turbine and governor are considered for this model, i.e., the generators are assumed to operate with constant T_m, Section 3.3.1.1.

The real power loads at buses #9 and #10 are 1500 and 2500 MW, respectively, which are modeled as constant-current loads. The corresponding reactive component of the loads are 100 and 250 MVAr (capacitive), which are modeled as constant impedance loads. Modeling of constant-current and constant-impedance load was described in Section 3.3.2.

The AC network beyond the PCC of the converter stations is modeled algebraically as mentioned in Section 3.3.3. The details of the generator parameters and AC transmission network of grid #1 can be found in [12]. The standard DAE model of the multi-machine AC system in the current injection framework was integrated with the dynamics of the converter transformers and phase reactors with proper adjustment of the reference frames as described in Section 3.5.

The second and third grids are isolated grids modeled by two aggregated generators $G5$ and $G6$, respectively. These generators were also modeled using similar assumptions as in $G1$–$G4$. The PCC bus of the converter stations #4 and #1 are buses #13 and #14, respectively, which are connected to the two grids through two short feeders (Fig. 3.31).

Under the nominal condition, each of $G1$, $G2$, and $G4$ generates 700 MW, and $G3$, $G5$, and $G6$ are left as slack. The tie-line transfer through the AC line connecting buses #9 and #10 is 200 MW, while converter stations #2, #3, and #4 operate in P–Q mode to maintain 900, 300, and 900 MW, respectively at upf at the PCC as shown in Fig. 3.31. Converter station #1 works in the V_{dc}–Q mode and keeps the DC link voltage constant at ±350 kV.

3.11.2 AC–MTDC Grid Load flow Solution

A sequential AC–MTDC grid load flow, Section 3.4.3, was done as described below.

- Step I – AC Load flow: First, the AC load flow is solved. There are three slack buses in the system shown in Fig. 3.31: $G3$, $G5$, and $G6$. Since stations #2, #3, and #4 are operating in Mode I, the corresponding PCC buses #17, #16, and #13 are treated as P–Q buses. For the PCC bus of station #1, i.e., bus #14, an initial guess of P_g is assumed on the basis of the real power balance in the MTDC grid neglecting losses. Table 3.2 summarizes the step-by-step load flow solution highlighting a few key components. It can be seen that the real power at bus #14 is assumed to be equal to equal to 3.0 p.u. (300 MW). The AC load flow converges after 18 iterations. The voltage profile of a few selected buses following the convergence of the AC load flow is shown in Table 3.2.

- As mentioned in Section 3.4.3, the voltages obtained from AC load flow solution for buses #5, #6, #9, and #10 are used for the MTDC grid load flow. Table 3.2 shows that these values are $1\angle 0°$, $1\angle 0°$, $0.979\angle -15.72°$, and $0.990\angle -1.18°$, respectively. The MTDC grid load flow converges in four iterations. Since the AC load flow was solved assuming exact value of P_g and Q_g at the PCC buses #13, #16, and #17, the converged voltage solution at these buses exactly matches with the MTDC grid load flow solution (Table 3.2). The real power at the PCC bus #14 was unknown, and therefore this value changes from $0.999\angle -0.39°$ to

Table 3.2 AC–MTDC grid load flow solution on 100 MVA base

AC–DC load flow iter.	1	2
$P_{g14}^{(k-1)}$	**3.0**	**3.3327**
No. of AC load flow iter.	18	18
V_5	$1\angle 0°$	$1\angle 0°$
V_6	$1\angle 0°$	$1\angle 0°$
V_9	$0.979\angle -15.72°$	$0.979\angle -15.72°$
V_{10}	$0.990\angle -1.18°$	$0.990\angle -1.18°$
V_{13}	$0.998\angle -1.18°$	$0.998\angle -1.18°$
V_{14}	$0.999\angle -0.39°$	$0.999\angle -0.44°$
V_{15}	$0.990\angle -21.28°$	$0.990\angle -21.28°$
V_{16}	$0.979\angle -15.31°$	$0.979\angle -15.31°$
V_{17}	$0.992\angle -25.46°$	$0.992\angle -25.46°$
No. of DC load flow iter.	4	4
V_{13}	$0.998\angle -1.18°$	$0.998\angle -1.18°$
V_{14}	$0.999\angle -0.44°$	$0.999\angle -0.44°$
V_{16}	$0.979\angle -15.31°$	$0.979\angle -15.31°$
V_{17}	$0.992\angle -25.46°$	$0.992\angle -25.46°$
$P_{g14}^{(k)}$	**3.3327**	**3.3327**
max$\lvert\Delta P_g^{(k)}\rvert$	**0.3327**	**0**

$0.999\angle -0.436°$ after the MTDC grid load flow converges. The real power at the PCC bus #14 at the end of the first iteration of the AC–MTDC grid load flow is given as $P_{g14}^{(k)} = 3.3327$. The maximum value of $|\Delta P_g^{(k)}| = 0.3327$ is more that the tolerance of 10^{-3} and therefore the AC–MTDC grid load flow iteration enters Step III of Section 3.4.3.

- The AC load flow is run again with the modified value of P_g at bus #14. After the convergence of the AC load flow, the final value of V_{14} becomes $0.999\angle -0.44°$, which is equal to the value obtained following the convergence of the MTDC grid load flow in iteration #1.

The AC–MTDC grid load flow solution is used to initialize the states of different components of the dynamic model as described in Section 3.5.1.

3.11.3 Small-signal Stability Analysis

After performing linearization, Section 3.6.3, the modal participation factor analysis was done to associate the oscillatory modes to different state variables of the combined AC–MTDC grid. Figure 3.32 shows the pole map of the 246th order test

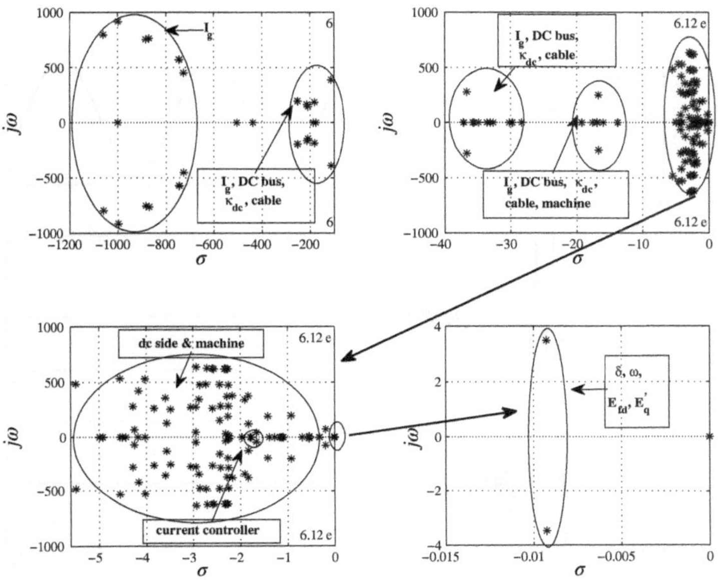

Figure 3.32 Case Study 3: Clustering of system poles based on participation factor analysis for the MTDC grid connected to a multi-machine AC system shown in Fig. 3.31. ©[36], 2011.

system. As expected, in addition to the modes associated with the MTDC grid shown in Section 3.9.3, the participation of generator states is more prominent as we move from left to right of the *s*-plane as shown in Fig. 3.32. There exists a poorly damped inter-area mode of 0.55 Hz frequency and 0.3 % damping where the states of generators, $G1$–$G4$ have the strongest participation as seen from the lower right subplot of Fig. 3.32. Refer to Sections 3.2.5 and 3.3 for a description of the MTDC grid states and AC grid states, respectively.

3.11.4 Nonlinear Simulation

3.11.4.1 AC Side Fault Nonlinear simulations in MATLAB®/SIMULINK® involving both AC and DC side faults were done. Figure 3.33 shows the dynamic behavior of the system following a three-phase self-clearing fault for about 5 cycles near bus #15 at $t = 0.01$ s.

In accordance with the small-signal stability analysis, the angular separation between $G1$ and $G3$ shows the presence of the poorly damped but stable inter-area mode (Fig. 3.33(a)). On the other hand, it can be seen from Fig. 3.33(b) and (c) that the positive pole DC link voltages of the stations #3 and #4 include higher frequency oscillatory modes. Careful observation of the converter station #3 DC link voltage reveals the presence of an 11-Hz mode immediately after the fault followed by the appearance of another mode with approximately 30 Hz frequency. These can primarily be attributed to the eigenvalues $-0.1948 \pm j73.6592$ and $-0.9259 \pm j196.7692$ having

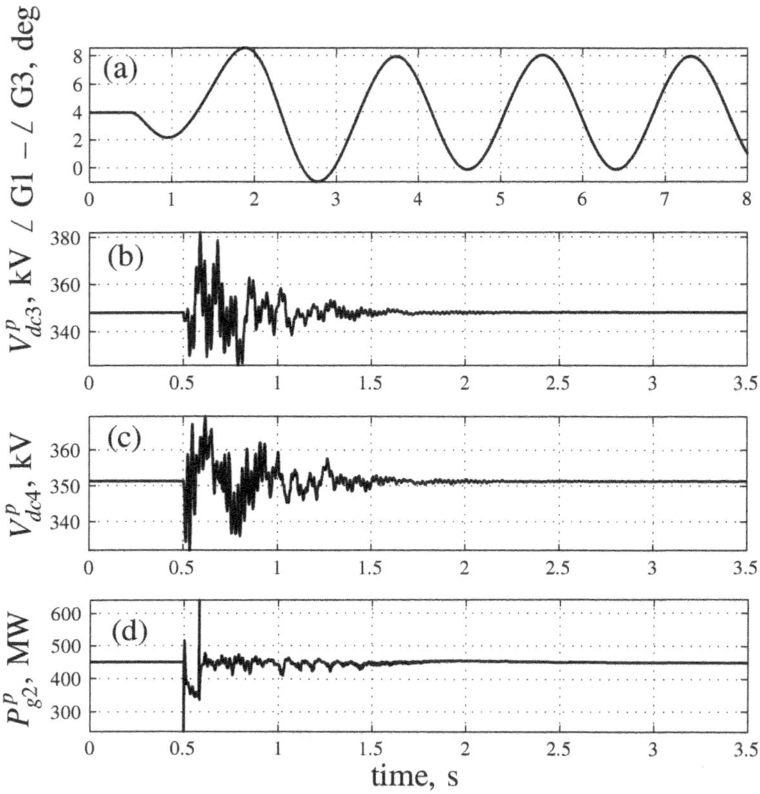

Figure 3.33 (a-d) Case Study 3: Dynamic performance of the combined AC–MTDC system (shown in Fig. 3.31) following a three-phase self-clearing fault at bus #15 for 80 ms starting at $t = 0.01$ s. ©[36], 2011.

very high participation from the DC network states as shown in Fig. 3.32. After the transient, all the parameters settle back to their pre-fault values, e.g., converter station #2 positive pole maintains around 450 MW (half of 900 MW) flow at bus #17 as shown in Fig. 3.33(d).

3.11.4.2 DC Cable Fault Next, a DC cable fault was considered at the mid-point of the negative pole cable connecting the converter stations #2 and #4. The fault was created at $t = 0.5$ s and cleared by opening the DC breakers at both ends of the faulted cable within 5.0 ms, Section 3.7.1. It can be seen from Fig. 3.34(a) that the poorly damped (but stable) inter-area mode is observable in the angular difference between $G4$ and $G2$. The DC link voltage of the positive pole at converter station #3, shown in (Fig. 3.34(b)) oscillates about its pre-fault value (348 kV) similar to the other DC side variables Fig. 3.34(c) and (d). The converter station #2 negative pole DC link voltage dips due to its vicinity to the fault on the cable connecting the converter stations #2

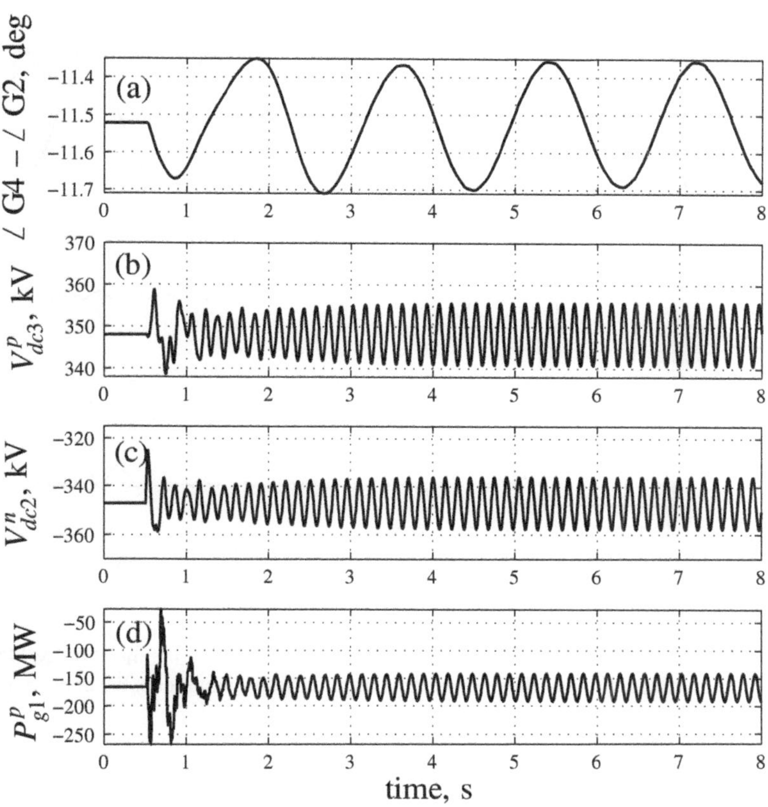

Figure 3.34 (a-d) Case Study 3: Dynamic performance of the combined AC–MTDC system (shown in Fig. 3.31) following a cable fault at the mid-point of the negative pole connecting converter stations #2 and #4. The fault occurs at $t = 0.5$ s and is cleared within 5 ms by isolating the faulted cable. ©[36], 2011.

Table 3.3 Critical modes of the MTDC grid connected to a multi-machine AC system

Pre-fault	Post-fault	Mode type
$-0.0093 \pm j3.4864$	$-0.0093 \pm j3.4864$	Inter-area
$-1.6650 \pm j42.3202$	$0.1975 \pm j40.2715$	DC network

and #4 as shown in Fig. 3.34(c). Also worth noting is the oscillations in the real power of the positive pole of station #1 about its initial value. The magnitude of the power is negative since the converter station is acting as a rectifier.

From the variations of several DC side quantities in Fig. 3.34(b-d), it is clear that the system approaches instability. This however, is not apparent from the angular difference between the generators G_4 and G_2 on the AC side shown in Fig. 3.34(a).

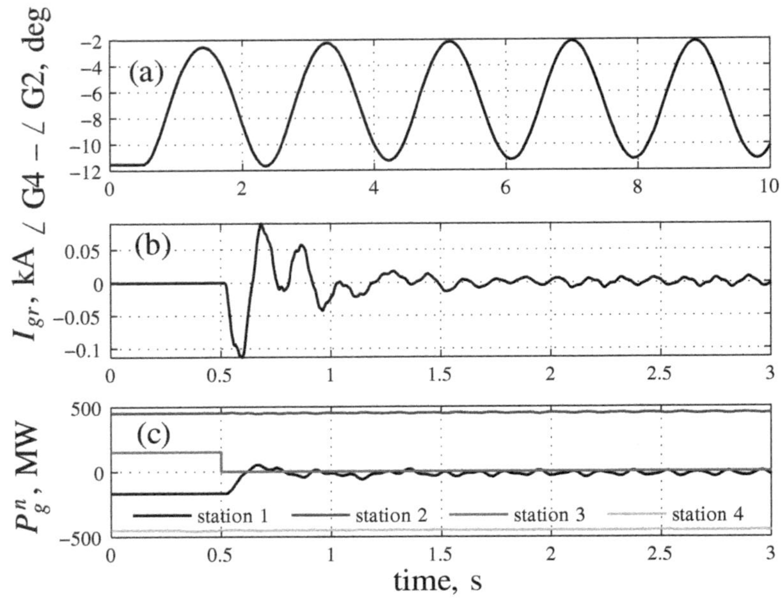

Figure 3.35 (a-c) Case Study 3: Dynamic performance of the combined AC–MTDC grid (shown in Fig. 3.31) following the outage of the negative pole converter at the converter station #3. ©[36], 2011.

It is to be noted that for a DC cable fault the oscillations between G_4 and G_2 on the AC side is much smaller than that in the case of AC side fault as shown in Fig. 3.33(a). Small-signal stability analysis was used to identify the critical modes associated with the AC and DC sides as shown in Table 3.3. It can be seen that outage of a DC cable has virtually no impact on the inter-area (0.55 Hz) mode which is reflected in similar responses of the angular differences between the generators as shown in Figs 3.33(a) and 3.34(a). However, the 7.0-Hz mode with high participation from the DC network states (refer to Fig. 3.32) becomes unstable, which is manifested in the responses in Fig. 3.34(b-d).

From the above results, it can be inferred that an averaged converter model along with the DC cable network should be taken into account while analyzing the stability and the dynamics of combined AC–MTDC systems.

3.11.4.3 Converter Outage In Case Study I, the converter outage was simulated for a system in which the surrounding AC grid was modeled by ideal voltage sources. For this system, the outage of the negative pole converter of the converter station #3 (inverter) was considered at $t = 0.5$ s by opening the AC side breaker.

The angular difference between the generators $G4$ and $G2$ in Fig. 3.35(a) shows stable but poorly damped oscillations around a reduced average (absolute) value because of less power flow through the AC tie-line connecting buses 9 and 10. The ground current before the outage of the converter is zero due to the balanced operation of the grid. Following the outage, the ground current shows oscillations (Fig. 3.35(b)). This current flows through the grounding resistor at the station #1 and returns through the cable charging capacitance. It can be seen from Fig. 3.35(c) that the power flow at the PCC of the negative pole of station #3 becomes zero while those of the stations #2 and #4 remain constant as they are in P–Q control. The negative pole converter of the station #1 (rectifier) operating in V_{dc}–Q mode reduces its power by about 150 MW (lost due to outage of converter in station #3). The positive pole converter powers, not shown here, remain at their respective pre-fault values, since the converter poles operate independently.

It should be noted that the station #1 absorbs the entire burden of the change in power going out of the MTDC grid. This will appear as a large loss of load to the grid $G6$, leading to over-frequency and other related operational challenges. This issue was mentioned in Section 1.5, and the concept of droop control was introduced in Section 2.4.5 for sharing the burden following a converter loss. This topic is the subject matter of the next chapter of this book.

CHAPTER 4

AUTONOMOUS POWER SHARING

4.1 Introduction

In a point-to-point HVDC link, typically one converter station controls the real power and the other station maintains the DC link voltage. As a natural extension, in an MTDC grid all converters except one could be operated in active power set point control and one converter station could be responsible for maintaining the DC voltage across the entire DC grid. The DC voltage controlling converter acts as the slack bus within the MTDC grid and deviates from an active power set point by the amount of loss that takes place within the MTDC grid. Due to a small percentage of loss in an MTDC grid this deviation is acceptable under normal operating conditions. However, if there is a converter outage the mismatch in active power is transferred to the DC voltage controlling converter. The DC voltage controlling converter either needs to increase or decrease its power depending on the nature of both the faulty converter and DC voltage controlling converter, i.e., rectifier or inverter. In an event of increase or decrease it may so happen that the DC voltage controlling converter needs to be loaded well beyond its rating (limited by the current controller limits mentioned in Section 2.4.1) or reverse the power direction that might have an impact on the surrounding AC systems. Therefore, following a converter outage in an MTDC

Multi-terminal Direct-Current Grids: Modeling, Analysis, and Control. First Edition.
Nilanjan Ray Chaudhuri, Balarko Chaudhuri, Rajat Majumder, and Amirnaser Yazdani.

grid, it is critical that the healthy converter stations share the power mismatch/burden in a desirable way. Similar to power-frequency droop used in AC generators, Section 3.3.1.5, a power-voltage droop in the DC link voltage control loops can ensure proper sharing according to the converter ratings.

The concept of droop control [21, 46, 47] has been proposed for power sharing and frequency support [48] through MTDC grids. A number of research papers have been published in recent years on control strategies for MTDC grids. In [49], a robust power control using DC voltage droop is demonstrated along with the system transient response under temporary and permanent DC faults. In [50], DC link voltage control is used on one converter station together with a DC voltage droop characteristic on the other converters. The impact of the proposed control structure on the stability of the AC and DC grid is investigated by simulations. Different operation modes of MTDC systems for large offshore wind farms are analyzed in [51]. While the offshore converters act as ideal voltage sources that absorb all the incoming power, the onshore converters inject the power to the AC grid and control the DC voltage using a droop control scheme. In [52], precise control of power flow in MTDC grids is demonstrated taking into account DC line voltage drop and power losses in DC lines and converters. In [53], the dynamic model of an MTDC grid is used to compare droop control, ratio control, priority control and voltage margin method for controlling the DC voltage at the converter stations. A case study on a four-bus MTDC system is presented to compare the different control strategies during the steady-state and a three-phase fault on the onshore system. Reference [54] provides a systematic comparison between the different voltage control strategies adopted for MTDC grids and discusses how all control strategies can be theoretically regarded as limiting cases of a voltage droop control. An advanced control combining voltage margin control, dead-band droop control, and undead-band droop control is demonstrated. In [55], various DC grid management strategies based on coordinated closed-loop DC voltage control and DC droop characteristics are proposed to ensure proper power sharing among various onshore AC networks, robust performance during wind speed and power variations, and large disturbances caused by the trip of one of the wind farms. In [56], the optimal power flow (OPF) set up for an AC–MTDC grid is presented to minimize the overall transmission loss with constant DC voltage control (master–slave control) and DC voltage droop control.

In [57], a distributed DC voltage control has been presented. Such an approach has some advantages over its concentrated slack bus counterpart. Here a number of converters can jointly control the DC system voltage. After a fault, a voltage droop controlled DC grid converges to a new operating point, which impacts the power flows in both the DC grid and the underlying AC grids. Reference [58] introduces a cascaded DC voltage control at every converter that allows a two-terminal VSC HVDC system to cope with converter outages. The model configures into a master–slave setup with the converters within an MTDC system taking over the DC voltage control in case the DC voltage controlling converter fails. It is shown that the model can be used to include a voltage droop control to share the power imbalance after a contingency in the DC system amongst the converters in the system. Reference [59] presented the influence of the converter droop settings and the MTDC network topology on the power sharing in MTDC grids. An analytical tool to evaluate the effect of the droop control settings on the steady-state voltage deviations and power sharing after a converter outage has been developed. An optimization algorithm is developed, considering two conflicting criteria. In [60], a dedicated droop control considering an offshore DC grid topology to implement different power flow scenarios has been proposed without any communication or master control. Operation of MTDC grids with special emphasis on wind spillage is studied. A process to synthesize spatially correlated winds is described and used to assess the value for different extensions and ratings. A resistance modulation device is suggested as a supplementary power flow controller, for significant wind spillage reduction.

In majority of the above papers, fixed values of droop constants, primarily based on the individual rating of the converters, have been used. However, this does not consider the actual loading condition, and hence the converter headroom (difference between the rated capacity and actual loading) available to share the additional burden.

In this chapter, an adaptive (variable) droop control scheme for appropriate power sharing taking into account the available headroom is described. This scheme can avoid possible overloading. The converters that are already operating close to the limits would not try to share the burden of the loss to the extent of the other converters with higher spare capacity or headroom. Similar to converter-fed AC networks (e.g. a microgrid [61, 62]), droop control impacts the stability of MTDC grids. For a fixed droop scheme a thorough stability analysis might not be essential as long as stability is ensured for the chosen fixed values through prior simulation. But for adaptive droop, the stability region needs to be ascertained *a priori* through small-signal stability analysis using the framework described in Chapter 3.

The amount of burden shared by a converter station following the outage is determined by the difference between the post- and pre-contingency power at PCC. Therefore, it is important to analyze the post-contingency condition. In this chapter, the post-contingency condition is ascertained using analytical derivation and the results are shown to be consistent with the transient simulations. Averaged models in MATLAB®/SIMULINK® [25] are used to perform the stability and small-signal stability analysis. Validation of this averaged model against switched model in EMTDC/PSCAD [26] are presented, which provides the required confidence in this analysis.

The topics in this chapter are organized as follows: First, the steady-state operating point of different converter stations is described. The concept of power sharing control in the context of AC systems and microgrids followed by MTDC grids is presented. Different techniques for power sharing are mentioned and the use of local feedback signal, and common signal-based droop control are described. Adaptive droop control is introduced and advantage of this scheme for autonomous power sharing is established through transient simulations on an MTDC grid with four bipolar converters and DC cable network with metallic return. Simulation results for both rectifier and inverter outages under two different scenarios are presented.

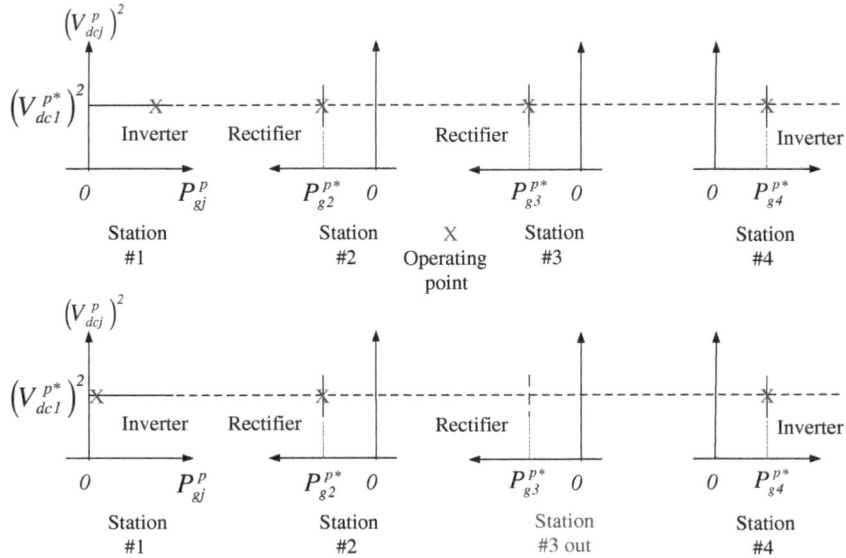

Figure 4.1 Steady-state operating characteristics of the positive pole converters in the four-terminal MTDC grid, Fig. 3.27, in V_{dcj}^2–P_{gj} plane. Upper subplot: Operating point under normal conditions. Lower subplot: Operating point following the outage of the converter station #3.

4.2 Steady-state Operating Characteristics

The steady-state operating characteristics of the MTDC grid can be represented on a two-dimensional plane using the control variables in different axes. Let us consider the four-terminal MTDC grid introduced in Section 3.10, Fig. 3.27, where converter station #1 operates in control Mode II (V_{dc}–Q) and converter stations #2 to #4 operate in the control Mode I ($P - Q$). Since in Mode II, the square of the DC link voltage is controlled (Sections 2.4.3 and 3.2.3.1), the converter steady-state operating characteristics is drawn using V_{dc}^2 as a variable. Similar characteristics can also be drawn simply using the DC voltage as reported in the literature.

The upper subplots in Fig. 4.1 show the steady-state operating characteristics of the positive pole converters in the four-terminal MTDC grid in the $V_{dcj}^2 - P_{gj}, j = 1, ..., 4$ plane corresponding to each converter station. Similar characteristics are applicable for the negative pole converters. For point-to-point VSC-HVDC links, the usual practice is to operate one of the converter stations in active power control mode to ensure the scheduled power exchange. The DC link voltage is maintained by the other converter station that consequently supplies the resistive losses in the DC link, thus acting as a slack converter station. In such a case, it is a common practice to draw these characteristics at any common DC node, which, however, is difficult for a meshed DC grid. Neglecting the system losses, the operating condition of the system can be represented by the point of intersection of the individual converter characteristics.

When the positive pole of converter station #3 is lost, the operating point gets shifted as shown in the lower subplot of Fig. 4.1. To maintain the real power balance in and out of the DC grid, the real power delivered by converter station #1 becomes zero. This impacts the AC grid connected to converter station #1 and was highlighted in Chapter 3. For the loss of a inverter station (e.g. converter station #4), the operating point for converter station #1 will move further right from the one shown in the upper subplot. This will result in an overloading of the converter station #1 and it might loose control over DC link voltage because of violation of current control limits mentioned in Section 2.4.3.

Therefore, the loss of converter stations can result in the overload of certain remaining converter stations (leading to hitting of current controller limits, Section 2.4.1) and also over-voltages in some nodes of the DC grid. The outage of converter station #1 will be the most critical. In the absence of any converter station controlling the DC link voltage, the whole DC grid might collapse. If the converter station controlling DC voltage is a rectifier, the outage will result in power deficit into the grid leading to the saturation of most or all the converter station. On the other hand, if it is an inverter, the outage will result in surplus power injection into the DC grid. This will lead to unacceptable over-voltage conditions.

To avoid these issues, sharing the burden of such a loss/power mismatch has to be appropriate. The concept of power sharing is common in interconnected AC power systems and small isolated systems like microgrids, which is reviewed in Section 4.3, followed by its application in MTDC grids.

4.3 Concept of Power Sharing

Sharing active power among synchronous generators based on frequency is common in the interconnected AC systems. Use of power frequency droop enables multiple synchronous generators to share the load [63]. Recently, sharing power by VSCs based on frequency signal has been reported extensively in the literature [61, 62, 64, 65, 66].

4.3.1 Power Sharing Among Synchronous Generators

All modern day generators have governors which are equipped with a power frequency droop characteristics, Section 3.3.1.5. The objective is to ensure that the generators that are operating in a synchronized manner can still be in synchronism in an event of a load switching (in or out) or outage of some generating units. The resulting power mismatch is shared appropriately by a group of generators with power frequency droop. A typical power frequency droop characteristics is shown in Fig. 4.2. Each generator maintains its rated angular speed (e.g., 50/60 Hz) at its rated power while the governor allows the speed to deviate from its rated speed in a linear manner when the delivered power is not equal to the rated power. This governor droop characteristics is described in equation (3.80). Different generators [63] can have different droop characteristics and can still operate in synchronism.

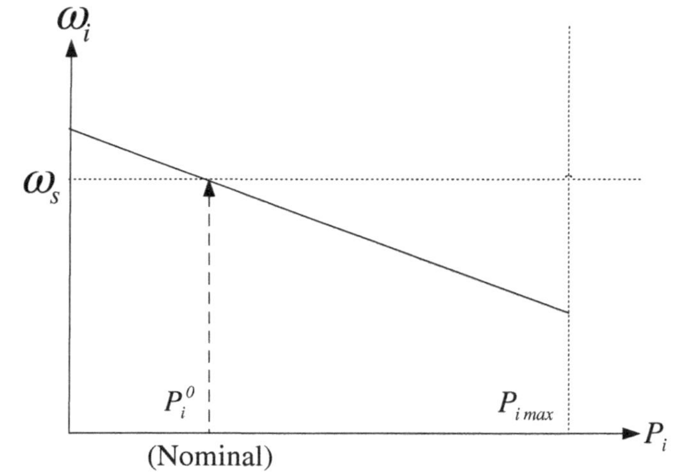

Figure 4.2 Typical power frequency droop characteristics.

4.3.2 Power Sharing in AC Microgrids

A microgrid has two major components, the static switch and the microsources. The static switch has the ability to autonomously island the microgrid from disturbances such as faults, IEEE 1547 events [67] or power quality events. Each microsource can seamlessly balance the power on the islanded microgrid using a power-frequency droop controller.

Within an AC microgrid each microsource regulates the voltage magnitude at the connection point and the injected active power. With this configuration, if a load switching takes place anywhere in the microgrid, the extra or surplus power gets balanced from the main grid, since every unit maintains a constant output power. This is similar to having a slack bus or DC voltage controlling converter. When the system operates in islanded mode the local power-frequency droop function ensures that the power is balanced within the island and each microsource participates in sharing any power imbalance. The loads within a microgrid receives power both from the local microsource and the main grid based on contractual agreement or specific situation. If the main grid gets disconnected due to IEEE 1547 events, voltage drops, faults, blackouts, etc., the microgrid can autonomously transfer to island operation. While regulating the output power, each source has a constant negative slope droop on the power-frequency plane. The slope of the power-frequency droop characteristic is chosen such that the frequency is allowed to drop by a given amount as the power output spans from zero to maximum as shown in Fig. 4.3. Figure 4.3 also shows the power set points P_{01} and P_{02} for two microsources. This is the amount of power injected by each source when connected to the grid, at system frequency (ω_s). The microsources need to increase their power output if the microgrid grid gets into islanded mode while importing power. The new operating point will be at a frequency

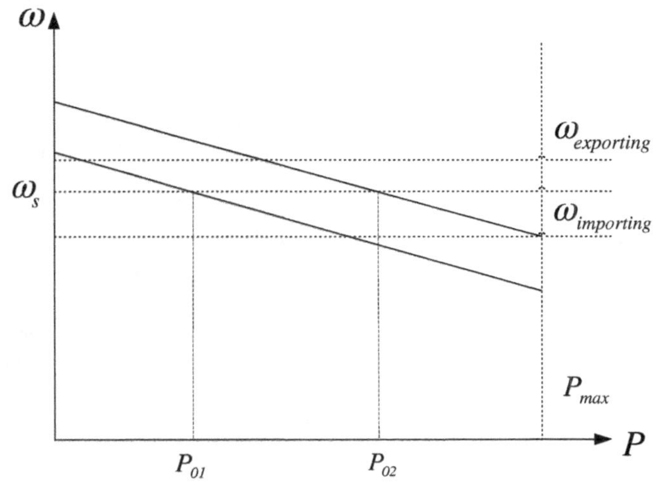

Figure 4.3 Power frequency droop of a microgrid.

that is lower than the nominal value. In this case both sources have increased their power output with unit 2 reaching its maximum power point. If the system transfers to islanded mode when exporting power to the grid, then the new frequency will be higher, corresponding to a lower power output from the sources with unit 1 at its zero power point. The characteristics shown in Fig. 4.3 are steady-state characteristics. They have a fixed slope in the region where the unit is operating within its power range. The slope becomes vertical as soon as any limit is reached. The droop is the locus where the steady-state points are constrained to come to rest, but during dynamics the trajectory will deviate from the characteristic.

The concept of power sharing control for the MTDC grid is discussed in Section 4.4.

4.4 Power Sharing in MTDC Grid

Two strategies have been proposed in the literature for autonomous power sharing control in an MTDC grid: (a) voltage margin control and (b) droop control.

4.4.1 Voltage Margin Control

As briefly mentioned in Section 1.4.1, in an LCC HVDC, a current margin control [9] is used to avoid adverse interactions between two converter stations [12]. As a dual to this strategy, the philosophy of voltage margin control was first proposed for VSC HVDC in [68]. The upper subplot in Fig. 4.4 shows the control characteristics of a point-to-point DC system with voltage margin control for the positive pole converter

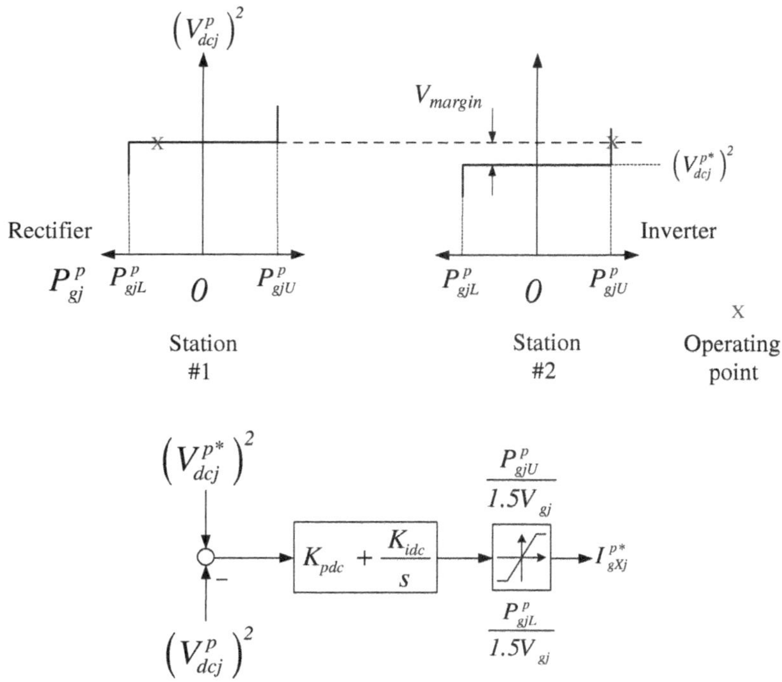

Figure 4.4 Upper subplot: control characteristics of a point-to-point DC system with voltage margin control for the positive pole converter stations. Lower subplot: implementation of voltage margin control for the positive pole converter stations.

stations. The difference between the references $\left(V^{p*}_{dcj}\right)^2$ for the rectifier and the inverter is called the voltage margin.

Let us assume that converter station #1 is operating as a rectifier and it maintains a constant DC voltage. When more power is drawn from the inverter side the limit P^p_{gjL} is violated. The DC voltage controller shown in the lower subplot hits the lower limit of the current reference and the converter station looses voltage control. This results in reduction in the DC voltage and the inverter station with lower DC voltage reference will assume voltage control. For the inverter station operating under DC voltage control, if the power increases beyond P^p_{gjU} the DC voltage rises and the rectifier will assume voltage control.

To facilitate a smooth transfer of the DC voltage control from one converter station to the other, a two-stage voltage margin control shown in Fig. 4.5 [68] is more effective. For each converter station two DC voltage controllers are used with references $\left(V^{p*}_{dcjU}\right)^2$ and $\left(V^{p*}_{dcjL}\right)^2$, respectively. The limits P^p_{giU}, P^p_{giM}, and P^p_{giL} are imposed using the limiting structure shown in Fig. 4.5. To appreciate the operational benefits of

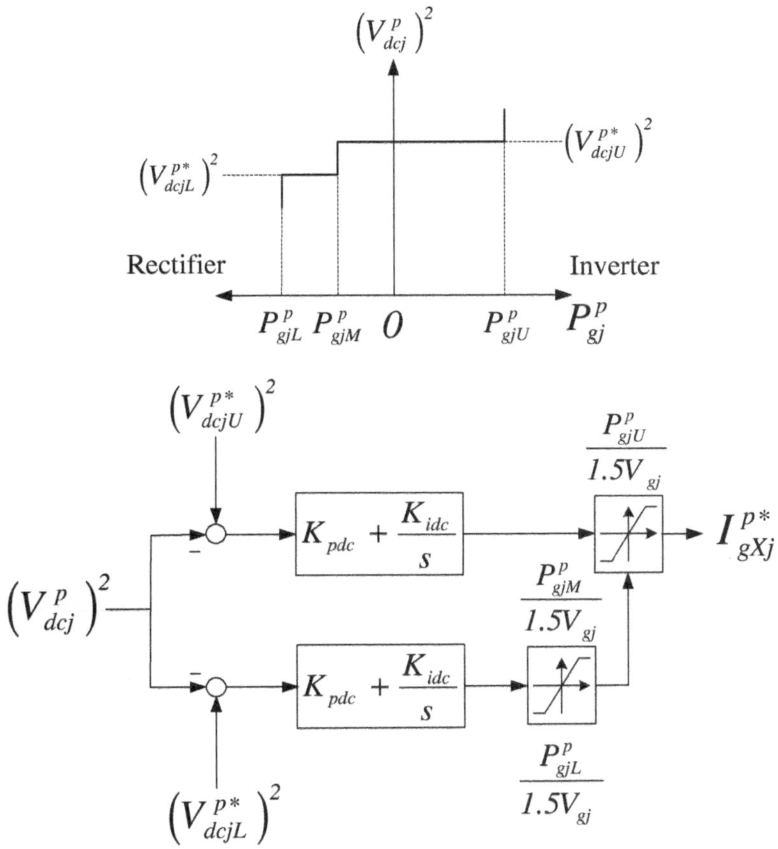

Figure 4.5 Upper subplot: characteristics of a two-stage DC voltage margin control. Lower subplot: implementation of a two-stage DC voltage margin control.

the two-stage voltage margin control, let us consider the four-terminal test network shown in the lower subplot of Fig. 3.27 as before.

Figure 4.6 shows the steady-state operating characteristics of the positive pole converter stations in the four-terminal MTDC grid, with a two-stage voltage margin control in V_{dcj}^2–P_{gj} plane. The converter station #1 operates with a single-stage voltage margin control and regulates the voltage under nominal condition. Stations #2 , #3, and #4 each work with a two-stage voltage margin control and maintain constant power in the vicinity of the nominal condition.

When converter station #1 is lost, the power balance necessitates more power output from the inverter station #4. The converter station #4 assumes DC voltage control mode and the operating point settles at a higher DC voltage as shown in Fig. 4.6. In this case, the voltage margin is defined as the difference between the successive voltage references corresponding to the converter stations that assume voltage control

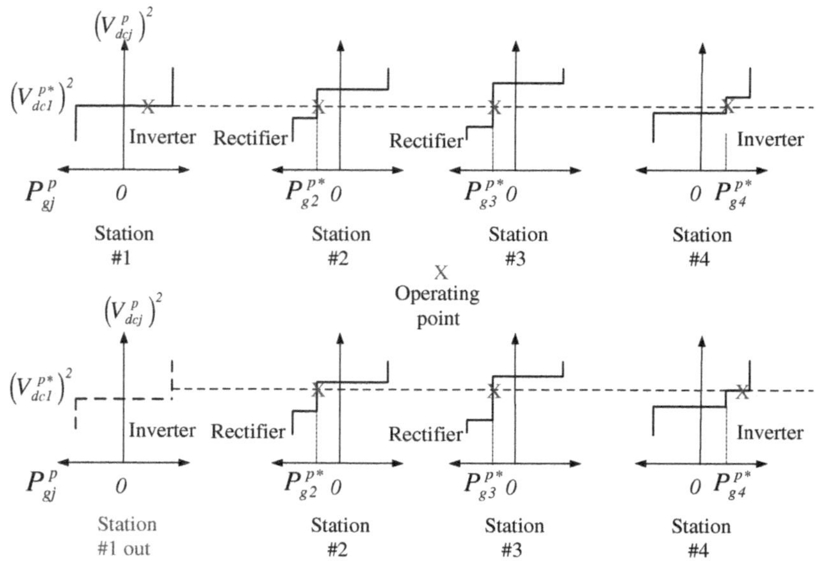

Figure 4.6 Steady-state operating characteristics of the positive pole converter stations in the four-terminal MTDC grid, Fig. 3.27, with a two-stage voltage margin control in V_{dcj}^2–P_{gj} plane. Upper subplot: Operating point under normal condition. Lower subplot: Operating point following the outage of the converter station #1.

one after another. The magnitude of the voltage margin should be chosen judiciously and can be formulated as an optimization problem. Too small voltage margin will result in unnecessary mode shift between converter stations whereas too large margin will lead to the under-utilization of the MTDC grid [69].

The settings used in voltage margin control is critical for an appropriate power sharing. It becomes more difficult to determine the voltage margins when the number of converter stations increase. Droop control can be used to solve this problem.

4.4.2 Droop Control

For an MTDC grid, it is desirable that following an outage of one or more converters, all the remaining ones should share the resulting power imbalance in certain appropriate proportion [21, 46, 47]. Thus all converter stations should operate in DC link voltage control mode (or as slack converters) rather than trying to strictly follow their respective active power references. However, set values of DC link voltage references at all converter stations could be conflicting unless they are modified properly depending on the reference and actual values of the active power and DC link voltage.

In Section 2.4.5, the concept of droop control was briefly introduced and is further elaborated in Section 4.3. The following modifications and simplifying assumptions

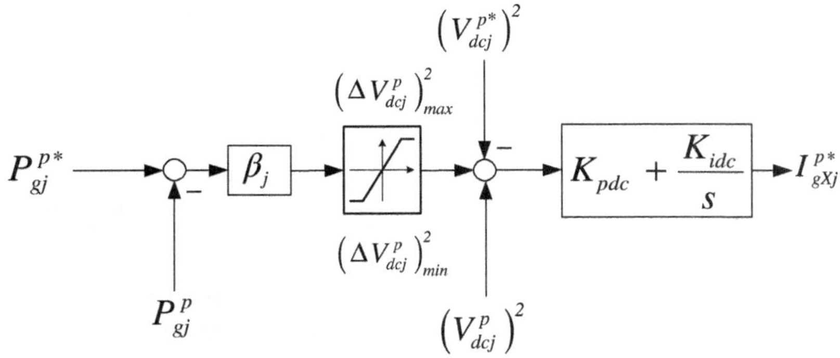

Figure 4.7 Active power-DC link voltage droop for sharing of power imbalance in the positive pole of the *j*th converter station using local DC bus voltage as feedback signal. ©[70], 2013.

on the droop controller shown in Fig. 2.33(a) are considered for analysis in this chapter and in Chapter 5:

- In the context of regulating power dispatch, the operators need access to the dispatch command P_g^*. This is not only important for the operation of the AC–MTDC grid, but also essential for a scenario where the converter stations will enter the power market. Therefore, the DC voltage will be drooped against the power error, i.e., the difference between reference and measured power—as opposed to only measured power.

- Additional limits on the allowable deviation of the DC voltage reference will be introduced.

- The feed forward compensation will not be included in the analysis.

- The delay time constant τ_P will be neglected assuming it to be sufficiently small.

- Subscript P will be dropped from the notation of the droop constant.

With the above-mentioned considerations, the active power-DC link voltage droop can be used to modify the reference values of DC link voltages V_{dcj}^2 at each converter station (under voltage control mode) as shown in Fig. 4.7. Using the modified voltage reference a PI controller[1] derives the X-axis current reference I_{gXj}^{p*} for the inner current control loop of the converters.

[1] As mentioned in Section 2.4.3, the compensator K_{Vdc} could be a PI controller designed through the method of symmetrical optimum [24]. Please note, $K_{Vdc}(s) = -K_{pdc} - \frac{K_{idc}}{s}$.

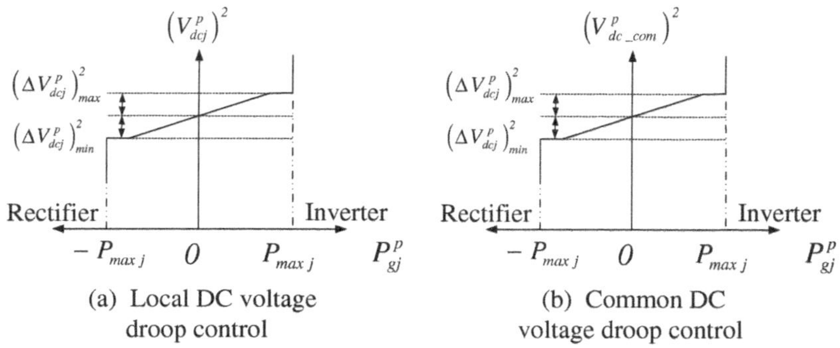

Figure 4.8 Droop control characteristics for the positive pole converter station: (a) local DC voltage droop control mode and (b) common DC voltage droop control mode.

Under steady-state condition the input to the PI compensator is zero. Assuming that the limits $\left(\Delta V_{dcj}^{p}\right)_{\max}^2$ and $\left(\Delta V_{dcj}^{p}\right)_{\min}^2$ are not violated we get

$$\left(V_{dcj}^{p}\right)^2 = \beta_j P_{gj}^{p} + \left[\left(V_{dcj}^{p*}\right)^2 - \beta_j P_{gj}^{p*}\right] \tag{4.1}$$

This leads to the droop characteristics shown in Fig. 4.8(a) whose slope is determined by the droop coefficient β_j. The section of the characteristics where the limits are hit is also highlighted. The limits on the current reference determine the value of P_{maxj} as described in Section 4.4.1 for voltage margin control. The values of the droop constants β_j at different converter stations determine how a power imbalance is shared among the converters. Identical values of β_j would result in equal sharing among the converters. For unequal β_j, the converter stations with higher values would have dominant contribution from active power control loop. Smaller β_j would ensure lesser deviation in DC link voltages, but it might adversely impact the overall stability as discussed later in Section 4.8.2. A special case is the DC link voltage control mode where $\beta_j = 0$ which results in the largest contribution to power sharing.

The problem with the local voltage feedback is that it results in a non-unique load flow solution for the MTDC grid. Analytical derivation of the post-contingency operation discussed in Section 4.6.1 will explain this. In case of parallel operation of multiple generators in AC systems or power sharing among multiple microsources in a microgrid, frequency acts as a common feedback signal that can very conveniently be used for droop characteristic, Section 3.3.1.5. However, in an MTDC grid, the quantity analogous to frequency is DC voltage, and unfortunately that is not uniform across the grid. Therefore, in order to implement a possible DC voltage–power droop characteristic, a common DC voltage signal $(V_{dc\_com})^2$ (at any node in the DC network or any converter station) needs to be chosen and communicated to all converters. This approach was proposed in [47] and modern DC power cables with embedded optical fiber communication channels can realize it. Figure 4.9 shows the droop control

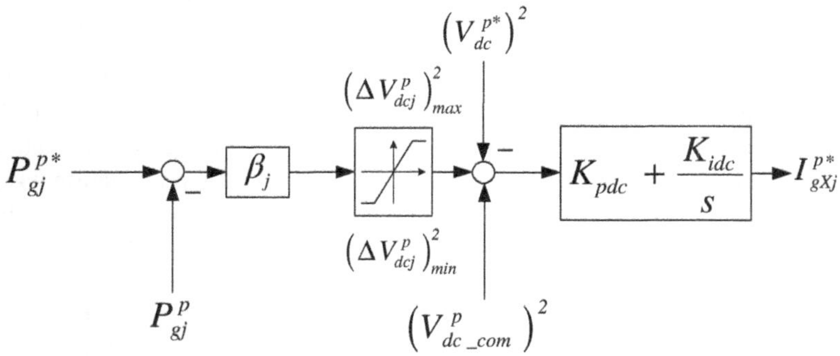

Figure 4.9 Active power–DC link voltage droop for sharing of power imbalance in the positive pole of the *j*th converter station. ©[70], 2013.

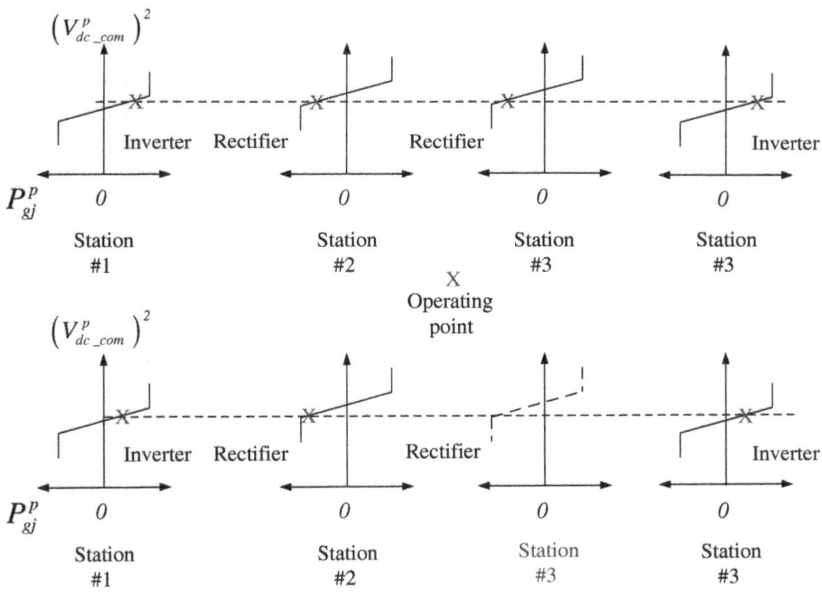

Figure 4.10 Steady-state operating characteristics of the positive pole converter stations in the four-terminal MTDC grid, Fig. 3.27, with common DC voltage droop control in V_{dcj}^2–P_{gj} plane. Upper subplot: Operating point under normal condition. Lower subplot: Operating point following the outage of the converter station #3.

scheme with a common DC voltage used as a feedback signal. The corresponding droop characteristics is shown in Fig. 4.8(b).

Fig. 4.10 shows the steady-state operating characteristics of the positive pole converter stations in the four-terminal MTDC grid shown in Fig. 3.27 with common DC voltage droop control in V_{dcj}^2–P_{gj} plane. The upper plots show the operating point under normal condition for converters with equal droop coefficient. For clarity of presentation the limits imposed by $\left(\Delta V_{dcj}^{p}\right)_{\max}^{2}$ and $\left(\Delta V_{dcj}^{p}\right)_{\min}^{2}$ are not shown. The operating point following the outage of the converter station #3 is shown in the lower plots. Due to the real power balance in the network, the rectifier (converter station #2) power increases and the inverter (converter station #1 and converter station #4) powers decrease in equal proportion. Unlike the post-outage condition shown in Fig. 4.1 where the converter station #1 assumes the entire burden, in this case the burden is distributed amongst the remaining converter stations.

Although a fixed or predefined values of the droop constants β_j can be chosen according to the ratings of the individual converter station [46, 21, 47], it does not consider the actual loading condition, and hence the headroom available for sharing the additional burden that arises due to a power imbalance situation. This problem with fixed droop can be avoided using an adaptive droop scheme described in Section 4.4.3.

4.4.2.1 Ratio and Priority Control A few papers [57]–[59] have proposed DC voltage droop with respect to the DC current instead of power. Reference [71] proposed "ratio control" approach in which the droop coefficient β_i of one converter station is kept constant and that of another converter station β_k is changed to maintain the DC current sharing between them at a desired value, Fig. 4.11(a).

The other approach is called priority control [72] in which one converter terminal controls the DC voltage. When the voltage control enters saturation and the voltage goes above a threshold $\left(V_{dcjU}^{p}\right)^{2}$ another terminal starts working in the droop control mode and carry power, Fig. 4.11(b).

The ratio and the priority control are essentially two different approaches of implementation of droop control.

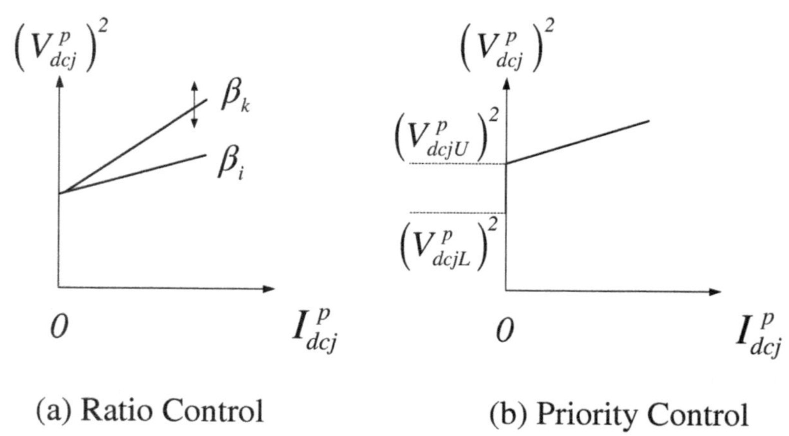

Figure 4.11 Characteristics of different operating modes for the positive pole converter station: (a) ratio control and (b) priority control.

4.4.3 Adaptive Droop Control

For a fixed droop scheme, shown in Fig. 4.9, the droop coefficients are usually dependent only on the respective converter ratings. Therefore, for an MTDC grid with equally rated converter stations the sharing of burden following a converter outage will be equally distributed. However, under a particular operating condition, all converter stations might not be equally loaded and hence some of them may not be able to participate in power sharing equitably. Therefore, it is desirable that the available converters should participate in power sharing depending on the headroom (difference between the rated capacity and present loading) available in addition to their ratings. To address this issue the contribution from the power control loop in Fig. 4.9 is modified as shown in Fig. 4.12. The idea is to calculate the droop coefficients based on a function of the normalized available headroom of each converter defined as [70]

$$\beta_j = \beta_j^0 \left(\frac{\mathfrak{R}_{base}}{\mathrm{H}_j} \right)^{\lambda} \tag{4.2}$$

where $\mathrm{H}_j = \mathfrak{R}_j - \left| P_{gj}^{p0} \right|$ is the available headroom of the jth converter of rating $\mathfrak{R}_j$, β_j^0 is the nominal droop which is in inverse proportion to the converter ratings, $\mathfrak{R}_{base}$ is the maximum value of all converter ratings, and λ is a user-defined positive constant. The above function is nonlinear in P_{gj}^{p} and results in higher droop values for lower available headroom. This ensures that the converters which are already operating very close to the operational limit would not try to share the burden of a lost converter

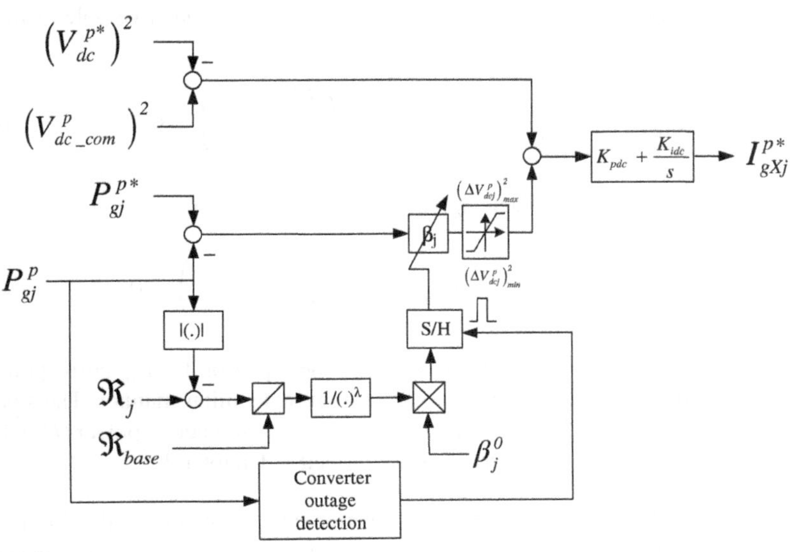

Figure 4.12 Block diagram of the proposed adaptive droop scheme in the positive pole of the jth converter station. ©[70], 2013.

(or due to loss of a cable) to the extent the burden is shared by the converters with higher spare capacity or headroom.

It is also anticipated that if the MTDC grid enters into power market in future, the converter station owners may want to use this spare capacity as an ancillary service and share the power mismatch appropriately. Therefore, due to both physical and market related limitations, adaptive droop constant will be an important aspect to have within HVDC control system intended for future MTDC grids.

The sequence of events followed to implement this approach is as follows

1. Use the droop coefficients $\beta_j^{(k)}$ calculated based on the pre-contingency steady-state of the kth converter outage (outage in the immediate past).

2. Keep calculating the normalized headroom-dependent droop coefficient $\beta_j^{(k+1)}$ online but do not use them until the next $(k+1)$th outage.

3. Upon detection of a converter outage, command the sample and hold function to invoke the droop values $\beta_j^{(k+1)}$ calculated just before the $(k+1)$th event.

4.5 AC–MTDC Grid Load flow Solution

It is important to identify the pre-contingency steady-state operating point of the MTDC grid, which could be done by solving an AC–MTDC load flow. The general sequential AC–MTDC grid load flow setup was described earlier in Section 3.4.3. Expressions for f_{15} and f_{16} are modified due to the presence of the droop control. The steady-state equation corresponding to droop control for the positive pole of the ith converter can be written as

$$f_{15i} = \left(V_{dci}^{p}\right)^2 - \left(V_{dci}^{p*}\right)^2 + 4\beta_i\left(\frac{P_{gi}^{*}}{2} - P_{gi}^{p}\right) = 0 \tag{4.3}$$

when a local DC voltage feedback signal is used, and

$$f_{15i} = \left(V_{dc\_com}^{p}\right)^2 - \left(V_{dc}^{p*}\right)^2 + 4\beta_i\left(\frac{P_{gi}^{*}}{2} - P_{gi}^{p}\right) = 0 \tag{4.4}$$

when a common DC voltage feedback is used. The equation for f_{16i} corresponding to the negative pole counterpart can also be written in a similar manner. The steady-state equations for f_{17i} and f_{18i} will depend on the control of reactive power (Q_g), PCC voltage (V_g), or a combination thereof as mentioned in Chapter 3.

Step I: AC Load flow Since the MTDC converter stations operate in the droop control mode, the real power is unknown at all PCC buses. Therefore, an initial guess of the real power $P_{gi}^{(0)} = P_{gi}^{*}$ is assumed. Relevant details regarding the steps to complete AC load flow were described in Section 3.4.1.

Step II: MTDC Load flow The MTDC load flow is done as described in Section 3.4.2. However, the following elements of the Jacobian differ:

$$\frac{\partial f_{15i}}{\partial P^{p}_{gi}} = -4\beta_i \tag{4.5}$$

For a local DC-voltage-feedback-based droop control:

$$\frac{\partial f_{15i}}{\partial V^{p}_{dci}} = 2V^{p}_{dci} \tag{4.6}$$

For a common DC-voltage-feedback-based droop control:

$$\frac{\partial f_{15i}}{\partial V^{p}_{dci}} = 2V^{p}_{dc\_com}, i = com \tag{4.7}$$

$$\frac{\partial f_{15i}}{\partial V^{p}_{dci}} = 0, i \neq com \tag{4.8}$$

where *com* is the common DC bus whose voltage is communicated to every other converter station as the feedback signal. The element corresponding to the negative pole can be obtained by replacing the superscript p by n.

Step III: Repetitions of Steps I and II This step remains identical to that described in Section 3.4.3.

4.6 Post-contingency Operation

After an event or a contingency like outage of one or more converters, the operating condition of the remaining converters and the MTDC grid could change drastically. It is important to understand the impact of different droop control philosophies on the so-called *burden* being shared by a converter station following a converter outage. Here, the burden is defined as the difference between the post- and pre-contingency power output of the converter. In this section, the post-contingency operating point in an MTDC network with n converter stations is derived analytically. The analysis is applicable to both the positive and negative pole converters and is broadly similar to [47] except for the adaptive droop considerations. Droop control based on both local and common DC voltage feedback is analyzed separately in Sections 4.6.1 and 4.6.2 for a fixed droop coefficient.

It should be noted that a fraction of the local or common DC link voltage can be used as a feedback signal without altering the principle laid out in Section 2.4.3. Without loss of generality, for the rest of this book $\left(\frac{V^{p}_{dcj}}{2}\right)^2$ is used as local feedback signal and $\left(\frac{V^{p*}_{dcj}}{2}\right)^2$ is used as the corresponding reference. Similarly, $\left(\frac{V^{p*}_{dc}}{2}\right)^2$ and

$\left(\frac{V^p_{dc\_com}}{2}\right)^2$ are used as the reference and feedback signals for the common DC-voltage-based droop control.

Let us assume P^p_{gj} is the real power at the PCC of the positive pole of the jth converter and P^{p*}_{gj} is the corresponding reference power. The reference setting of the converters are chosen such that $\sum_j P^{p*}_{gj} = 0$.

Let us denote the variables in the pre-contingency condition with superscript "0". In this condition the total real power loss in the positive pole DC cable network is $P^{p0}_{loss(dc)} = -\sum_j P^{p0}_{gj}$. We assume that the reference quantities are not changed following the contingency.

4.6.1 Local DC Link Voltage Feedback

From Fig. 4.7, assuming that DC feedback and the reference signals are given by $\left(\frac{V^p_{dcj}}{2}\right)^2$ and $\left(\frac{V^{p*}_{dcj}}{2}\right)^2$, respectively, the input to the PI compensator is zero under steady-state condition. Under pre-contingency condition, without any limit violation we get

$$4\left(P^{p*}_{gj} - P^{p0}_{gj}\right)\beta_j + \left(V^{p0}_{dcj}\right)^2 - \left(V^{p*}_{dcj}\right)^2 = 0 \tag{4.9}$$

Summing over all converter stations

$$\sum_j^n \left\{4\left(P^{p*}_{gj} - P^{p0}_{gj}\right)\beta_j + \left(V^{p0}_{dcj}\right)^2 - \left(V^{p*}_{dcj}\right)^2\right\} = 0 \tag{4.10}$$

Assuming equal droop coefficients for all converter stations

$$\beta_j = \beta, \quad \forall j \tag{4.11}$$

$$4\beta P^{p0}_{loss(dc)} + \sum_j^n \left\{\left(V^{p0}_{dcj}\right)^2 - \left(V^{p*}_{dcj}\right)^2\right\} = 0 \tag{4.12}$$

Following the outage of the nth converter station

$$\sum_j^{n-1} P^{p*}_{gj} = -P^{p*}_{gn}, \sum_j^{n-1} P^p_{gj} = -P^p_{loss(dc)} \tag{4.13}$$

Equation (4.10) can be rewritten for the remaining converters as

$$4\beta\left\{P^p_{loss(dc)} - P^{p*}_{gn}\right\} + \sum_j^{n-1}\left\{\left(V^p_{dcj}\right)^2 - \left(V^{p*}_{dcj}\right)^2\right\} = 0 \tag{4.14}$$

From (4.12) and (4.14) we can see that the converter DC voltages for the pre-disturbance and post-contingency condition can have a non-unique combination of

$\left(V^p_{dcj}\right)^2$ and $\left(V^{p*}_{dcj}\right)^2$ that satisfies the equality. As a result there is a possibility of multiple operating conditions. It will be shown in the next section that a common DC voltage feedback can solve this problem.

4.6.2 Common DC Link Voltage Feedback

Assuming $\left(\frac{V^{p*}_{dc}}{2}\right)^2$ and $\left(\frac{V^{p}_{dc\_com}}{2}\right)^2$ are used as the reference and feedback signals, respectively, it can be seen from Fig. 4.9 that under pre-contingency steady-state,

$$P^{p*}_{gj} - P^{p0}_{gj} + \frac{\left(V^{p0}_{dc\_com}\right)^2 - \left(V^{p*}_{dc}\right)^2}{4\beta_j} = 0 \tag{4.15}$$

$$\sum_j (P^{p*}_{gj} - P^{p0}_{gj}) + \frac{\left(V^{p0}_{dc\_com}\right)^2 - \left(V^{p*}_{dc}\right)^2}{4} \sum_j \frac{1}{\beta_j} = 0 \tag{4.16}$$

From (4.15) and (4.16) we get,

$$\left(V^{p0}_{dc\_com}\right)^2 - \left(V^{p*}_{dc}\right)^2 = -4\frac{P^{p0}_{loss(dc)}}{\sum_j \frac{1}{\beta_j}} \tag{4.17}$$

$$P^{p0}_{gj} = P^{p*}_{gj} + \frac{-P^{p0}_{loss(dc)}}{\beta_j \sum_j \frac{1}{\beta_j}} \tag{4.18}$$

For a fixed droop scheme it is usual to choose $\beta_j \mathfrak{R}_j = \beta_i \mathfrak{R}_i, \forall i, j$ where $\mathfrak{R}_j$ is the rating of the jth converter. Modifying (4.18) we get

$$P^{p0}_{gj} = P^{p*}_{gj} + \frac{-P^{p0}_{loss(dc)}}{\frac{1}{\mathfrak{R}_j} \sum_j \mathfrak{R}_j} \tag{4.19}$$

It is clear from (4.19) that in the case of equal droop coefficients (i.e., equal converter ratings) the converter real powers deviate from their reference equally by the amount $\frac{P^p_{loss(dc)}}{n}$. Following the outage of the nth converter:

$$\sum_j^{n-1} P^{p*}_{gj} = -P^{p*}_{gn}, \sum_j^{n-1} P^p_{gj} = -P^p_{loss(dc)} \tag{4.20}$$

where $P^{p}_{loss(dc)}$ is the total DC transmission loss in the post-outage condition. The post-contingency steady-state operating point gets modified to:

$$\left(V^{p}_{dc\_com}\right)^2 - \left(V^{p*}_{dc}\right)^2 = -4\frac{(P^{p}_{loss(dc)} - P^{p*}_{gn})}{\sum\limits_{j}^{n-1}\frac{1}{\beta_j}} \tag{4.21}$$

$$P^{p}_{gj} = P^{p*}_{gj} + \frac{(-P^{p}_{loss(dc)} + P^{p*}_{gn})}{\frac{1}{\mathfrak{R}_j}\sum\limits_{j}^{n-1}\mathfrak{R}_j} \tag{4.22}$$

Note that for converters with equal ratings the difference between the pre-outage and the post-outage power for the jth converter is

$$\Delta P^{p}_{gj} = P^{p}_{gj} - P^{p0}_{gj} = \left[\frac{P^{p0}_{loss(dc)}}{n} - \frac{P^{p}_{loss(dc)}}{(n-1)}\right] + \frac{P^{p*}_{gn}}{(n-1)} \tag{4.23}$$

Neglecting the bracketted terms, we can say that the remaining $n-1$ converters will share the burden of the nth converters equally amongst themselves.

4.6.3 Adaptive Droop Control

For the adaptive droop scheme mentioned in Section 4.4.3 the available headroom (H_j) of the jth converter is taken into consideration. Note that $\mathrm{H}_j = \mathfrak{R}_j - \left|P^{p0}_{gj}\right|$, not $\mathrm{H}_j = \mathfrak{R}_j - \left|P^{p}_{gj}\right|$ since it depends on the pre-contingency operating point as mentioned in Section 4.4.3. After some algebraic manipulations and neglecting the loss terms as in (4.23) it can be shown that the difference between the pre-outage and the post-outage power for the positive pole of the jth converter is

$$\Delta P^{p}_{gj} \approx \frac{P^{p*}_{gn}}{\left(\frac{1}{\mathfrak{R}_j}\right)\left(\frac{\mathfrak{R}_{base}}{\mathrm{H}_j}\right)^{\lambda}\sum\limits_{j}^{n-1}\mathfrak{R}_j\left(\frac{\mathrm{H}_j}{\mathfrak{R}_{base}}\right)^{\lambda}} \tag{4.24}$$

From the above derivations it can be concluded that the adaptive droop will ensure sharing of power mismatch amongst the converters based on their respective available headroom. It is to be noted that when all converter ratings are equal, i.e.,$\mathfrak{R}_j = \mathfrak{R}_{base}$, $\forall j$, putting $\lambda = 0$ in (4.24) results in (4.23) (neglecting loss terms) which corresponds to the fixed droop situation. Post-contingency steady-state operating points observed in the transient simulations (in Section 4.8.3) are shown to correspond with (4.23) and (4.24) for fixed and adaptive droop, respectively.

4.7 Linear Model

Once the pre-contingency steady-state operating point is established through AC–MTDC load flow, a linear model around that operating point could be derived by modifying the state-space representation described in Section 3.2.5. In this case, only common DC-voltage-feedback-based droop control is considered. The same per unit (p.u.) system mentioned in Section 3.2.6.1 is followed. The modified state equations are

$$\varsigma_1 = \dot{I}^{p*}_{gXi} = -\frac{R^p_i + K_{pI} + 4K_{pI}K_{pdc}V_{gi}\beta_i}{L^p_i} I^p_{gXi} + \frac{K_{iI}\kappa_{IX}}{L^p_i} + \frac{K_{pI}K_{idc}}{L^p_i}\kappa_{dr}$$
$$+4\frac{K_{pI}K_{pdc}\beta_i}{L^p_i}P^{p*}_{gi} + \frac{K_{pI}K_{pdc}}{L^p_i}\left(V^p_{dc\_com}\right)^2 - \frac{K_{pI}K_{pdc}}{L^p_i}\left(V^{p*}_{dc}\right)^2 \tag{4.25}$$

$$\varsigma_3 = \dot{\kappa}_{IX} = K_{pdc}\left\{4\left(P^{p*}_{gi} - V_{gi}I^p_{gXi}\right)\beta_i + \left(V^p_{dc\_com}\right)^2 - \left(V^{p*}_{dc}\right)^2\right\}$$
$$+K_{idc}\kappa_{dr} - I^p_{gXi} \tag{4.26}$$

$$\varsigma_{10} = \dot{\kappa}_{dr} = 4\left(P^{p*}_{gi} - V_{gi}I^p_{gXi}\right)\beta_i + \left(V^p_{dc\_com}\right)^2 - \left(V^{p*}_{dc}\right)^2 \tag{4.27}$$

The expressions for ς_2 and ς_4 are the same as that in $P - Q$ control.

The linear model for the positive pole converter in the ith converter station can be derived as follows:

- The state variables, control variables, and algebraic variables are

$$\Delta x = \left[\Delta I^p_{gXi}\Delta I^p_{gYi}\Delta\kappa_{IX}\Delta\kappa_{IY}\Delta\kappa_{dr}\right]^T$$
$$\Delta u = \left[\Delta P^{p*}_{gi}\Delta V^{p*}_{dci}\Delta Q^{p*}_{gi}\right]^T$$
$$\Delta z = \left[\Delta V_{gi}\Delta V^p_{dc\_com}\right]^T$$

- The nonzero elements of $\left[\frac{\partial\varsigma}{\partial x}\right]$ are given by

$$\frac{\partial\varsigma_1}{\partial I^p_{gXi}} = -\frac{R^p_i + K_{pI} + 4K_{pI}K_{pdc}V_{gi}\beta_i}{L^p_i}$$
$$\frac{\partial\varsigma_1}{\partial\kappa_{IX}} = \frac{K_{iI}}{L^p_i}, \frac{\partial\varsigma_1}{\partial\kappa_{dr}} = \frac{K_{pI}K_{idc}}{L^p_i} \tag{4.28}$$

$$\frac{\partial\varsigma_3}{\partial I^p_{gXi}} = -1, \frac{\partial\varsigma_3}{\partial\kappa_{dr}} = K_{idc}, \frac{\partial\varsigma_{10}}{\partial I^p_{gXi}} = -4V_{gi}\beta_i \tag{4.29}$$

- The nonzero elements of $\left[\frac{\partial \varsigma}{\partial u}\right]$ are given by

$$\frac{\partial \varsigma_1}{\partial P_{gi}^{p*}} = \frac{4K_{pI}K_{pdc}\beta_i}{L_i^p}, \frac{\partial \varsigma_1}{\partial V_{dc}^{p*}} = -\frac{2K_{pI}K_{pdc}}{L_i^p}V_{dc}^{p*} \tag{4.30}$$

$$\frac{\partial \varsigma_3}{\partial P_{gi}^{p*}} = 4K_{pdc}\beta_i, \frac{\partial \varsigma_3}{\partial V_{dc}^{p*}} = -2K_{pdc}V_{dc}^{p*} \tag{4.31}$$

$$\frac{\partial \varsigma_{10}}{\partial P_{gi}^{p*}} = 4\beta_i, \frac{\partial \varsigma_{10}}{\partial V_{dc}^{p*}} = -2V_{dc}^{p*} \tag{4.32}$$

- The nonzero elements of $\left[\frac{\partial \varsigma}{\partial z}\right]$ are given by

$$\frac{\partial \varsigma_1}{\partial V_{gi}} = -\frac{4K_{pI}K_{pdc}\beta_i}{L_i^p}I_{gXi}^p, \frac{\partial \varsigma_1}{\partial V_{dc\_com}} = \frac{2K_{pI}K_{pdc}}{L_i^p}V_{dc\_com} \tag{4.33}$$

$$\frac{\partial \varsigma_3}{\partial V_{gi}} = -4K_{pdc}\beta_i I_{gXi}^p, \frac{\partial \varsigma_3}{\partial V_{dc\_com}} = 2K_{pdc}V_{dc\_com} \tag{4.34}$$

$$\frac{\partial \varsigma_{10}}{\partial V_{gi}} = -4\beta_i I_{gXi}^p, \frac{\partial \varsigma_{10}}{\partial V_{dc\_com}} = 2V_{dc\_com} \tag{4.35}$$

Expressions related to the output equations can also be derived in a similar manner.

As before, the linear model for the negative pole converter can be derived using a similar approach. The linear model of the DC network and the elements corresponding to other control modes were described in Section 3.6.

4.8 Case Study

4.8.1 Study Network

In Chapter 3, a study network consisting of four bipole converter stations connected with 500-km cables (shown in Fig. 3.27) was used. Converter stations #2 and #3 were considered to be acting as rectifiers while #1 and #4 as inverters under nominal condition. For the studies without droop control, converter station #1 operates in V_{dc}–Q control mode and keeps the DC link voltage at ±350 kV with unity power factor at the PCC while the rest operate in *P–Q* control mode.

Figure 4.13 shows the test system used here, which is essentially the same as in Fig. 3.27, but the control at each converter station is changed due to the introduction of droop control. Unlike Fig. 3.27, the real power flow at all converter stations cannot be highlighted in Fig. 4.13. The DC link voltage of converter station #1 is chosen as the common reference for the droop scheme. The steady-state operating condition is determined by solving the AC–DC load flow equations [9] mentioned in Sections 3.4 and 4.5. The averaged model of the converter stations were used in developing the nonlinear model in MATLAB®/SIMULINK® [25] platform. The cables were represented by a single pi-section model.

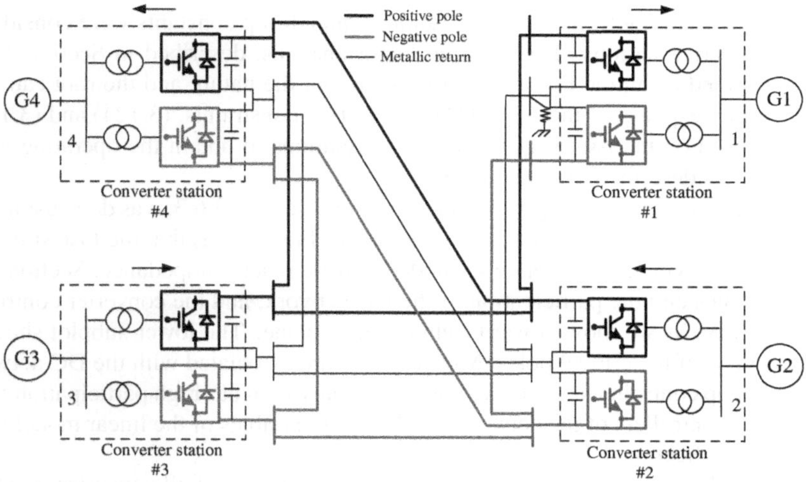

Figure 4.13 Bipole MTDC grid with four converter stations under droop control. ©[70], 2013.

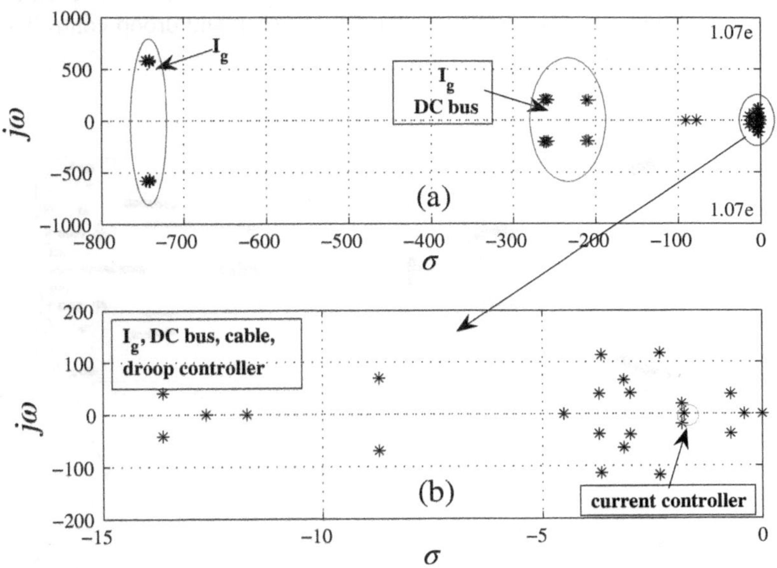

Figure 4.14 Clustering of system poles based on participation factors following the outage of negative pole of converter station #2. ©[70], 2013.

4.8.2 Small-signal Stability Analysis

For small-signal stability analysis, the study network in Section 4.8.1 was linearized (as in Sections 3.6 and 4.7) around an operating condition where the negative pole converter of station #2 was out of operation. Equal droop constants were considered for all converter stations. Participation factor analysis, described in Section 3.6.3, was preformed using the linear model to figure out the nature and the root cause of the dynamic response of the AC–MTDC grid. Two constraints, (3.124) and (3.125), were imposed on the AC–MTDC load flow equations to obtain the operating point (x_0, z_0, u_0) as described in Section 3.4.2.5.

An eigenvalue and participation factor analysis, Section 3.6.3, was done using this linear model. It can be seen from the pole map of Fig. 4.14(a) that the fastest modes are associated with the AC network (modeled by tie reactor impedance, Sections 2.2 and 2.3) states and the participation of the DC network, and the converter controller states increase as we move toward right on the s-plane. The lower subplot shows a zoomed view of the critical poles. Most of these are associated with the DC network and the droop controller states except one pole pair which has high participation from the current controllers of the converters. The state-variables of the linear model were mentioned in Section 4.7.

Figure 4.15 shows the locus of the system poles when β_j of all converters are kept equal and varied from 50% to 1%. The arrows indicate the movement of poles due to such variation. It can be seen that the pole clusters having higher participation from the AC side current move toward right when β_j is reduced. However, the pole clusters with higher participation from the DC network and droop controller states show a mixed movement (Fig. 4.15(b)).

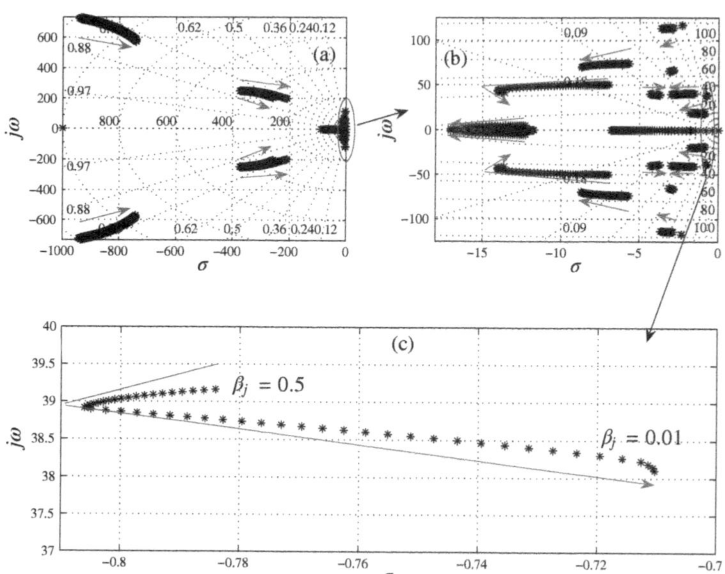

Figure 4.15 Root locus of the system with change of droop constants (β_j) in the operating condition with the outage of the negative pole converter at converter station #2. ©[70], 2013.

The movement of the most critical pole is zoomed in Fig. 4.15(c). This pole pair moves toward the left half of the s-plane when β_j is decreased from 0.5. However, there is a point of inflexion beyond which further decrement results in movement toward the right half of s-plane. A low value of β_j is desirable since it reduces the steady-state deviation in the DC link voltage and stops it from violating the limits at the output of the droop controller as shown in Fig. 4.12. On the other hand, in the proposed adaptive droop scheme a lower (higher) headroom might result in higher (lower) droop values. To avoid instability due to large/small droop values the stability analysis as above is important for a range of possible operating conditions. Only one typical operating condition is presented here to indicate the need of upper and lower bounds on the droop constants from stability considerations.

4.8.3 Nonlinear Simulation

4.8.3.1 Validation Against Switched Model The averaged model developed in MATLAB®/SIMULINK® was validated against a switched model in EMTDC/PSCAD following a converter outage. Figure 4.16 shows the power sharing amongst the converter stations following the outage of the negative pole converter in station #3. A fixed droop scheme (shown in Fig. 4.9) with $\beta = 0.1$ is considered for all converters where stations #1 and #4 (#2 and #3) export (import) about 900 MW each. As expected, the remaining negative pole converters share the loss of 450 MW and the resultant unbalance in the metallic return currents in equal proportion while the positive pole power is maintained at the pre-disturbance value (Fig. 4.16). The loss of one rectifier has resulted in an *increase* in power share by the other rectifying unit

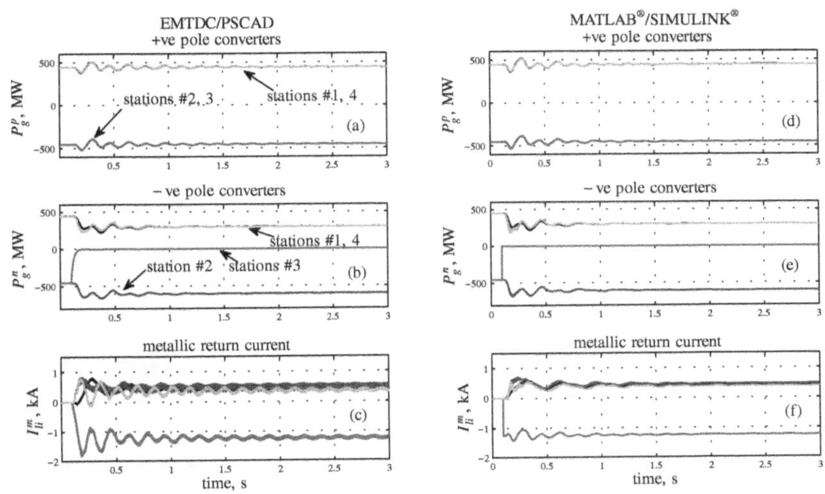

Figure 4.16 (a-f) Validation of the averaged model in MATLAB®/SIMULINK® with the switched model in EMTDC/PSCAD for autonomous power sharing following the outage of the negative pole converter of the converter station #3. ©[70], 2013.

(in converter station #2) from −450 MW to −600 MW and a similar *decrease* for both the inverter units (in converter stations #1 and #3). These responses are to be compared against those in Figs 3.23 and 3.25 where the burden of such an outage was taken only by the converter under V_{dc}–Q control mode.

It can be seen that there is a very close match between these results except that the switching ripple is absent in the averaged model. Moreover, the converter outage is slower in the EMTDC/PSCAD [26] model since the AC breaker opening considers natural current zero in each phase while the averaged model considers almost instantaneous switching, refer to Section 3.9.2.2. This validates the small-signal stability analysis with the averaged model reported in the previous section. The dynamic responses from the switched model are reported throughout the rest of this chapter.

4.8.3.2 Problems with Local Voltage Feedback As mentioned in Section 4.4.2, due to the non-unique load flow solution with the local-voltage-based-droop control, a common DC bus voltage reference signal needs to be communicated to all the converter locations. Several tests were done to analyze this problem and one of them is presented here to establish the point. As shown in Fig. 4.17 the negative pole converters share the power change equally for equal droop coefficients following the negative pole converter outage at station #2. However, the power in positive pole converters drift unacceptably to reach a different post-contingency operating point from their nominal values. On the contrary, common reference signal-based droop control does not suffer from this issue and thus it is justifiable to use this architecture also for the proposed adaptive (variable) droop scheme.

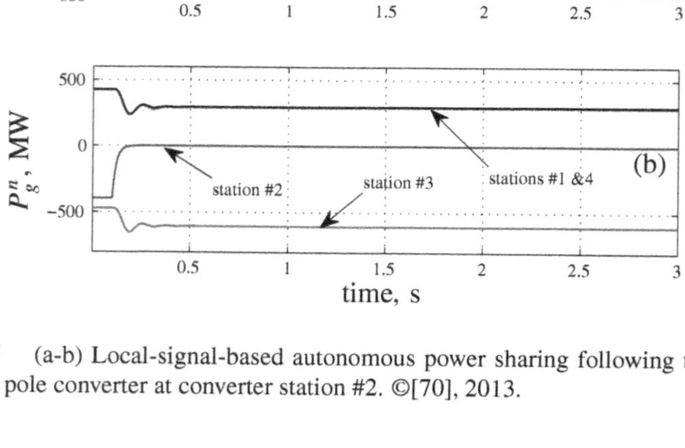

Figure 4.17 (a-b) Local-signal-based autonomous power sharing following the outage of the negative pole converter at converter station #2. ©[70], 2013.

4.8.3.3 *Fixed vs Adaptive Droop* In this section the performance of the fixed and adaptive droop scheme is compared in the backdrop of no power sharing control (no droop) wherein only converter station #1 is under DC link voltage control (slack). For the fixed droop, $\beta = 1\%$ is chosen while $\lambda = 2$ is used for adaptive droop. Unequal loading conditions at the four converter stations are considered to demonstrate the benefit of adaptive droop. Both rectifier and inverter outages are simulated under a number of scenarios out of which two representative cases are presented below.

Inverter Outage: In the first case each converter pole in stations #1 and #4 exports about 420 MW and 200 MW, respectively, from the DC grid, while each of those in converter stations #2 and #3 import around 310 MW into it. The outage of the negative pole converter in the inverter station #4 is considered. It can be seen from Fig. 4.18(a) that in the absence of droop control the slack converter station (#1) operating in V_{dc}–Q mode increases its power to 600 MW to compensate the loss of the inverter.

This results in the overloading of the unit whereas the inverter in stations #2 (Fig. 4.18(b)) and #3 (not shown) maintains the pre-contingency power.

With a fixed droop, the burden is shared equally by the remaining converters in the negative pole as shown in Fig. 4.19.

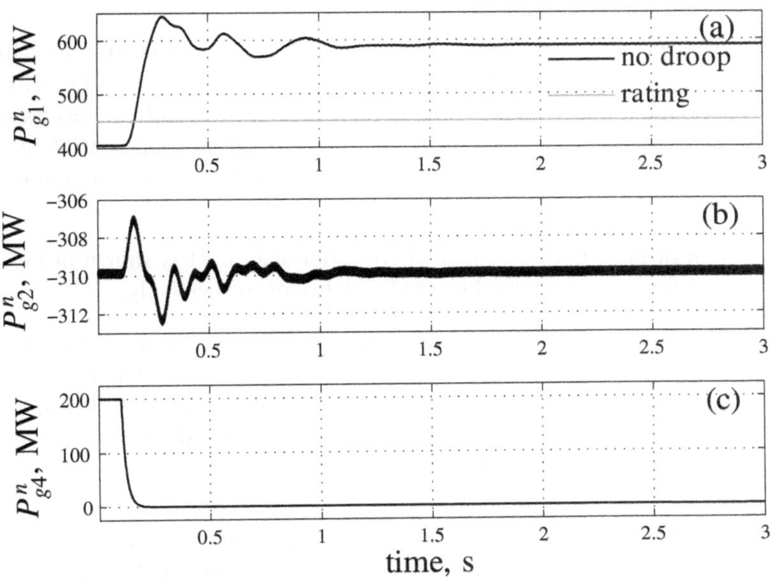

Figure 4.18 (a-c) Power sharing without droop scheme following the outage of the negative pole converter (inverter) of the converter station #4. ©[70], 2013.

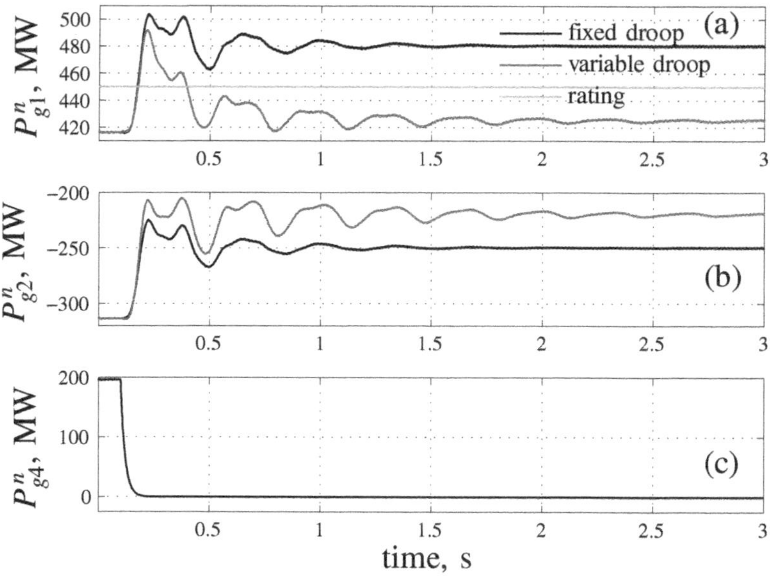

Figure 4.19 (a-c) Comparison of power sharing for fixed vs adaptive droop scheme following the outage of the negative pole converter (inverter) of the converter station #4. ©[70], 2013.

However, this does not account for the available converter headroom, which in this particular case is only 30 MW for each converter in station #1. Therefore, the inverting unit at that station shares about 66 MW power and violates the power rating of 450 MW as shown in Fig. 4.19(a). Adaptive droop control ensures a higher droop value for lower headroom and prevents converter station #1 from overloading, Fig. 4.19(a).

Variation of the DC link voltage at converter station #1 is shown in Figs 4.20 and 4.21 for no droop, variable and fixed droops, respectively.

The DC bus voltage of the slack converter (no droop) is maintained constant at ±350 kV with very little overshoot in its transient response. This is larger for the common bus voltage for fixed and adaptive droop control. It can be seen from Fig. 4.21 that the adaptive droop scheme does not deteriorate the DC link dynamics any further than the fixed droop scheme.

Rectifier Outage: In this case, converter pole in stations #2 and #3 act as rectifiers importing about 200 MW and 420 MW, respectively, into the MTDC grid, while those in converter stations #1 and #4 work as inverters carrying around 310 MW each. The outage of the negative pole converter in the rectifying station #2 was considered.

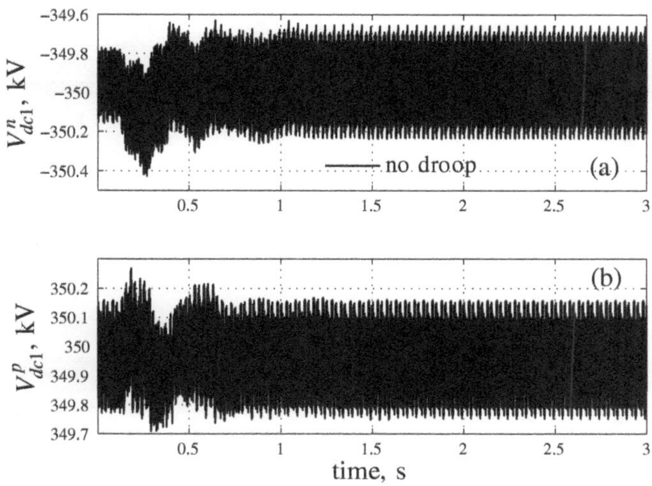

Figure 4.20 (a-b) DC link voltage of the converter in $V_{dc} - Q$ control mode following the outage of the negative pole converter (inverter) of the converter station #4. ©[70], 2013.

Dynamic response of the system with no droop are shown in Fig. 4.22. Although no converter has violated their power ratings, the large drop in converter station #1 power by about 200 MW might not be acceptable for the AC side utility connected to this converter station. It can be seen from Fig. 4.23 that the adaptive droop scheme is effective for the rectifier outage scenario too.

Table 4.1 Post-contingency steady-state operating points with fixed and adaptive droop control for rectifier and inverter outages

	Inverter outage				Rectifier outage			
	Pre	H_j	Fixed	Variable	Pre	H_j	Fixed	Variable
P^n_{g1}	417.2	32.8	**483.9**	422.8	307.6	142.4	241.0	209.5
P^n_{g2}	−312.8	137.2	−246.1	−215.6	−202.4	247.6	N/A	N/A
P^n_{g3}	−312.8	137.2	−246.1	−215.6	−422.4	27.6	**−489.0**	−426.1
P^n_{g4}	197.2	252.8	N/A	N/A	307.6	142.4	241.0	209.5

It is to be noted that the post-outage steady-state operating points in above simulation results agree with the analytical values shown in Table 4.1, which were obtained from the formulation in Section 4.6.

Current Controller Limits: Note that, in practice there will be limits on $I^{p*}_{gX}(I^{n*}_{gX})$ and $I^{p*}_{gY}(I^{n*}_{gY})$ of the current controller, as described in Sections 2.4.1 and 3.2.3.2.

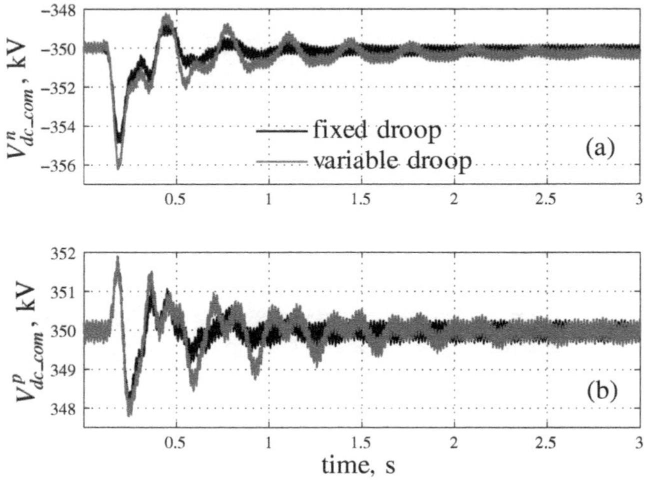

Figure 4.21 (a-b) Comparison of common bus DC link voltage for fixed vs adaptive droop scheme following the outage of the negative pole converter (inverter) of the converter station #4. ©[70], 2013.

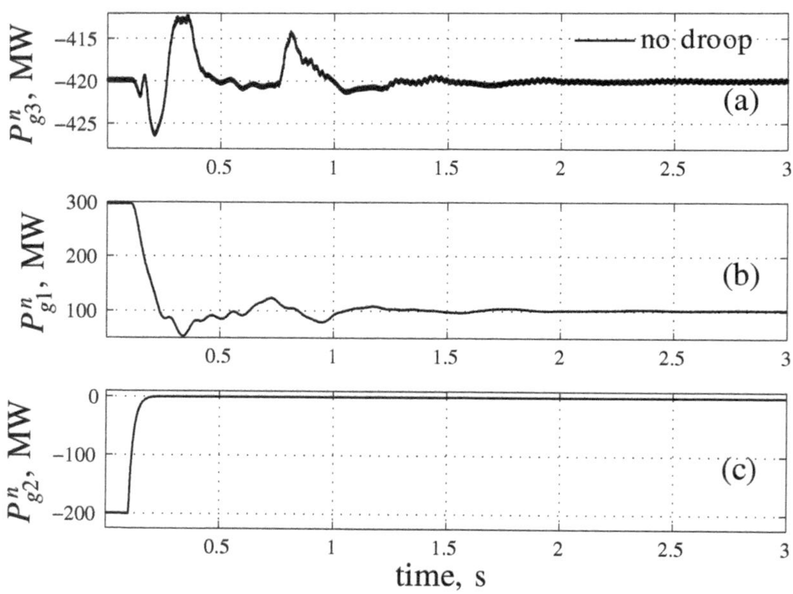

Figure 4.22 (a-c) Power sharing without droop scheme following the outage of the negative pole converter (rectifier) of the converter station #2. ©[70], 2013.

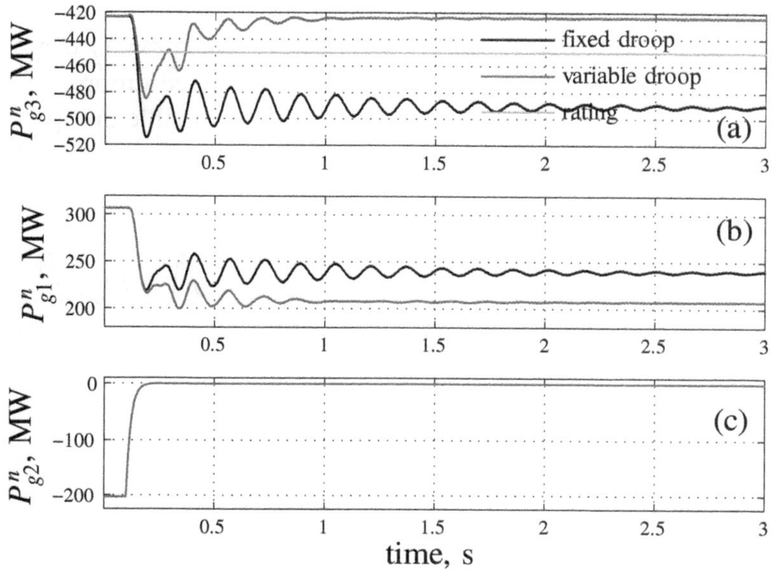

Figure 4.23 (a-c) Comparison of power sharing for fixed vs adaptive droop scheme following the outage of the negative pole converter (rectifier) of the converter station #2. ©[70], 2013.

In the case studies shown above, these limits were ignored for illustrative purposes to validate the post-contingency operating condition obtained from the analytical expression through time domain nonlinear simulation as shown in Table 4.1.

The simulation results with the current controller limits in place are shown in Figs 4.24 and 4.25 for inverter outage condition. The current limiting strategy is based on the principle mentioned in Section 3.2.3.2, i.e., limiting the magnitude of the components in $X - Y$ frame and re-computing the values of $I^{p*}_{gX}(I^{n*}_{gX})$ and $I^{p*}_{gY}(I^{n*}_{gY})$ while keeping the phase intact.

Figure 4.24 shows the response of the converter real powers following the outage of the negative pole of converter station #4 under the operating condition described in Section 4.8.3.3. Due to repeated violation of current controller limits with fixed droop control, the power sharing amongst the converter stations are jeopardized and the dynamic behavior is poorer in Fig. 4.24(a) and (b). On the other hand, the adaptive droop scheme leads to a desirable system performance.

As shown in Fig. 4.25, there is larger variation in DC link voltage when a fixed droop control is employed while the proposed strategy results in much tighter DC link voltage control.

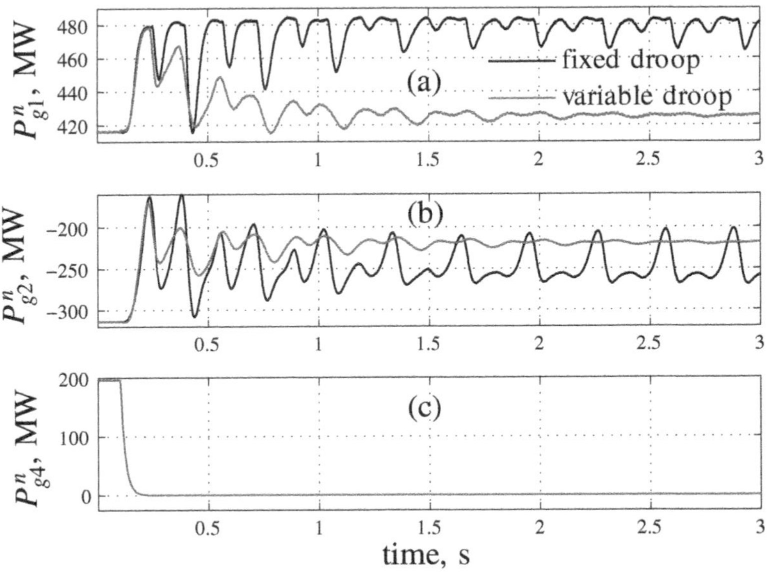

Figure 4.24 (a-c) Comparison of power sharing for fixed vs adaptive droop scheme following the outage of the negative pole converter (inverter) of the converter station #4. ©[70], 2013. Converter current control limits are in place.

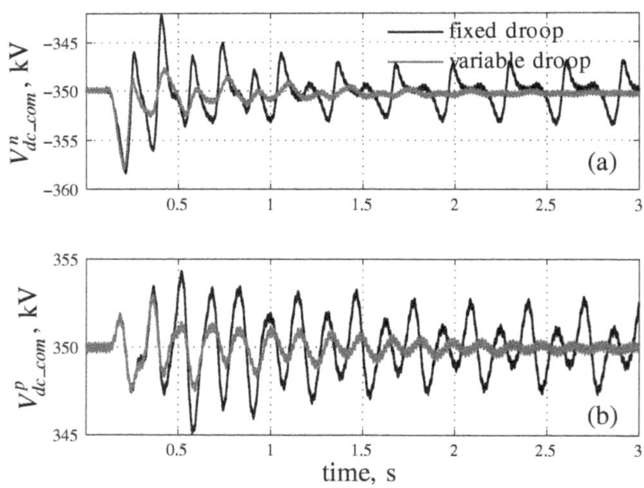

Figure 4.25 (a-b) Comparison of common DC link voltage for fixed vs adaptive droop scheme following the outage of the negative pole converter (inverter) of the converter station #4. Converter current control limits are in place. ©[70], 2013.

The common DC-link-voltage-feedback-based droop control is very attractive for system operation. However, it might face certain challenges under conditions where significant latency is present in the communication channel or the communication channel is lost (in the worst case). Research on developing a fall back option (e.g., using local-signal-based droop) during such extreme conditions should be considered.

CHAPTER 5

FREQUENCY SUPPORT

5.1 Introduction

Wind turbine generators (WTGs) are usually interfaced with the electric power grids through power electronic converters. Due to fast power reference control of the converters, WTGs do not inherently contribute to the system inertia unlike conventional synchronous power plants. In future, WTGs are likely to replace a significant fraction of the conventional power plant fleet which could reduce the overall system inertia drastically. Moreover, asynchronous interconnections through DC links act as "fire breaks" contributing further to the reduction in overall system inertia. With growing penetration of both asynchronous generation and interconnection, it is envisaged that the inertia of future systems could become low. This would result in unacceptably high rate of change and absolute variation in system frequency following a supply–demand imbalance condition.

To resolve this problem, mechanisms for extracting inertial support from WTGs have been proposed and demonstrated [73, 74]. These methods rely on detected changes in grid frequency to control the power reference of the WTGs. For instance, if the grid frequency goes down, the wind turbines are made to slow down so that

Multi-terminal Direct-Current Grids: Modeling, Analysis, and Control. First Edition.
Nilanjan Ray Chaudhuri, Balarko Chaudhuri, Rajat Majumder, and Amirnaser Yazdani.

they can release the stored kinetic energy in the blades in order to provide the inertial support. The mechanism is relatively straightforward for onshore wind farms and offshore wind farms with AC connection to the onshore AC system. For remote offshore wind farms connected through a DC link, extraction of inertial support is more complicated [51] and further research is required.

A multi-terminal DC (MTDC) grid with only autonomous power sharing control (as described earlier in Chapter 4) at the converter stations is not very effective in allowing exchange of primary frequency support among the host AC systems. Although this “fire-break” feature is helpful to arrest the propagation of fault level, this depletes the overall inertia which could have serious consequences for future systems with little individual inertia. Exchange of inertial and primary frequency support through an MTDC grid is, therefore, essential and is the subject matter of this chapter.

Disturbances in the AC (e.g., loss of generation, increase in load) and the DC grid (e.g., converter loss) lead to deviation of frequency in the AC system. The rate of change and the extent of frequency deviation depends on the magnitude of the disturbance, inertia of the system, and governor response characteristics [63]. One particular issue is how would an MTDC grid react to loss/outage of one or more converters and the resulting power imbalance. Sharing the burden of such a loss/power mismatch has to be appropriate in order to minimize the impact on the neighboring AC system. Very few papers have considered the problem of system frequency support through MTDC grids. In [48], a frequency support strategy based on local DC bus voltage–frequency droop is proposed.

In this chapter, frequency support for the surrounding AC systems interconnected through an asymmetric bipole MTDC grid with metallic return network is analyzed. The basic idea is to modify the standard autonomous power sharing (droop) control (described earlier in Chapter 4) by a frequency droop control loop to share the power mismatch in order to minimize the deviation in the AC system frequency. The topics are organized a follows: First, the fundamentals of frequency control is reviewed followed by brief discussions on the primary frequency support from wind farms and participation in automatic generation control (AGC). Next, the modification of the power-voltage droop control loop for frequency support is described. An analytical formulation is presented to compute the post-contingency frequency in an AC–MTDC grid in the presence of frequency droop control. The effectiveness of the frequency support is established through nonlinear simulation of an MTDC grid with four converter stations connected to three isolated AC systems. Small-signal Stability analysis is shown to establish the sensitivity of the eigenvalues with respect to different droop coefficients, control parameters, and change in operating conditions involving converter outage. It is demonstrated that the frequency droop control for the converters in an MTDC grid is effective in reducing the deviation in the frequency of the AC system following disturbances on both AC and DC sides. The results of the analytical formulation and the simulation are compared to gain more insight into the interaction of the AC and MTDC system.

5.2 Fundamentals of Frequency Control

In electric power systems, the total power generated must match with the power consumed by the loads plus the losses in the network on an instantaneous basis in order to maintain the grid frequency. This balance must be preserved in the face of continuous fluctuations in loads and potentially significant intermittence in renewable generation. The system frequency increases if the generation is more than the demand/load, and reduces if the load exceeds the generation. Certain generators (e.g., gas power plants) are commanded to quickly ramp up or down their power output to compensate for these fluctuations so that supply-demand balance and hence, the grid frequency, is restored [63]. Frequency sensitive loads also play a part in this balancing process. Collective response of all the components (e.g., generators, loads etc.,) to an unbalance between supply-demand is referred to as *system frequency response*.

Based on the underlying mechanism and speed of action, the system frequency response following a large disturbance is divided into the following categories: inertial, primary frequency response, and secondary frequency response or automatic generation control (AGC). Each category refers to the method for providing frequency support with conventional generators. Inertial response exploits stored kinetic energy (inertia) within the rotor of the synchronous generators. Primary frequency support involves automatic control of generator governors to control the mechanical power input (Sections 3.3.1.5 and 4.3.1). Sometimes, inertial control is considered within the primary frequency response. Control of power output of the generators in response to the system operators' command constitutes the secondary/AGC response. For further details and exact definitions of inertial, primary, and secondary response, the readers should refer to [74]. A typical time variation of system frequency following a sudden loss of generation is shown in Fig. 5.1. The inertial response immediately follows the generation loss followed by primary frequency response typically within 20–30 seconds. Within 5–10 minutes the secondary response restores the system frequency back to its nominal value of 50/60 Hz.

The extent of inertial response is determined by the physical inertia and rotational speed of the generator rotors which determines the kinetic energy stored in them. In response to an increase in electrical load or loss of generation elsewhere, the remaining generators decelerate and release stored kinetic energy in the process. Immediately after the disturbance this helps restore the generation-demand imbalance to an extent reducing the rate and extent of change frequency. It is to be noted that inertial response comes naturally from all the generators without requiring any control loop. Within seconds time frame, the generator governors respond to the system frequency change and adjusts the mechanical power input to the generators offering primary frequency support. Only the generators with active governor control contributes to the primary frequency response. Participating generators could be made to share the burden of primary frequency support in certain proportion by appropriate choice of

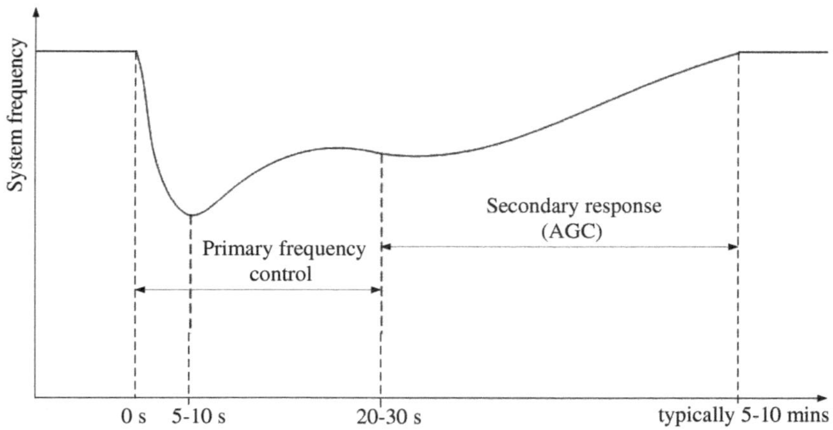

Figure 5.1 Primary and secondary (AGC) frequency response.

power-frequency droop [63]. The principle of droop control in synchronous generators was described in Sections 3.3.1.5 and 4.3.1. Finally, within 5–10 min time frame the power plants adjust their power levels in response to commands issued by the systems operator to restore the frequency to its nominal value.

5.3 Inertial and Primary Frequency Support from Wind Farms

Unlike the synchronous generators, wind turbines do not inherently provide inertial contribution. However, a wind turbine could be made to contribute inertial support by controlling the power output to compensate for deviations in grid frequency. If the frequency drops, the control loop will act to increase the electrical torque which slows the turbine down and extracts the kinetic energy stored in the blades. The inertial support is provided over a short time scale typically to avoid a longer recovery phase and prevent the turbines from stalling in the worst case. However, sometimes wind turbines are capable of providing more effective inertial support than a conventional synchronous generator with equivalent per unit of spinning inertia. This is because wind turbines with converter interface to the grid can change their speed of rotation over a much wider range compared to synchronous generators and hence release more kinetic energy. Also actuation of the torque command signal is much faster with power electronic converters than what could be achieved in conventional synchronous generators. Inertial response from wind farms is now required by several European transmission system operators to help regulate the grid frequency.

For example, the Irish grid code requires that wind plants have active power curtailment capabilities, and outlines specific active power generation set-points as a function of available power in the event of a frequency deviation. This code further specifies a minimum response rate for individual turbines of 1% of rated power per second [75]. Denmark's TSO, Energinet requires that wind farms are able to track a reserve power offset and reference power levels generated by the system operator [76]. Red Electrica, the Spanish system operator requires that wind farms respond to frequency deviations with proportional control of active power output within a range of percentages of rated power [77]. In Canada, Hydro-Quebec requires that wind farms larger than 10 MW should be able to modify their active power output for at least 10 s in response to grid frequency deviations greater than 0.5 Hz [78], [79, 80].

There are multiple approaches to extract additional power from the rotating turbine blades and to handle the recovery phase. As perviously mentioned, one approach is to quickly increase the torque in response to reduction in the grid frequency, which causes the turbine to slow down and release the kinetic energy stored in the blades. Another approach is to allow the turbine to accelerate initially before increasing the torque command [81]. The benefit is that the turbine power and rotor speed do not fall below normal operation speed throughout the course of inertial response which eliminates the recovery period and prevents the turbine from stalling. However, a temporary over-speed can be caused when operating at rated power which may not be desirable. Moreover, inertial support could be delayed due to the initial speed-up process resulting in relatively larger frequency drop than the other approach.

The conventional synchronous generators automatically provide inertial support and have governors for primary control. For wind turbines, which do not have governors, the distinction between inertial and primary responses is less clear. To provide primary frequency support, the wind turbines need to maintain a reserve by operating at a suboptimal tip-speed ratio [73, 82]. Such derated operation is expensive unless there is adequate incentive to participate in the frequency response market. Pitch control is also used for derated operation and contribute to the primary response [73]. Further research is required to address the persistent over-speeding of the turbines while under above-rated wind speeds. Also, the effect of sudden and frequent torque changes the structural loading of the turbine and hence its lifetime is to be carefully considered as well.

Inertial and primary control are effective when implemented at the wind farm level. Intelligent and distributed response of an entire wind farm could potentially minimize grid frequency deviation and reduce the recovery phase. Primary frequency control with varying levels of wind penetration suggests that the wind farms' contribution to frequency support could greatly increase grid robustness and reduce the maximum frequency deviation [83]. In the simulation the grid was represented as an electrical bus to analyze the response of the wind farm following outage of a conventional generator. While these simulation results are promising, they are limited by their use of a number of turbines much smaller than a typical wind farm and simplified models of turbines and turbine wake interactions.

5.4 Wind Farms in Secondary Frequency Control (AGC)

A secondary (AGC) response consists of the power plants raising and lowering power output in response to commands from the system operators. Conventional generators typically receive an absolute power reference and are usually capable of ramping the power output upto their rated capacity. Similarly, the wind farms participating in AGC receives an absolute power command signal or a "delta control" command which specifies the percentage of the available wind power to be captured. The delta control ensures that there is a percentage of available wind power that is kept as overhead in case the the system operator requires more power. The system operator often specifies a "power gradient limiter" which is a lower limit on the rate of change of wind farm power output to meet the desired set-point. Wind turbines respond to a change in power demand by pitch angle and rotor speed control although it involves suboptimal operation. Combined torque-pitch controllers have also been used to respond very quickly to a change in power reference command.

Conventional power plants with synchronous generators automatically contribute to inertial support while governor control action is required to participate in primary and secondary (AGC) frequency response. For wind farms, separate control is required to derive inertial support and make them contribute to inertial, primary and secondary frequency control as described above. With AC connection to the wind farm and AC interconnection between multiple systems, frequency support could be readily exchanged or shared. However, for DC connected wind farms and DC interconnection between AC systems, exchanging frequency support is not automatic as the DC acts as a "fire-break" electrically decoupling the AC systems and the wind farms from the main grid. An MTDC grid would interconnect several onshore AC systems and remote offshore wind farms in future with each AC system likely to have relatively low inertia. Hence, it is essential to devise a mechanism for exchanging frequency support across the MTDC grid which is described in the next section. This approach is generic and applies to the exchange of inertial, primary and secondary (AGC) frequency response from conventional power plants and AC and DC connected wind farms. For remote offshore wind farms connected via DC link, the offshore DC converter would usually act as an AC voltage source to transfer the incoming power from wind farms on to the MTDC grid (Section 2.4.6 and Mode IV in Section 3.2.3.1). Hence, the modification to DC voltage droop control scheme, described in the next section, do not apply to such offshore converters for wind farms.

5.5 Modified Droop Control for Frequency Support

In Section 2.4.5, the concept of active power-DC link voltage droop control was introduced. In Chapter 4, appropriate modifications and simplification were introduced to the droop control loop for autonomous power sharing. The objective there was to ensure that all the healthy converter stations share the burden following a contingency (e.g., loss of one converter) in a desirable way. The proposed droop control reduces the change in power exchange with the surrounding AC systems and the frequency deviation also gets distributed amongst different AC grids connected to different converter stations. However, such a droop control strategy could still result in potentially unacceptable frequency deviations in the AC systems due to the following reasons:

- After a disturbance in the AC side, e.g., step change in a generator power reference, the frequency of that system deviates from the nominal value and recovers due to governor action. The autonomous power sharing loop does not aid in reducing the frequency deviation, since there is no change in the net power exchange with the concerned AC grid.

- Change in the real power reference of a converter station results in appropriate distribution of the power deviation amongst different converter stations. Depending on how many stations are connected to each AC grid and the governor droop coefficients of different AC systems, this power sharing may result in larger frequency deviations in some AC systems.

- Following the outage of a converter in one pole, those in the unaffected pole do not participate in power sharing. Therefore, frequency deviation in certain AC systems could be relatively large.

- Autonomous power sharing control alone can not appropriately share the frequency support (as an ancillary service) among AC systems interconnected through the MTDC grid.

Thus there is a need to modify the power sharing among different converter stations in order to provide frequency support for the host AC systems. One way to achieve this is to update the real power reference of each converter station with a supplementary power-frequency droop control [84] as shown in Fig. 5.2. The AC system frequency $f_{(k)}$ at the PCC of the kth converter is used as a feedback signal to derive the frequency error. The frequency error is multiplied by the frequency droop constant β_{fk} to modify the power reference set-point. It was mentioned in Sections 3.2 and 3.7 that the dynamic variation of Θ_{gk} with respect to time gives the deviation in angular frequency from the nominal condition, as the AC network current balance

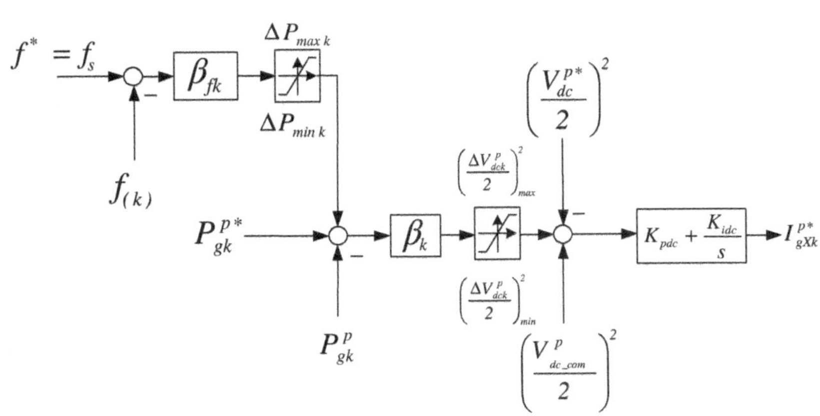

Figure 5.2 Active power-DC link voltage droop for sharing of power imbalance in the positive pole of the kth converter station and frequency support of the adjacent AC system. ©[84], 2013.

equations are solved at each solution time step. Therefore, $f_{(k)}$ can be computed as $f_{(k)} = \frac{1}{2\pi}\left(\frac{d\Theta_{gk}}{dt} + \omega_s\right)$.

Here, P^{p*}_{gk} is considered positive, if the converter station is acting as an inverter injecting power into the AC system. Upon reduction (increase) in system frequency, the real power injection into the AC system is commanded to increase (reduce) by changing the reference accordingly (Fig. 5.2). If the converter station acts as a rectifier exporting power out of the AC system, P^{p*}_{gk} is negative. A decrease (increase) in system frequency in that case would result in a reduction (increase) in power export command from the AC system which helps restore the frequency. It is to be noted that both the converter poles participate in the frequency support. The absolute values of the droop constants used for autonomous power sharing (β_k) and frequency support (β_{fk}) are critical for satisfactory performance and stability as demonstrated later in this chapter.

The frequency droop control loop acts to reduce the frequency deviation in the system where the disturbance has taken place. This is achieved by importing frequency support from the remaining (unaffected) AC systems. As a result, the frequency of those AC systems would deviate from their nominal values (which would not have happened without the frequency droop control loop). Thus, rather than having a large and potentially unacceptable variation in the frequency of the affected system, there would be relatively small (but hopefully within acceptable limits) variations of frequency in all the AC systems. The extent of variation of frequency depends on the nature and size of the disturbance and also the inertia of the systems.

It is essential to ensure that the variation of the frequency in the unaffected AC systems does not violate the acceptable limits while trying to support the affected system. Hard limits are therefore, placed on the incremental power change ΔP_{min}, ΔP_{max} caused by the frequency deviations as shown in Fig. 5.2. These limits are chosen based on the available reserve capacity within a particular AC system.

5.6 AC–MTDC Load Flow Solution

It is important to determine the pre-contingency steady-state operating point of the AC–MTDC grid which could be done by solving a AC–MTDC load flow. The general sequential AC–MTDC grid load flow setup was described in Section 3.4.3 and the necessary modifications for incorporating the droop control for autonomous power sharing was elaborated in Section 4.5. Inclusion of the frequency droop control loop in f_{15} for the kth converter station leads to

$$\begin{aligned} f_{15k} = \left(V^{p}_{dc\_com}\right)^2 - \left(V^{p*}_{dc}\right)^2 + \\ 4\beta_k\left(\frac{P^{*}_{gk}}{2} - P^{p}_{gk} + \beta_{fk}\left(f_s - f_{(k)}\right)\right) = 0 \end{aligned} \tag{5.1}$$

Since the pre-disturbance frequency $f_{(k)}$ is assumed to be equal to the rated value f_s, the expression in (5.1) becomes identical to (4.4). Therefore, no further change is needed to the elements of Jacobian in addition to those mentioned in Section 4.5. Post contingency frequency is derived analytically in the next section.

5.7 Post-Contingency Operation

It is crucial to derive an analytical formulation to calculate the steady state frequency of the AC systems in the AC–MTDC grid under post-contingency condition. It can throw significant insight into the frequency support contribution of an area to the other. Figure 5.3 shows the jth AC system of the interconnected AC–MTDC grid. In this context, let us define the following symbols:

$f_s(=\frac{\omega_s}{2\pi})$ nominal frequency of the AC system

$f_{(k)}$ frequency at the PCC of the kth converter station

ℓ number of AC areas connected to the MTDC grid

n total number of converter stations in the MTDC grid

f_j post-contingency steady state frequency in the jth AC area

P^0_{ej} pre-disturbance total real power load in the jth AC area

P_{ej} post-contingency total real power load in the jth AC area

N_j total number of generators in jth area

P^0_k mechanical power input of the kth generator connected to the jth AC area before the disturbance

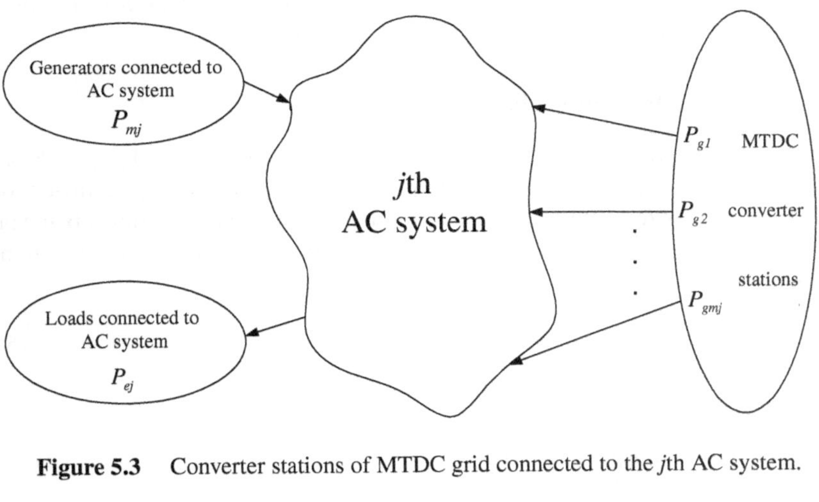

Figure 5.3 Converter stations of MTDC grid connected to the jth AC system.

R_{gok} droop coefficient of the kth governor

m_j total number of converter stations connected to jth area.

P^{p0}_{gk}, P^{n0}_{gk} real power injected by the positive and negative pole of the kth converter station before the disturbance

In this analysis we assume that the reference quantities remain unchanged following the disturbance.

5.7.1 Analysis for AC System

For the kth generator equipped with a governor (please see Section 3.3.1.5) in the jth AC area, the mechanical power input P_k under the post-contingency condition can be expressed as

$$P_k = P^0_k - \left(\frac{f_j - f_s}{f_s}\right)\frac{1}{R_{gok}} \tag{5.2}$$

The total mechanical power input to all generators in this area under the post-contingency operating condition is

$$\sum_k^{N_j} P_k = \sum_k^{N_j}\left\{P^0_k - \left(\frac{f_j - f_s}{f_s}\right)\frac{1}{R_{gok}}\right\} \tag{5.3}$$

Regrouping the terms appropriately, we can write

$$\begin{aligned} P_{mj} &= \sum_k^{N_j} P^0_k - \left(\frac{f_j - f_s}{f_s}\right)\sum_k^{N_j}\frac{1}{R_{gok}} \\ P_{mj} &= P^0_{mj} - \left(\frac{f_j - f_s}{f_s}\right)\frac{1}{\beta_{gj}} \end{aligned} \tag{5.4}$$

where, P^0_{mj} and P_{mj} are the total mechanical power input to generators in the jth area under pre and post contingency condition and β_{gj} is the equivalent governor droop coefficient of the aggregated generators in the jth area.

5.7.2 Analysis for Converter Station

Let us denote the frequency at the PCC of the kth converter station by $f_{(k)}$. Prior to a disturbance, $f_{(k)} = f_s$. Therefore, the output of the frequency droop feedback loop shown in Fig. 5.2 is zero. With similar assumptions mentioned in Section 4.6, the pre-disturbance power output of the positive pole converter at the kth converter station is given by:

$$P^{p0}_{gk} = P^{p*}_{gk} + \frac{-P^{p0}_{loss(dc)}}{\beta_k \sum_k^n \frac{1}{\beta_k}} \tag{5.5}$$

where, $\sum_k^n P^{p*}_{gk} = 0$ and $\sum_k^n P^{p0}_{gk} = -P^{p0}_{loss(dc)}$.

5.7.2.1 *AC Side Disturbances* After a disturbance in the AC system, the frequency deviates from f_s. If no converter outage occurs in the positive pole of the converter stations, the post-disturbance power output of the positive pole converter at the kth converter station is given by

$$P^p_{gk} = P^{p*}_{gk} + \frac{\left(V^p_{dc\_com}\right)^2 - \left(V^{p*}_{dc}\right)^2}{4\beta_k} + \left(f_s - f_{(k)}\right)\beta_{fk} \tag{5.6}$$

Summing over all converter stations in the MTDC grid and deriving the expression for $\frac{\left(V^p_{dc\_com}\right)^2 - \left(V^{p*}_{dc}\right)^2}{4}$ we get

$$\sum_k^n \left\{P^{p*}_{gk} - P^p_{gk} - \left(f_{(k)} - f_s\right)\beta_{fk}\right\} + \frac{\left(V^p_{dc\_com}\right)^2 - \left(V^{p*}_{dc}\right)^2}{4}\sum_k^n \frac{1}{\beta_k} = 0 \tag{5.7}$$

$$\frac{\left(V^p_{dc\_com}\right)^2 - \left(V^{p*}_{dc}\right)^2}{4} = \frac{-P^p_{loss(dc)} + \sum_k^n \left(f_{(k)} - f_s\right)\beta_{fk}}{\sum_k^n \frac{1}{\beta_k}} \tag{5.8}$$

where, $\sum_k^n P^p_{gk} = -P^p_{loss(dc)}$. Substituting (5.8) in (5.6):

$$P^p_{gk} = P^{p*}_{gk} + \frac{-P^p_{loss(dc)} + \sum_k^n \left(f_{(k)} - f_s\right)\beta_{fk}}{\beta_k \sum_k^n \frac{1}{\beta_k}} + \left(f_s - f_{(k)}\right)\beta_{fk} \tag{5.9}$$

Similar expression can be derived for the negative pole converter.

5.7.2.2 *Converter Outage* Following the disturbances due to the outage of the positive pole of the nth converter station the post-disturbance power output of the positive pole converter at the kth converter station can be obtained as described in the previous section

$$P^{pout}_{gk} = P^{p*}_{gk} + \frac{P^{p*}_{gn} - P^{pout}_{loss(dc)} + \sum_k^{n-1} \left(f_{(k)} - f_s\right)\beta_{fk}}{\beta_k \sum_k^{n-1} \frac{1}{\beta_k}} + \left(f_s - f_{(k)}\right)\beta_{fk} \tag{5.10}$$

where, $\sum_k^{n-1} P^p_{gk} = -P^{pout}_{loss(dc)}$, and $\sum_k^{n-1} P^{p*}_{gk} = -P^{p*}_{gn}$.

The power output for the negative pole counterpart is identical to that described in the previous section.

5.7.3 Analysis for AC System Connected to Converter Stations

Let us now analyze the post-contingency operating condition of the jth AC area connected to m_j converter stations. For this analysis the losses in the AC–MTDC system will be neglected. Following a disturbance in the AC system the total real power injected into the jth AC system from the MTDC grid is

$$P_{dcj} = \sum_k^{m_j} P^p_{gk} + \sum_k^{m_j} P^n_{gk} = 2\sum_k^{m_j} P^p_{gk} \tag{5.11}$$

The power balance equation leads to

$$P_{mj} - P_{ej} + P_{dcj} = 0 \tag{5.12}$$

Substituting values of P_{mj} and P_{dcj} from (5.4) and (5.9) and neglecting losses, we get

$$P^0_{mj} - \left(\frac{f_j - f_s}{f_s}\right)\frac{1}{\beta_{gj}} - P_{ej} + 2\sum_k^{m_j}\left\{P^{p*}_{gk} + \frac{\sum_k^n \left(f_{(k)} - f_s\right)\beta_{fk}}{\beta_k \sum_k^n \frac{1}{\beta_k}} + \left(f_s - f_{(k)}\right)\beta_{fk}\right\} = 0 \tag{5.13}$$

Grouping appropriate terms from (5.13) we get

$$\underbrace{P^0_{mj} - P_{ej} + \frac{1}{\beta_{gj}} + 2\sum_k^{m_j} P^{p*}_{gk} + 2f_s\left\{\sum_k^{m_j}\beta_{fk} - \sum_k^{m_j}\frac{\sum_k^n \beta_{fk}}{\beta_k\sum_k^n \frac{1}{\beta_k}}\right\}}_{\Gamma_j} +$$

$$\underbrace{\left\{-\frac{1}{f_s\beta_{gj}} - 2\left(\sum_k^{m_j}\beta_{fk} - \sum_k^{m_j}\frac{\sum_k^n \beta_{fk}}{\beta_k\sum_k^n \frac{1}{\beta_k}}\right)\right\}}_{w_{jj}} f_j + \underbrace{2\sum_k^{m_j}\frac{\sum_{k=m_j+1}^n f_{(k)}\beta_{fk}}{\beta_k\sum_k^n \frac{1}{\beta_k}}}_{\sum_{r=1, \neq j}^{l} w_{jr}f_r} = 0 \tag{5.14}$$

Following the outage of the positive pole of the nth converter station, substituting values of P_{mj} and P_{dcj} from (5.4), (5.9), and (5.10) in (5.12) and neglecting losses, we get

$$\underbrace{P^0_{mj} - P_{ej} + \frac{1}{\beta_{gj}} + \sum_k^{m_j^p} P^{p*}_{gk} + \sum_k^{m_j^p} \frac{P^{p*}_{gn}}{\beta_k \sum_k^{n-1} \frac{1}{\beta_k}} + \sum_k^{m_j} P^{n*}_{gk}}_{\Gamma_{j1}} +$$

$$\underbrace{f_s \left[\left\{\sum_k^{m_j^p} \beta_{fk} - \sum_k^{m_j^p} \frac{\sum_k^{n-1} \beta_{fk}}{\beta_k \sum_k^{n-1} \frac{1}{\beta_k}}\right\} + \left\{\sum_k^{m_j} \beta_{fk} - \sum_k^{m_j} \frac{\sum_k^{n} \beta_{fk}}{\beta_k \sum_k^{n} \frac{1}{\beta_k}}\right\}\right]}_{\Gamma_{j2}} +$$

$$\underbrace{f_j \left\{-\frac{1}{f_s \beta_{gj}} - \left(\sum_k^{m_j^p} \beta_{fk} - \sum_k^{m_j^p} \frac{\sum_k^{n-1} \beta_{fk}}{\beta_k \sum_k^{n-1} \frac{1}{\beta_k}}\right) - \left(\sum_k^{m_j} \beta_{fk} - \sum_k^{m_j} \frac{\sum_k^{n} \beta_{fk}}{\beta_k \sum_k^{n} \frac{1}{\beta_k}}\right)\right\}}_{w_{jj}} \tag{5.15}$$

$$+ \underbrace{\left\{\sum_k^{m_j^p} \frac{\sum_{k=m_j^p+1}^{n-1} f_{(k)} \beta_{fk}}{\beta_k \sum_k^{n-1} \frac{1}{\beta_k}} + \sum_k^{m_j} \frac{\sum_{k=m_j+1}^{n} f_{(k)} \beta_{fk}}{\beta_k \sum_k^{n} \frac{1}{\beta_k}}\right\}}_{\sum_{r=1, \neq j}^{l} w_{jr} f_r} = 0$$

where, $m_j^p = m_j$ for the AC systems where no converter loss has happened and $m_j^p = m_j - 1$ for the AC system that lost the converter pole connected to it. Please note that $f_{(k)} = f_j, k = 1(1)m_j$.

5.7.4 Analysis of AC–MTDC Grid

To determine the post contingency steady state frequency of l AC systems connected through the MTDC grid, (5.14) or (5.15) can be used depending on whether it is an AC side disturbance or a converter outage. There are l linear equations with l unknown frequencies:

$$\sum_r^l w_{jr} f_r + \Gamma_j = 0, j = 1(1)l \tag{5.16}$$

The steady state frequency can be obtained by

$$[f]_{l\times 1} = -[w]^{-1}_{l\times l}[\Gamma]_{l\times 1} \tag{5.17}$$

We define matrix w as the "Inertial Coupling" matrix since the elements of this matrix determines the interaction between the steady state frequency of one AC system with another. The diagonal elements w_{jj} depend on the aggregated governor droop coefficients β_{gj} in the AC area as well as the droop coefficients β_k and β_{fk} of the converter stations. On the other hand, the off-diagonal elements are only functions of the converter droop coefficients β_k and β_{fk}. Therefore, the inertial contribution from one AC system to the other can be modified by changing β_k and β_{fk} of the "Inertial Coupling" matrix. When β_k is equal for all converter stations, the off-diagonal elements become function of only β_{fk}. The diagonal elements depend on both β_{fk} and β_{gj}.

5.8 Case Study

A case study is reported in this section to illustrate the benefits of using a combined droop control for both autonomous power sharing and frequency support (as shown in Fig. 5.2).

5.8.1 Study Network

Figure 5.4 shows the study network consisting three AC systems asynchronously interconnected through the MTDC grid. One AC system is modelled in details with four generators ($G1$–$G4$) while other two AC grids are modelled as a single equivalent generator. Each generator is equipped with their governors.

Referring to the list of symbols in Section 5.7, for this test system, $\ell = 3$, $n = 4$ $f_s = 60$ Hz.

For AC area #1: $m_1 = 2$, $N_1 = 4$, $P^0_{m1} = 2800$ MW, $P^0_{e1} = 4000$ MW, $\beta_{g1} = 0.00471$, $P^{p*}_{g2} = P^{n*}_{g2} = 450$ MW, $P^{p*}_{g3} = P^{n*}_{g3} = 150$ MW

For AC area #2: $m_2 = 1$, $N_2 = 1$, $P^0_{m2} = 900$ MW, $P^0_{e2} = 0$ MW, $\beta_{g2} = 0.01885$, $P^{p*}_{g4} = P^{n*}_{g4} = -450$ MW

For AC area #3: $m_3 = 1$, $N_3 = 1$, $P^0_{m3} = 300$ MW, $P^0_{e3} = 0$ MW, $\beta_{g3} = 0.01885$, $P^{p*}_{g1} = P^{n*}_{g1} = -150$ MW

The same system as in Section 3.11 has been used with the following modifications. The generators ($G1$–$G6$) are equipped with turbine and governor control, Section 3.3.1.5, in addition to the IEEE DC1A exciters (Section 3.3.1).

The MTDC grid consisting of four bipole converter stations are connected with 500 km cables as shown in Fig. 5.4. The DC cables were represented by a single pi-section model. Converter stations #2 and #3 are acting as inverters while #1 and #4 as rectifiers under nominal condition. The DC link voltage of station #1 is chosen as the common reference for the droop scheme. The feedback signal used in the frequency

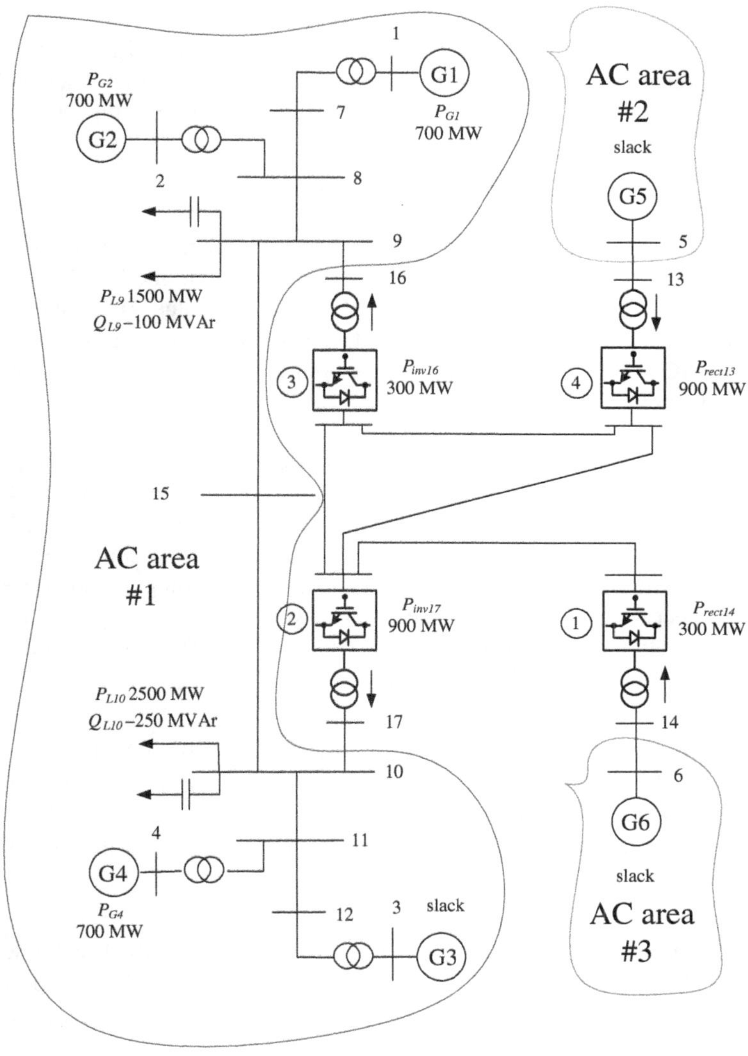

Figure 5.4 Asymmetric bipole MTDC grid (single line diagram) connected to a multi-machine AC system. ©[84], 2013.

droop control loop is derived from the speed of the generator nearest to the converter bus.

Under the nominal condition, each of $G1$, $G2$, and $G4$ generates 700 MW and the rest were left as slack. The tie-line transfer through the AC line connecting buses #9 and #10 is around 200 MW. All converter stations operate in droop control mode while maintaining unity power factor at the PCC. The isolated AC systems represented by equivalent generators, $G5$ and $G6$ could be wind farms in which case the control of the interface converters would be different. The converter would be controlled to act as a stiff voltage source with constant voltage magnitude and frequency and thus allowing the incoming power from the wind farms to transfer on to the DC grid.

5.8.2 AC–MTDC Grid Load flow Solution

The load flow solution of the combined AC–MTDC systems, Section 3.4 is used to calculate the initial conditions of the states of the dynamic model. As the study system in Section 5.8.1 includes a multi-machine interconnected AC system (unlike the study system in Section 3.10.1 where all the AC systems were equivalent voltage sources), the AC–MTDC load flow was carried out following a sequential approach as elaborated below:

1. First, the AC system was considered for the load flow, where $G3$, $G5$, and $G6$ were assumed to be slack generators and the converter buses #13, #14, #16, and #17 as load buses. Since the real power at these buses were not known apriori as the converter stations were working under droop control, they were initialized with their respective reference values. Assuming a load convention with respect to the AC systems (i.e., real power is positive (negative) if it is a load (generation)), the real power at the concerned buses were initialized as 900, 300, −300, and −900 MW, respectively.

2. The voltage magnitudes and angles of buses #5, #6, #9, and #10 (denoted in complex polar form as V_5, V_6, V_9, and V_{10}, respectively, in Table 5.1), calculated from the AC load flow in Step 1, were used as initial guess for the MTDC grid load flow (in fact voltages at bus #5 and #6 were $1\angle 0$ since they were slack buses).

3. The MTDC grid load flow solution in Step 2 gave the modified real powers at buses #13, #14, #16, and #17 (denoted as P_{g13}, P_{g14}, P_{g16}, and P_{g17}, respectively, in Table 5.1). With these updated values, the AC load flow was run again.

4. Steps 1–3 were repeated until the maximum value of ($|\Delta P_g^{(k)}|$) at the converter PCC buses (denoted as max $|\Delta P_g^{(k)}|$ in Table 5.1) fell below a certain tolerance ϵ (assumed 10^{-10} in this work). Here, $|\Delta P_g^{(k)}| = |P_g^{(k)} - P_g^{(k-1)}|$ and k is the number of iterations.

Three iterations of sequential AC–MTDC grid load flow leading toward convergence are shown in Table 5.1 for the study system in Fig. 5.4. There is virtually no difference between the bus powers and voltages from iteration 2 to 3 as the maximum value of the power mismatch at the converter PCC buses ($max|\Delta P_g^{(k)}|$) was already at a low value of 1.57E−06.

Table 5.1 Iterations leading to convergence of sequential AC–MTDC grid load flow

AC–DC load flow iter.	1	2	3
P_{13}	9.0	9.0341	9.0341
P_{14}	3.0	3.0341	3.0341
P_{16}	−3.0	−2.9659	−2.9659
P_{17}	−9.0	−8.9659	−8.9659
V_5	$1\angle 0$	$1\angle 0$	$1\angle 0$
V_6	$1\angle 0$	$1\angle 0$	$1\angle 0$
V_9	$0.98\angle -15.72$	$0.98\angle -16.17$	$0.98\angle -16.17$
V_{10}	$0.99\angle -26.66$	$0.99\angle -26.89$	$0.99\angle -26.89$
$max\lvert\Delta P_g^{(k)}\rvert$	0.0341	1.57E−06	7.31E−11
No of AC load flow iter.	15	9	5
No of DC load flow iter.	4	4	4

5.8.3 Small-signal Stability Analysis

For small-signal stability analysis, the AC–MTDC grid model described by differential-algebraic equations (DAEs) is linearized around the nominal operating condition described in Section 5.8.1. Participation factor analysis (Section 3.6.3), is done to analyze the nature and the root cause of the dynamic response of the MTDC grid. It was found that the poles in the far left of the s-plane has more participation from the states of the converter transformer and phase reactor (i.e., AC side states). As we move toward the right of the s-plane, the participation from the DC network, machine and droop controller states increase. To establish a stability zone for the droop controller gains β (pu kV^2/MW), β_f (MW/Hz), K_{pdc} (pu MW/kV^3) and K_{idc} (pu MW-s/kV^3) the sensitivity of the eigenvalue movement w.r.t these parameters are investigated (for the pu calculations, the base quantities used were: MVA = 608, kV = 120.09).

For $\beta = 0.1$ and $\beta_f = 100$, the value of K_{pdc} was increased from 0.2 to 1.0. The integral coefficient K_{idc} was kept 10 times that of K_{pdc}. It can be seen from the root locus plot of Fig. 5.5 that there is a mixed movement of the poles - some toward further left and some toward right. One of the DC network modes have moved into the right half of the s-plane, making the system unstable. Therefore, it can be concluded that increasing the PI controller gains can reduce system stability.

To find out the implication of variation in β on the system stability the values of K_{pdc}, K_{idc}, and β_f were fixed at 1.0, 10.0, and 100.0, respectively. When β is reduced from 0.5 to 0.1 the system becomes unstable as shown in Fig. 5.6.

It is clear from this exercise that increase in β can increase the system stability as opposed to K_{pdc} and K_{idc}. Also worth noting is the movement of the AC network poles in the same direction of the critical poles of the system toward the right half of the s-plane.

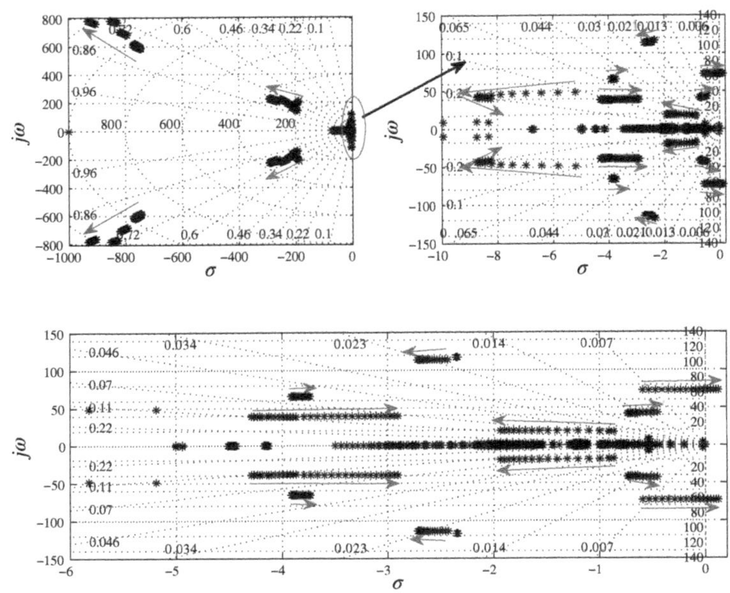

Figure 5.5 Root locus of system poles with increase of K_{pdc} from 0.2 to 1.0, keeping β and β_f constant at 0.1 and 100, respectively. The value of K_{idc} is kept 10 times that of K_{pdc}. ©[84], 2013.

These studies provide some insight into the root-cause of the system dynamic behavior and stability. It was also observed that β_f has a low sensitivity toward the movement of the system eigenvalues. After establishing the stability zone at nominal condition, the following values were considered for the droop controller $\beta = 0.1$, $\beta_f = 100.0$, $K_{pdc} = 0.5$ and $K_{idc} = 5.0$. However, it was found that following the outage of one pole of converter station #3 (($N - 1$) scenario) the system becomes unstable as shown in Fig. 5.7.

This highlights the importance of considering different contingency scenarios during the design process. Also, it was found that the operating condition alters the sensitivity of the eigen-value movements. Finally, the following values were chosen for the droop control: $\beta = 0.1$, $\beta_f = 100.0$, $K_{pdc} = 0.1$ and $K_{idc} = 1.0$.

5.8.4 Nonlinear Simulation

5.8.4.1 AC Side Disturbances A sudden loss of infeed to the MTDC grid is one of the major concerns for the system operators as this will impact other AC grids connected to it. To emulate such a disturbance step changes in the mechanical power input of the generators and loads are considered. The analytical expression to obtain frequency following AC side disturbances is described in Section 5.7.2.1. According to (5.17) the frequency of the three AC systems shown in Fig. 3.31 can be expressed

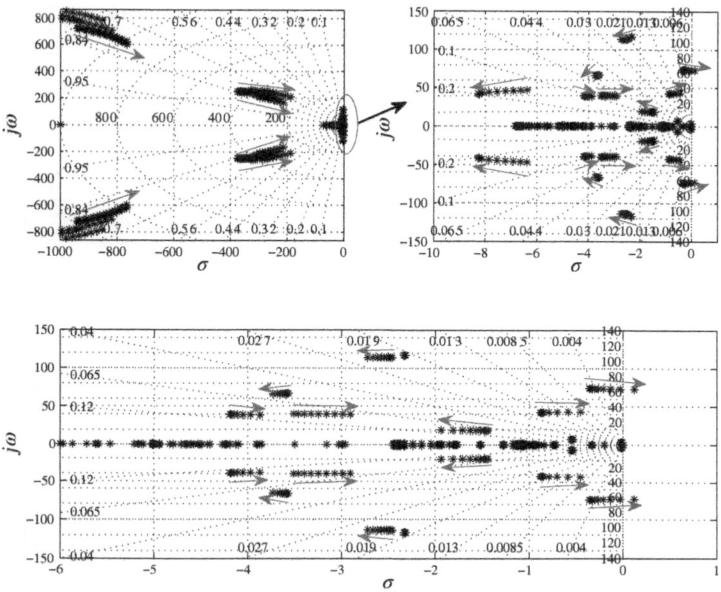

Figure 5.6 Root locus of system poles with reduction of β from 0.5 to 0.1, keeping K_{pdc}, K_{idc} and β_f constant at 0.1, 10.0, and 100.0, respectively. ©[84], 2013.

as

$$\underbrace{\begin{bmatrix} \left(-\frac{1}{\beta_{g1}} - 2\beta_f\right) & \beta_f & \beta_f \\ \beta_f & \left(-\frac{1}{\beta_{g2}} - \frac{3\beta_f}{2}\right) & \frac{\beta_f}{2} \\ \beta_f & \frac{\beta_f}{2} & \left(-\frac{1}{\beta_{g3}} - \frac{3\beta_f}{2}\right) \end{bmatrix}}_{w} \begin{bmatrix} f_1 \\ f_2 \\ f_3 \end{bmatrix} = - \begin{bmatrix} \Gamma_1 \\ \Gamma_2 \\ \Gamma_3 \end{bmatrix} \tag{5.18}$$

Substituting appropriate values of droop coefficients we get the "Inertial Coupling" matrix as

$$w = \begin{bmatrix} -412.2016 & 100 & 100 \\ 100 & -203.0504 & 50 \\ 100 & 50 & -203.0504 \end{bmatrix} \tag{5.19}$$

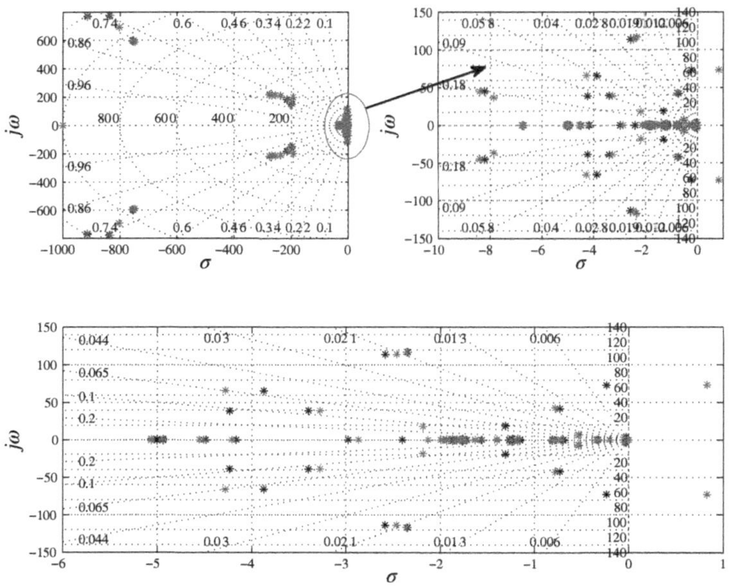

Figure 5.7 System pole locations following the outage of the negative pole of converter station #3. The pole map in grey presents the converter outage condition while its black counterpart show that of nominal condition. ©[84], 2013.

In the absence of any frequency droop (i.e., $\beta_f = 0$), the "Inertial Coupling" matrix becomes

$$w_{nd} = \begin{bmatrix} -212.2016 & 0 & 0 \\ 0 & -53.0504 & 0 \\ 0 & 0 & -53.0504 \end{bmatrix} \tag{5.20}$$

The cross-coupling terms are absent in this matrix. In the following sections the impact of this will be demonstrated through nonlinear time-domain simulations and verified with the analytical formulation.

Step Reduction in Mechanical Power Input of $G1$ Figure 5.8 shows the variation in the system frequency due to a step reduction in the power reference of $G1$ which is one of the four generators within the interconnected AC system by 20% .

As a result of the loss of generation, the frequency of the interconnected 4-machine AC system drops below 59.5 Hz. This can be explained following the equations described in Section 3.3.1.5. From (3.78), under steady state, $\Delta P_i = \Delta P_{Ti}$. Since the power reference is reduced, $\Delta P_{Ti} < 0$ as observed from (3.79). From (3.80), it can be seen that the frequency reduces under steady state condition. During dynamic condition the generator frequencies are different

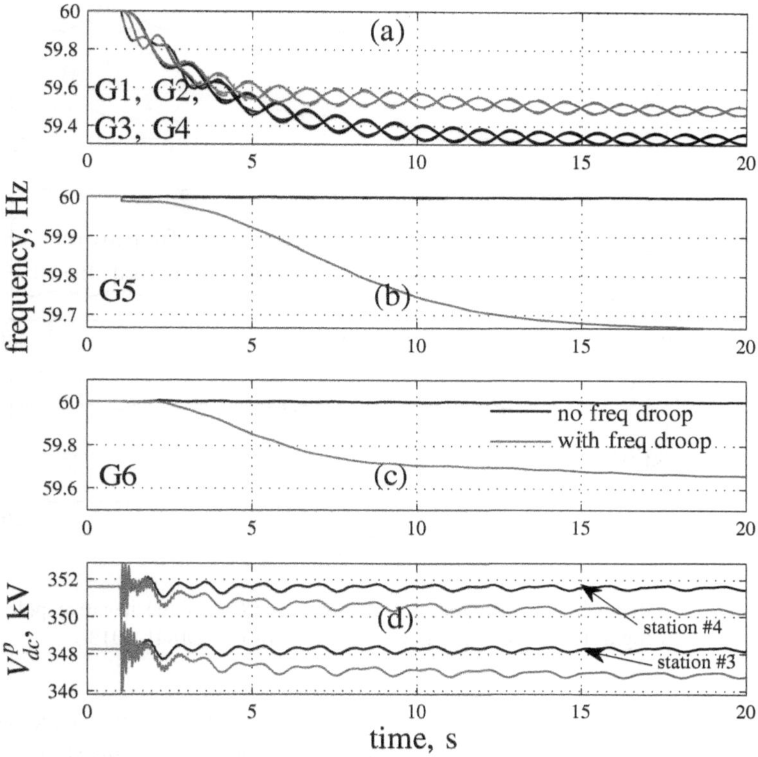

Figure 5.8 (a-d) AC system frequency and MTDC DC link voltage variation following a step reduction of power reference of $G1$. ©[84], 2013.

from each other—varying according to the modeshapes of the generator speeds. The frequency of the two isolated power systems corresponding to $G5$ and $G6$ remain unchanged as they are electrically decoupled from the 4-machine AC system by the MTDC network.

Without frequency droop, the real power at the point of common coupling of all converter stations oscillate about the pre-disturbance average value as shown in Fig. 5.9.

Only negative pole real powers are shown due to symmetry—i.e., identical behavior is observed in the positive pole converters. In the presence of frequency droop the total real power injection into the 4-machine AC system increases (Fig. 5.9(b)). This compensates for the reduction in the power of $G1$ and prevents the frequency from falling below 59.5 Hz as shown in Fig. 5.8(a).

Closer look reveals that this frequency support comes at the expense of increase in real power drawn by the two other converter stations #1 and #4 from the two isolated AC systems. This results in reduction of frequencies of those power

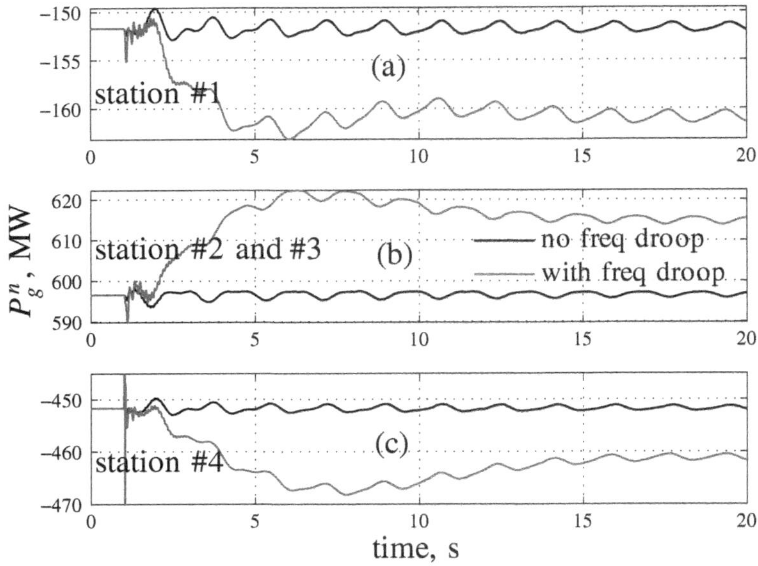

Figure 5.9 (a-c) Dynamic behavior of real power at PCC of the negative pole MTDC converters following a step reduction of power reference of $G1$. ©[84], 2013.

systems, as shown in Fig. 5.8(b) and (c). However, the frequency excursions for the isolated AC systems are within the allowable tolerance. This study confirms that the presence of the frequency droop control loop ensures effective exchange of frequency support across the MTDC grid in order to limit the frequency deviations in all the host AC systems.

- Analytical Formulation:
  Without any frequency droop control, the post disturbance frequency for the three areas can be obtained after substituting appropriate values in (5.18):

$$\begin{bmatrix} f_1 \\ f_2 \\ f_3 \end{bmatrix} = -\left[w_{nd}\right]^{-1} \begin{bmatrix} 12592 \\ 3183 \\ 3183 \end{bmatrix} = \begin{bmatrix} 59.34 \\ 60.00 \\ 60.00 \end{bmatrix} \tag{5.21}$$

  With frequency droop we get:

$$\begin{bmatrix} f_1 \\ f_2 \\ f_3 \end{bmatrix} = -[w]^{-1} \begin{bmatrix} 12592 \\ 3183 \\ 3183 \end{bmatrix} = \begin{bmatrix} 59.50 \\ 59.67 \\ 59.67 \end{bmatrix} \tag{5.22}$$

  The results from the analysis closely match the simulation results of Fig. 5.8.

Step Reduction in Mechanical Power Input of *G*5 A step reduction in mechanical power input in one of the isolated AC systems is considered next. As shown in Fig. 5.10, the effect of frequency support through the MTDC grid is more pronounced in this case due to comparatively low inertia than the interconnected 4-machine AC system. In the absence of frequency droop, the frequency of this isolated AC system goes well below 58.0 Hz while the frequency droop control limits this deviation to only 1 Hz variation below the nominal 60 Hz frequency. As before, such tighter regulation is achieved at the expense of slight (less that 0.5 Hz) change in frequency of the two other AC systems. It is to be noted that the resulting change in the DC link voltages for both the disturbances are within the acceptable (within ±1%) limits.

It is also interesting to note that a disturbance in the interconnected 4-machine 2-area AC system excited the low frequency inter-area oscillations. These oscillations are visible in the dynamic response of the generators *G*1–*G*4 (Fig. 5.8(a)).

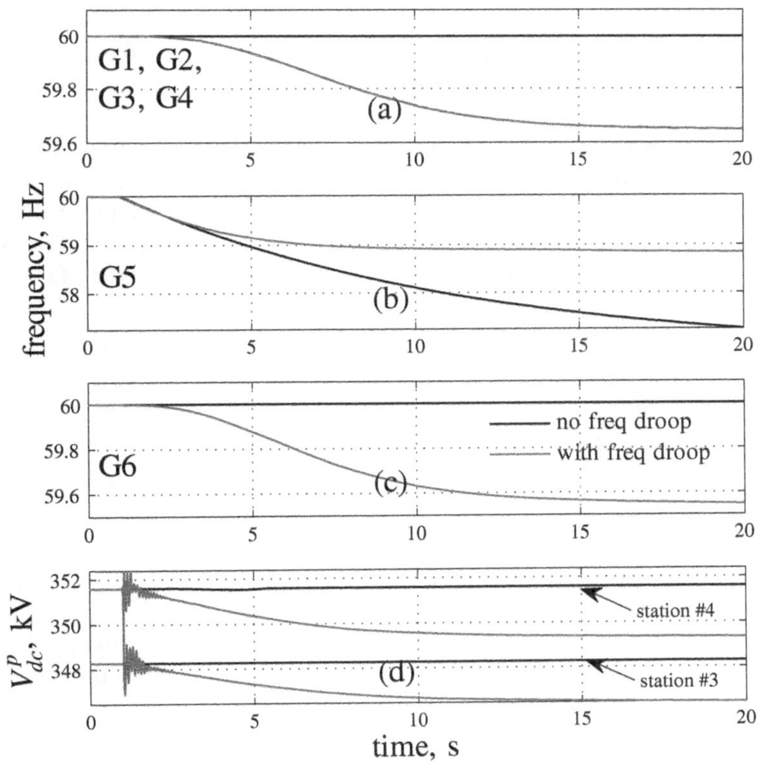

Figure 5.10 (a-d) AC system frequency and MTDC DC link voltage variation following a step reduction of power reference of *G*5 by 20%. ©[84], 2013.

However, as expected, a similar disturbance at the isolated AC system *G5* did not trigger such oscillations as shown in Fig. 5.10(a).

- Analytical Formulation:
  With frequency droop control, the post disturbance frequency for the three areas can be obtained after substituting appropriate values in (5.18):

$$\begin{bmatrix} f_1 \\ f_2 \\ f_3 \end{bmatrix} = -[w]^{-1} \begin{bmatrix} 12732 \\ 3003 \\ 3183 \end{bmatrix} = \begin{bmatrix} 59.58 \\ 58.78 \\ 59.49 \end{bmatrix} \tag{5.23}$$

without any droop control:

$$\begin{bmatrix} f_1 \\ f_2 \\ f_3 \end{bmatrix} = \begin{bmatrix} 60.00 \\ 56.60 \\ 60.00 \end{bmatrix} \tag{5.24}$$

The results from the analysis closely match the the simulation results of Fig. 5.10.

Step Increase in Load at Bus 10 The previous studies emulated a sudden loss of generation within the interconnected and the isolated AC systems. Sudden change in demand within the interconnected AC system is considered next. A 1% step increase in the 1400 MW load connected at bus 10 is simulated. The dynamic variation of frequency in the host AC networks surrounding the MTDC grid is shown in Fig. 5.11.

As expected the frequency deviation in the 4-machine interconnected AC system is more when there is no frequency droop. However, with droop control frequency support is provided from the remote AC systems (*G5*, *G6*) through the MTDC grid which results in relatively less frequency variation.

The variation of real power at the PCC of the negative pole MTDC converter is shown in Fig. 5.12. From Fig. 5.12(b) it can be seen that in the presence of frequency droop, the total real power injection into the 4-machine AC system increases to cater for the increase in load. This excess power for the 4-machine system is provided by the two isolated power systems through converter stations #1 and #4 as seen in Fig. 5.9(a) and (c). This results in slight reduction of frequencies (but within allowable limits) of those power systems, as shown in Fig. 5.11(b) and (c).

Above simulation results involving drop of AC system frequency due to either loss of generation or increase of load confirms the effectiveness of using a frequency droop control loop to exchange frequency support among the host AC systems interconnected through an MTDC grid.

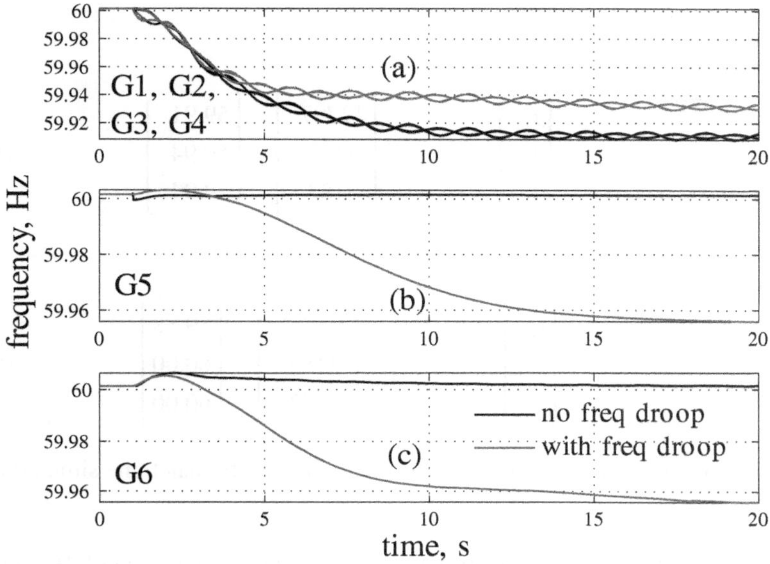

Figure 5.11 (a-c) AC system frequency variation following a step increase in load at bus 10. ©[84], 2013.

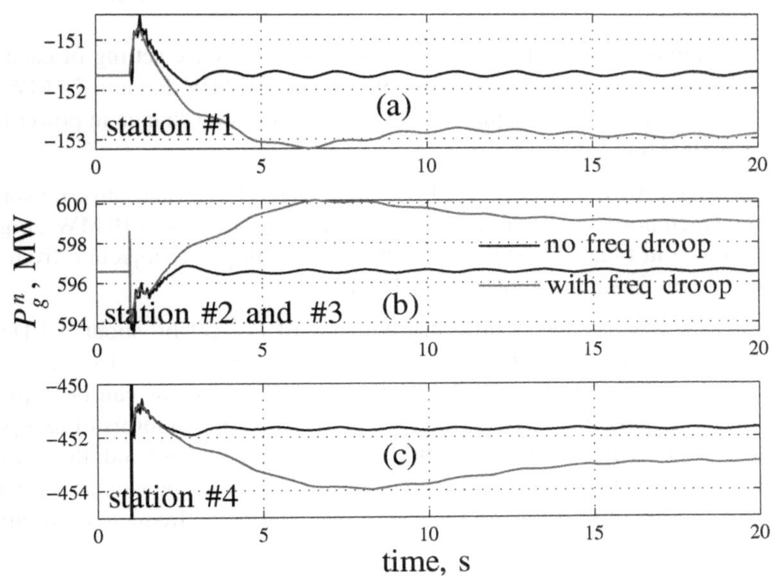

Figure 5.12 (a-c) Dynamic behavior of real power at PCC of the negative pole MTDC converters following a step increase in load at bus 10. ©[84], 2013.

- Analytical Formulation:
  With frequency droop:

$$\begin{bmatrix} f_1 \\ f_2 \\ f_3 \end{bmatrix} = -[w]^{-1} \begin{bmatrix} 12707 \\ 3183 \\ 3183 \end{bmatrix} = \begin{bmatrix} 59.91 \\ 59.94 \\ 59.94 \end{bmatrix} \tag{5.25}$$

  No frequency droop:

$$\begin{bmatrix} f_1 \\ f_2 \\ f_3 \end{bmatrix} = -\left[w_{nd}\right]^{-1} \begin{bmatrix} 12707 \\ 3183 \\ 3183 \end{bmatrix} = \begin{bmatrix} 59.88 \\ 60.00 \\ 60.00 \end{bmatrix} \tag{5.26}$$

As before, results from analytical formulation closely match the simulation results.

5.8.4.2 Converter Station Disturbances After studying the effect of AC side disturbances, the impact of disturbances within the MTDC grid is investigated. In this section, the performance of the frequency droop control loop is evaluated in the face of sudden change in converter power reference or outage of a converter.

Step Reduction in P_g^* of Station #3 The real power reference setting of each pole of converter station #3 is reduced by 50% from 150 MW down to 75 MW. The variation in AC system frequency as a result of this step change in power reference is shown in Fig. 5.13.

As expected, without frequency droop, each pole of the four converter stations change their power by one fourth of the total change (i.e., about 19 MW), Fig. 5.14. It is shown in Fig. 5.14(b) that the net reduction in power injection from each pole of stations #2 and #3 is around 38 MW.

These changes result in a drop in frequency of *G1*–*G4* by about 0.3 Hz and increase in frequency of *G5* and *G6* by about 0.6 Hz (Fig. 5.13). The cause of increase/decrease of frequency of the respective systems can be explained by the corresponding decrease/increase of system loads in terms of real power exchange at the PCC of the converter stations. On the other hand, the introduction of frequency droop reduces the change in converter powers at all stations (Fig. 5.14). As observed from Fig. 5.13, this reduces the frequency variation in all the AC systems.

Outage of Negative Pole of Station #3 A more severe disturbance in the form of outage of one pole of a converter station is considered in this section. In this particular case, outage of the negative pole of converter station #3 was studied which showed unbalanced operation in the DC side. This is clear from the non-zero metallic return current shown in Fig. 5.15(d).

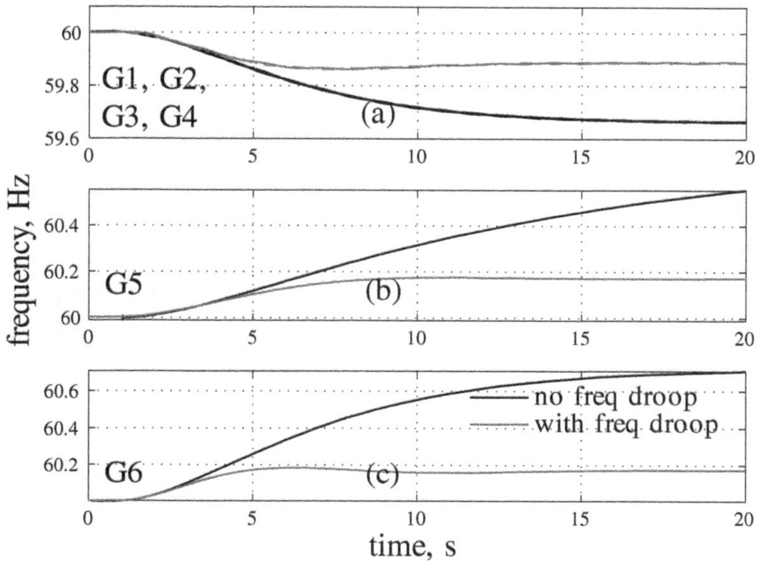

Figure 5.13 (a-c) AC system frequency variation following a step reduction of real power reference of converter station #3. ©[84], 2013.

With only $P - V_{dc}$ droop, as the station #3 negative pole power becomes zero (Fig. 5.16(c)), the rest of the negative poles share the lost power (i.e., 150 MW) equally amongst themselves resulting in 50 MW deviation (shown in Fig. 5.16) for each.

It is to be noted that the aggregate P_g of stations #2 and #3 negative poles are separately shown in Fig. 5.16(b) which changes by about 100 MW. It is also worth noting that the positive pole control operates independently and the converter powers do not change from the pre-disturbance values in the absence of power-frequency droop shown in Fig. 5.17.

Similar to the previous case, the changes in the converter real powers resulted in deviation of AC system frequencies as shown in Fig. 5.15. Frequency of the 4-machine AC system reduces to 59.6 Hz while that of $G5$ and $G6$ increases to 60.8 and 60.9 Hz, respectively. It is interesting to observe the converter real power response with power-frequency droop for the positive pole (Fig. 5.17) and the negative pole (Fig. 5.16), respectively. Station #1 reduces the net deviation in power output from 50 to around 15 MW. To be more specific, under post-disturbance condition P_g^p is approximately -170 MW and P_g^n is about -115 MW totalling about -285 MW—which is a 15 MW deviation from the pre-disturbance -300 MW output. This trend is clearly visible for the other converter stations as well.

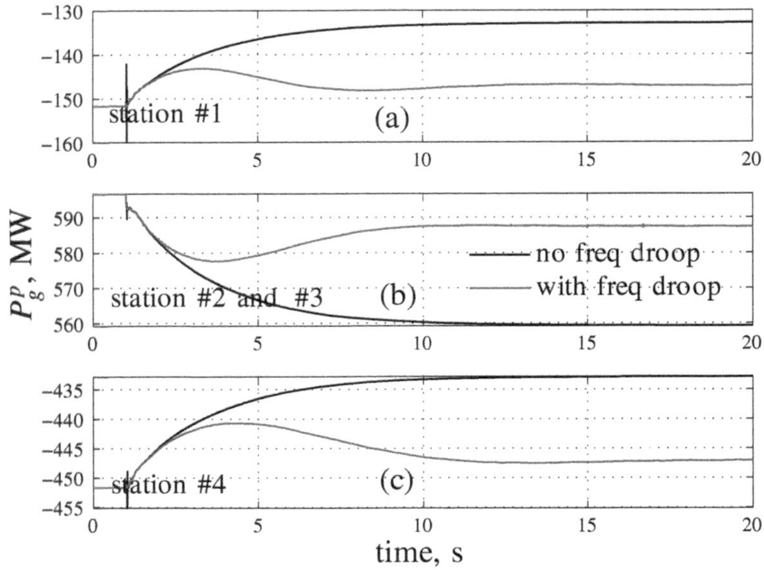

Figure 5.14 (a-c) Dynamic behavior of real power at PCC of the positive pole MTDC converters following a step reduction of real power reference of converter station #3. ©[84], 2013.

As a result the individual AC systems experience reduced change in system loads. Therefore, the frequency deviation of the 4-machine AC system is reduced to 0.2Hz while that of *G*5 and *G*6 are less than 0.3 Hz (Fig. 5.15).

The above case study substantiates the fact that use of power-voltage droop for autonomous power sharing alone is not sufficient to reduce the frequency deviations in host AC systems especially, following a large disturbance. An additional power-frequency droop control is required to modify the voltage reference for each converter station. The power-frequency droop control loop acts to reduce the frequency deviation by importing frequency support from the remaining unaffected AC systems. As a result, the frequency of those AC systems would deviate from their nominal value (which would not have happened without the frequency droop control loop). Thus, rather than having a large and potentially unacceptable variation in frequency of the affected system, there is relatively small (but hopefully within acceptable limits) variation of frequency in all the AC systems.

- Analytical Formulation:
  For this condition $m_1^n = 1$, $m_2^n = 1$ and $m_3^n = 1$. Using the expressions

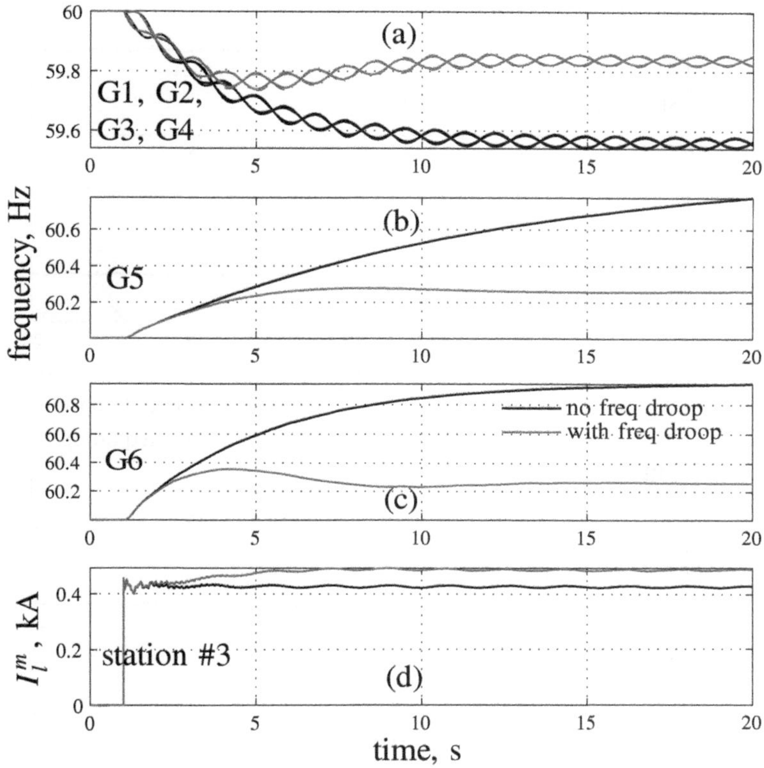

Figure 5.15 (a-d) AC system frequency and metallic return current variation following the outage of negative pole converter of station #3. ©[84], 2013.

derived in Section 5.7.2.2 we can write:

$$\underbrace{\begin{bmatrix} \left(-\frac{1}{\beta_{g1}}-\frac{5\beta_f}{3}\right) & \frac{5\beta_f}{6} & \frac{5\beta_f}{6} \\ \frac{5\beta_f}{6} & \left(-\frac{1}{\beta_{g1}}-\frac{17\beta_f}{12}\right) & \frac{7\beta_f}{12} \\ \frac{5\beta_f}{6} & \frac{7\beta_f}{12} & \left(-\frac{1}{\beta_{g1}}-\frac{17\beta_f}{12}\right) \end{bmatrix}}_{w} \begin{bmatrix} f_1 \\ f_2 \\ f_3 \end{bmatrix} = -\begin{bmatrix} \Gamma_1 \\ \Gamma_2 \\ \Gamma_3 \end{bmatrix} \tag{5.27}$$

The "Inertial Coupling" matrices considering frequency droop control is:

$$w = \begin{bmatrix} -378.8683 & 83.3333 & 83.3333 \\ 83.3333 & -194.7171 & 58.3333 \\ 83.3333 & 58.3333 & -194.7171 \end{bmatrix} \tag{5.28}$$

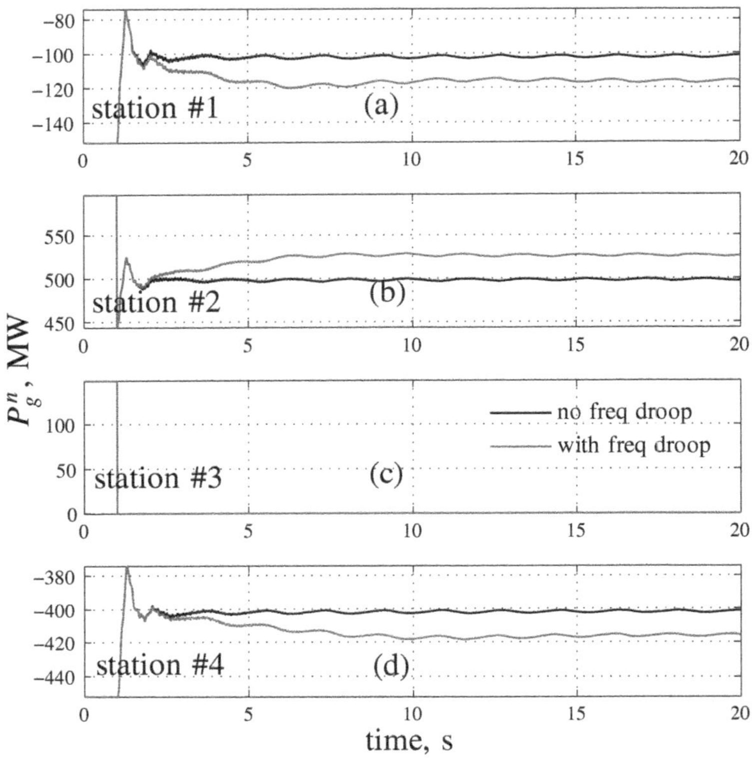

Figure 5.16 (a-d) Dynamic behavior of real power at PCC of the negative pole MTDC converters following the outage of negative pole converter of station #3. ©[84], 2013.

Which leads to:

$$\begin{bmatrix} f_1 \\ f_2 \\ f_3 \end{bmatrix} = -[w]^{-1} \begin{bmatrix} 12632 \\ 3233 \\ 3233 \end{bmatrix} = \begin{bmatrix} 59.86 \\ 60.28 \\ 60.28 \end{bmatrix} \tag{5.29}$$

The analysis can also be done without frequency droop control by equating β_f to zero:

$$w_{nd} = \begin{bmatrix} -212.2016 & 0 & 0 \\ 0 & -53.0504 & 0 \\ 0 & 0 & -53.0504 \end{bmatrix} \tag{5.30}$$

$$\begin{bmatrix} f_1 \\ f_2 \\ f_3 \end{bmatrix} = -\left[w_{nd}\right]^{-1} \begin{bmatrix} 12632 \\ 3233 \\ 3233 \end{bmatrix} = \begin{bmatrix} 59.53 \\ 60.94 \\ 60.94 \end{bmatrix} \tag{5.31}$$

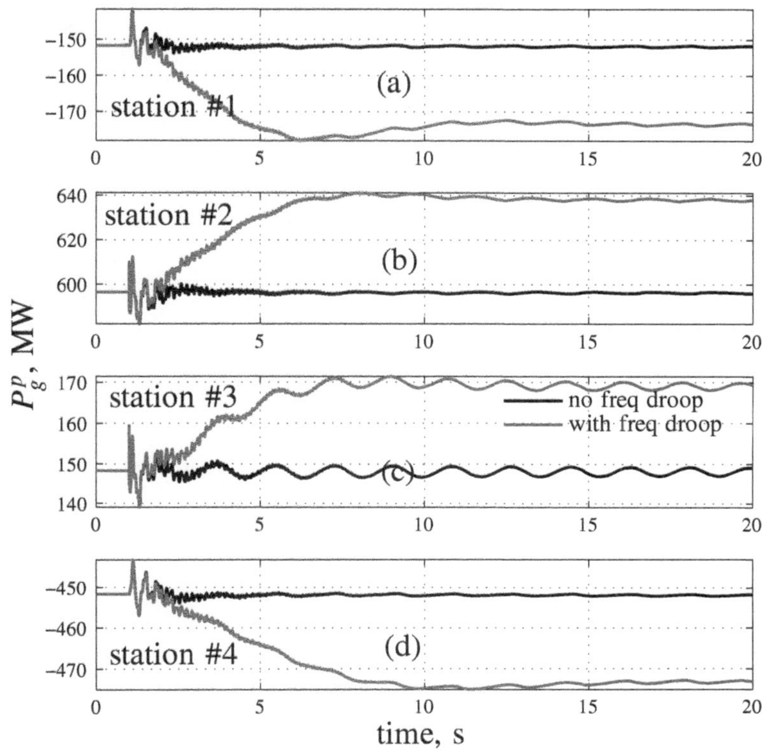

Figure 5.17 (a-d) Dynamic behavior of real power at PCC of the positive pole MTDC converters following the outage of negative pole converter of station #3. ©[84], 2013.

The results from the analysis closely match the the simulation results of Fig. 5.15.

It is to be noted that the nature of frequency variation with frequency droop control is different when there is AC side disturbance compared to the DC side disturbance. For the AC side disturbances, the variation of frequency occurs in the same direction for all AC systems. For the DC side disturbances frequency of the affected AC system varies in the opposite direction to that of the unaffected system. This is also confirmed by the results obtained from the analytical formulation.

CHAPTER 6

PROTECTION OF MTDC GRIDS

6.1 Introduction

Protection of a VSC DC grid in the event of a fault in the DC cable network is a research problem till date. It was mentioned in Section 2.4.2.4 that regardless of the control strategy, conventional VSCs cannot limit or interrupt the current that flows following a DC-side fault. DC circuit breakers are yet to be commercially available for high power ratings. Therefore, the only way to clear a DC-side fault is to open the circuit breakers on the AC side. For a point-to-point VSC HVDC link, this is not a major problem as the loss-of-infeed is not unacceptably high even if the entire link (e.g., 500 MW) is taken out of service by opening the AC breakers at both ends. However, for a DC grid with a capacity in the range of tens of GWs, opening all the AC circuit breakers to clear a DC-side fault would result in unacceptably large loss-of-infeed. Thus, one cannot rely on the AC circuit breakers to clear a fault within the DC grid.

There are essentially two ways of clearing DC grid faults without causing a large loss-of-infeed. One option is to use DC breakers (as and when they are commercially available) to isolate only the faulty cable while continuing to operate the rest of the DC grid as usual. Another option is to use converter topologies which are capable

Multi-terminal Direct-Current Grids: Modeling, Analysis, and Control. First Edition.
Nilanjan Ray Chaudhuri, Balarko Chaudhuri, Rajat Majumder, and Amirnaser Yazdani.

of interrupting the DC fault current. Such topologies are usually less efficient and a proper balance between fault-blocking capability and efficiency is required.

Detection of a DC fault and locating the faulty cable is another big challenge. In a DC grid, the rate of rise of fault current is much higher than an AC system where the reactance of lines limits the rise of current. Hence, the DC grid protection system would have to act much faster (typically within 1–2 ms) before the fault currents reach a level which is difficult to clear. Moreover, locating a fault within the DC grid is difficult as traditional distance protection using impedance relays is not effective given that DC cables have very low resistance and virtually zero series reactance. So the protection system is likely to have to rely on advanced signal processing techniques (e.g., Wavelets) to detect and locate the DC fault. Ensuring sensitivity and selectivity with so little reaction time is a major challenge for DC grid protection systems and a standard methodology is yet to be developed. In this chapter, we have summarized the challenges, the state-of-the-art, and the ongoing research and development work in the area of DC grid protection and fault current interruption using the DC circuit breakers or fault-blocking converters.

6.2 Converter Station Protection

Different components of a converter station have their own protection zones as shown in Fig. 6.1 for an asymmetric bipolar configuration with switchgears and sensors. The switchgears will include AC circuit breakers, DC circuit breakers, and DC isolators. Multiple current sensors will be used at different locations as highlighted in Fig. 6.1. State-of-the art fiber optic sensing technology can be used for current measurement. These sensors have extremely low latency in measurement (6 kHz bandwidth) [85].

Different zones of protection are also highlighted in Fig. 6.1. The measured signals will be processed by relays that determine the faulted component and command appropriate breaker tripping. Different protection philosophies can be used to achieve selective isolation—which determine the characteristics of the relay. Well-proven existing schemes can be used for the AC side protection. For example, the converter transformers are protected using differential protection [86]. On the DC-side, the DC bus protection can be based on traditional bus differential protection [86].

The most challenging part of the DC grid protection is protection against DC cable faults. The response to such DC cable faults is described in Section 6.3.

6.3 DC Cable Fault Response

Most VSC converter topologies are defenseless against DC-side faults. Primarily, two types of cable faults can happen: cable-to-cable fault and cable-to-ground fault. Analysis of the fault response for the converter topologies mentioned in Section 2.3 is essential in determining the challenges involved in DC grid protection. In this section the fault response of two widely used VSC converter topologies: two-level and half-bridge MMC are analyzed.

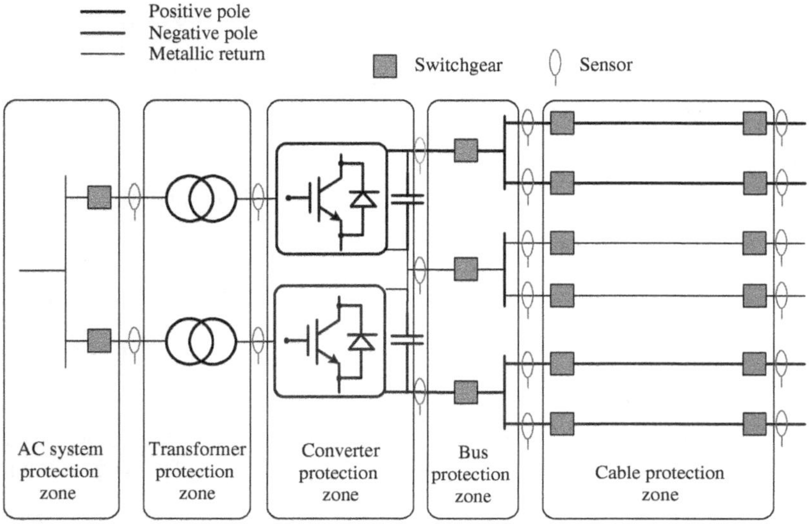

Figure 6.1 Zones of protection, switchgears, and sensors in an asymmetric bipolar MTDC converter station.

6.3.1 Fault Response of Two-level VSC

The fault response of a two-level VSC, Section 2.3.1, following a cable-to-cable fault was reported in [87, 88]. Although this type of fault has a low probability of occurrence, it is the most critical fault from the protection standpoint. The IGBTs are blocked during such faults and the antiparallel diodes carry the fault current. Figure 6.2 shows a two-level VSC with a DC fault between two cables. The DC cable up to the fault point is assumed to be modeled by a single pi-section. The cable resistance ($\frac{R_c}{2}$) and inductance ($\frac{L_c}{2}$) up to the fault point are shown and the capacitance (C_{dc}) is assumed to be the equivalent capacitance of the cable pi-section and the converter DC bus (Fig. 3.1). The resistance R of the tie reactor is neglected.

As described in [87, 88], the response following the DC-side fault can be expressed as a superposition of the natural response and the forced response of the system. Further, the natural response can be divided into two sequential phases. These responses are summarized briefly. Interested readers should refer to [87, 88] for further details.

1. Natural Response—capacitor discharge phase ($t_0 \leq t \leq t_1$): This is the natural response of the circuit immediately after the fault at $t = t_0$. Figure 6.3(a) shows the equivalent circuit of this phase. Assuming $R_c < 2\sqrt{\frac{L_c}{C_{dc}}}$, the expressions for the cable current and the DC bus voltage are computed for the second-order $R - L - C$ circuit.

$$I_l = C_{dc}\frac{dV_{dc}}{dt} = -\frac{I_0\omega_0}{\omega_n}e^{-\gamma t}\sin(\omega_n t - \xi) + \frac{V_0}{\omega_n L_c}e^{-\gamma t}\sin\omega_n t \tag{6.1}$$

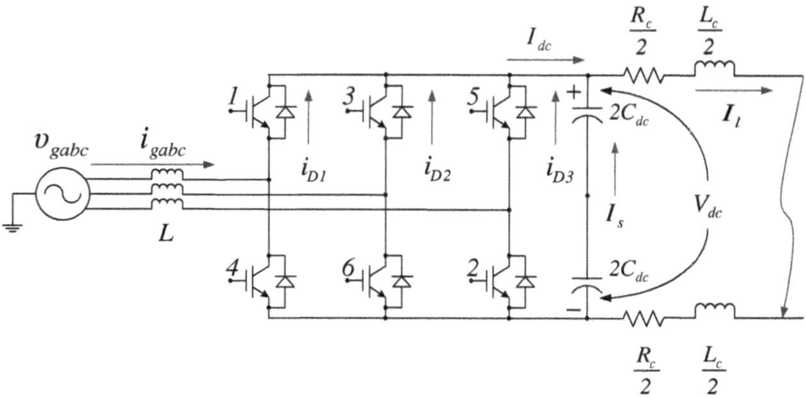

Figure 6.2 A two-level VSC HVDC converter station following a cable-to-cable DC fault. The IGBTs are blocked and the antiparallel diodes conduct to feed the fault current from the AC side.

$$V_{dc} = \frac{V_0\omega_0}{\omega_n} e^{-\gamma t} \sin(\omega_n t + \xi) - \frac{I_0}{\omega_n C_{dc}} e^{-\gamma t} \sin \omega_n t \tag{6.2}$$

where $\gamma = \frac{R_c}{2L_c}, \omega_n = \sqrt{\frac{1}{L_c C_{dc}} - \left(\frac{R_c}{2L_c}\right)^2}, \omega_0 = \sqrt{\gamma^2 + \omega_n^2},$
$\xi = \tan^{-1}\left(\frac{\omega_n}{\gamma}\right), I_l(t_0) = I_0, V_{dc}(t_0) = V_0$

From (6.2) it can be seen that the DC bus voltages reduces to zero at $t_1 = t_0 + \frac{(\pi-\lambda)}{\omega_n}$, where $\lambda = \tan^{-1}\left(\frac{V_0\omega_0 C_{dc} \sin\xi}{V_0\omega_0 C_{dc} \cos\xi - I_0}\right)$

2. Natural response—diode freewheel phase ($t_1 < t$): After V_{dc} becomes zero, the cable current I_l is transferred to the antiparallel diodes of the VSC bridge. This results in a freewheeling response through those diodes as shown in Fig. 6.3(b). The simple $R - L$ circuit equation can be solved with $I_s = 0$.

$$I_l = I_1 e^{-(\frac{R_c}{L_c})t} \tag{6.3}$$

where $I_l(t_1) = I_1$.
Each diode carries one-third of the cable current:

$$i_{Di} = \frac{I_l}{3}, i = 1, 2, ..., 6 \tag{6.4}$$

3. Forced response—grid current feeding: The IGBTs are blocked when the current through them rises above a threshold. Under this condition, the response is determined by the equivalent $R - L - C$ circuit shown in Fig. 6.3(c). Assuming

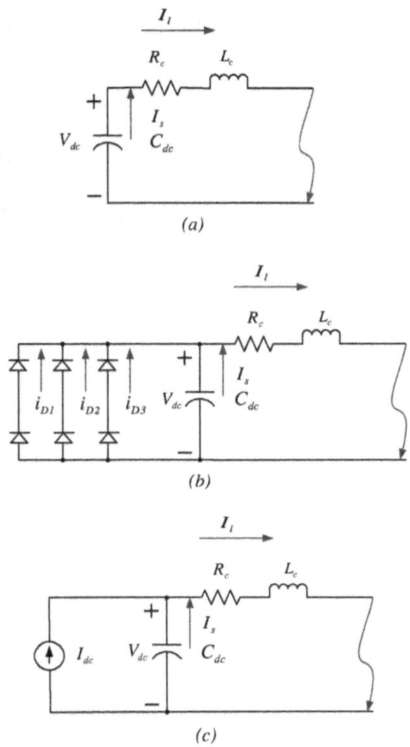

Figure 6.3 Equivalent circuits for different phases of response of a two-level VSC following a cable-to-cable fault. (a) Natural response: capacitor discharge phase, (b) natural response: diode freewheel phase, and (c) forced response: grid current feeding [87, 88].

a worst-case scenario for the phase a where the fault occurs at the zero crossing of the phase a grid voltage, we can write

$$v_{ga} = V_g \sin(\omega_g t + \theta_{g0}), t_1 \leq t \tag{6.5}$$

where $\theta_{g0} = \omega_g(t_1 - t_0)$

$$i_{ga} = I_g \left[\sin(\omega_g t + \theta_{g0} - \delta) - \sin\left(\theta_{g0} - \delta\right) e^{-t/\tau}\right] + I_{a1} e^{-t/\tau}, t_1 \leq t \tag{6.6}$$

where $\delta = \tan^{-1}\left(\omega_g \tau\right)$,

$\tau = \frac{(L_c+L)}{R_c}$,

$I_{a1} = i_{ga}(t_1)$

The current from the converter (I_{dc}) under this condition (Fig. 6.3(a)) is the positive half-cycle of current from the grid side:

$$I_{dc} = i_{D1}(i_{ga} > 0) + i_{D2}(i_{gb} > 0) + i_{D3}(i_{gc} > 0) \tag{6.7}$$

The expression for the cable current and the DC bus voltage can be written as

$$I_l = C_1 \sin(\omega_g t + \lambda) + C_2 e^{-t/\tau} + \frac{C_3 \omega_0 e^{-\gamma t}}{\omega_n} \sin(\omega_n t + \theta_{g0}) + \frac{C_4 e^{-\gamma t}}{\omega_n} \sin \omega_n t \tag{6.8}$$

$$V_{dc} = R_c I_l + L_c \frac{dI_l}{dt} \tag{6.9}$$

where $C_1 = I_m \sqrt{\left(1 - \omega_g^2 L_c C_{dc}\right)^2 + \left(R_c C_{dc} \omega_g\right)^2}$,

$\lambda = \theta_{g0} - \delta - \sigma$,

$\sigma = \tan^{-1}\left(\frac{R_c C_{dc} \omega_g}{1 - \omega_g^2 L_c C_{dc}}\right)$,

$C_2 = I_m \left(\frac{\tau^2}{\tau^2 - R_c C_{dc} \tau + L_c C_{dc}}\right)$,

$I_m = I_{a1} - I_g \sin\left(\theta_{g0} - \delta\right)$
$C_3 = -C_1 \sin \lambda - C_2$,

$C_4 = \frac{C_2}{\tau} + \omega_g C_1 \cos \lambda$

Cable-to-ground fault is more probable but less harmful compared to the cable-to-cable fault. Similar analysis can be done for the ground fault which is not covered here. Interested readers can refer to [87, 88].

6.3.1.1 Analysis The cable current rises very sharply during the capacitor discharge phase, but the most critical phase for the system is the free-wheeling phase. The thermal inertia of IGBTs is much lower compared to the cables. Therefore, they are blocked as soon as current through them rises above a threshold. However, the diode current jumps to dangerously high value of $\frac{I_l}{3}$ when the freewheeling period begins which might be beyond their surge current rating. Therefore, appropriate device protection measures should be considered. The steady-state current and voltage are governed by the forced response which is not as severe as the freewheeling condition.

With considerable fault resistance R_f, the condition $R_c + R_f > 2\sqrt{\frac{L_c}{C_{dc}}}$ might satisfy which produces a first-order response. In this condition, the diode freewheeling phase does not arise.

When a DC cable fault occurs in an MTDC grid, the converter stations with two-level VSC topology will go through the phases of response mentioned here. This class of VSC converters cannot block or control the current during such faults. Since the fault current in the cable rises extremely fast, it is very challenging for the relaying and protection system of such grids.

6.3.2 Fault Response of Half-bridge MMC

The half-bridge MMC topology is gaining popularity among the major HVDC manufacturers. This topology was described in detail in Section 2.3.3. Unlike the two-level VSC, this topology does not have a large DC bus capacitor. Therefore, no converter capacitor discharge phase is present when a DC cable-to-cable short circuit occurs.

As described in Section 2.3.3, the half-bridge converters in Fig. 2.15 are also called *sub-modules*. In this chapter, these sub-modules are represented as blocks marked "SM" in figures, for the sake of clarity. In the rest of the chapter, the sub-module capacitance is denoted by C_{SM} and the voltage across the DC bus of each sub-module is denoted by V_c. Figure 6.4 shows a converter station with half-bridge MMC topology where a DC cable-to-cable short circuit has happened. Similar to the two-level topology, here the IGBTs are blocked and the antiparallel diodes experience a surge in current. This current can be limited by properly designing the arm inductors L_{arm}. The diodes usually have poor surge current capability. To protect them press-pack thyristors are triggered [89] as shown in Fig. 6.6(a).

Let us consider the AC-side voltages and currents of the converter are (Fig. 6.4)

$$v_{ga} = V_g \cos\left(\omega_g t + \phi + \rho\right), \quad (6.10)$$

$$v_{a1} = V_{a1} \cos\left(\omega_g t + \phi\right) \quad (6.11)$$

$$i_{ga} = I_g \cos \omega_g t \quad (6.12)$$

$$(6.13)$$

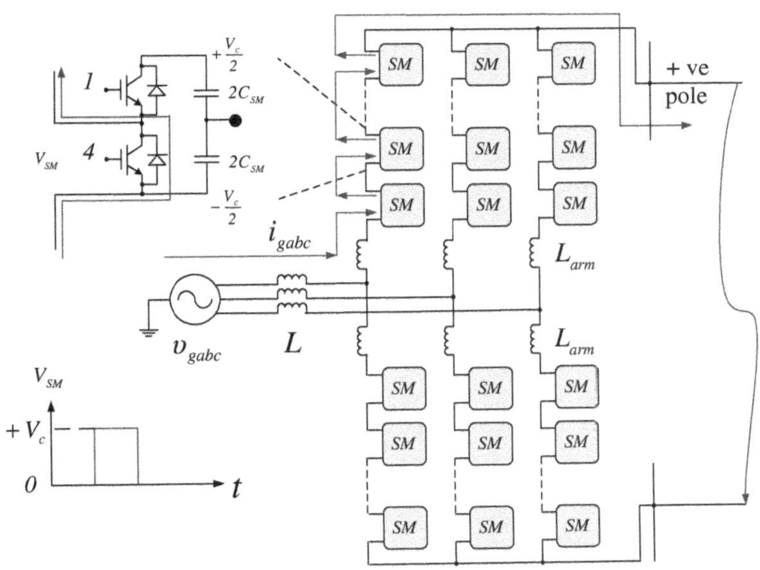

Figure 6.4 A converter station with half-bridge MMC configuration. Following a cable-to-cable DC fault the antiparallel diodes conduct to feed the fault current from the AC side.

Here, v_a is the voltage behind the phase reactor L produced by the converter arm for phase a. The positive/negative arm voltages v_{ap}/v_{an} and corresponding currents i_{ap}/i_{an} in the a-phase can be expressed as

$$v_{ap} = \frac{V_{dc}}{2} - v_{a1}, v_{an} = \frac{V_{dc}}{2} + v_{a1} \tag{6.14}$$

$$i_{ap} = \frac{i_{ga}}{2} + \frac{I_{dc}}{3}, i_{an} = -\frac{i_{ga}}{2} + \frac{I_{dc}}{3} \tag{6.15}$$

The equivalent circuits corresponding to different phases of response were described in [90] and are shown in Fig. 6.5.

1. Phase I: before IGBT turn off ($t_0 \le t \le t_1$): Let us assume the fault occurs at time t_0 and the IGBTs are turned off at time t_1. This time interval is of the order of tens of microseconds. Thus, the AC voltages and currents shown in Fig. 6.5(a) can be assumed to be equal to their respective values at time t_0. The current I_{dc}

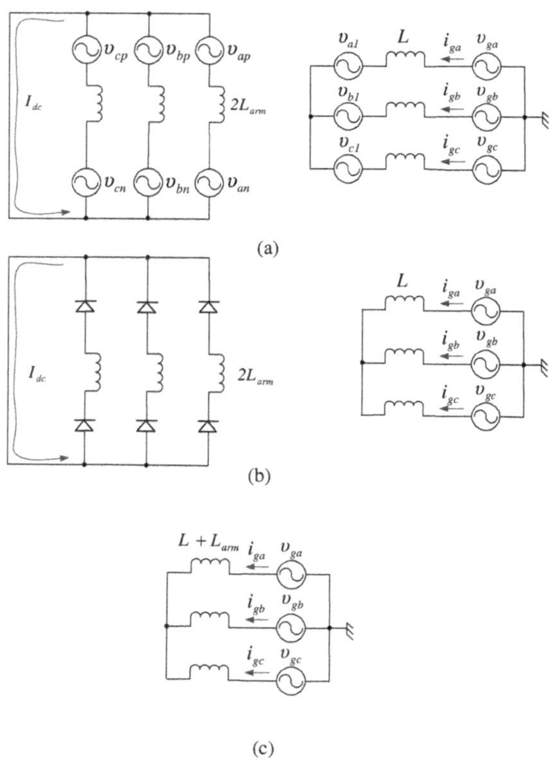

Figure 6.5 Equivalent circuits for different phases of response of a half-bridge MMC following a cable-to-cable fault. (a) Phase I: before IGBT turn off, (b) phase II: diode freewheel phase, and (c) phase III: grid current feeding [90].

rises very fast and increases the arm currents. The phase-a arm currents at the end of this interval are

$$i_{ap}(t_1) = \frac{i_{ga}(t_0)}{2} + \frac{I_{dc}(t_0)}{3} + \frac{V_{dc}(t_0)}{2L_{arm}}\left(t_1 - t_0\right), \tag{6.16}$$

$$i_{an}(t_1) = -\frac{i_{ga}(t_0)}{2} + \frac{I_{dc}(t_0)}{3} + \frac{V_{dc}(t_0)}{2L_{arm}}\left(t_1 - t_0\right) \tag{6.17}$$

2. Phase II: diode freewheel phase ($t_1 \leq t \leq t_2$): At time t_1 the IGBTs are blocked and the current starts freewheeling through the antiparallel diodes as shown in Fig. 6.5(b). One arm current will increase while the other reduces to zero. In the AC side the entire AC voltage gets applied across the inductor L which results in increase in the AC current. The time interval in which the current in one arm becomes zero is given by

$$t_2 - t_1 = \frac{i_{ga}(t_2) - i_{ga}(t_0)}{L} \tag{6.18}$$

where $i_{ga}(t_2) = 2\left(\frac{I_{dc}(t_0)}{3} + \frac{V_{dc}}{2L_{arm}}\left(t_1 - t_0\right)\right)$

3. Phase III: grid current feeding ($t_1 < t$): This phase is similar to the grid current feeding phase of the two-level converter (Section 6.3.1).

It can be seen that the arm inductor has a direct impact on the diode current. Reference [90] derived a relationship to design L_{arm} to limit the peak current to $i_{D\_\max}$ for known values of of DC bus voltage V_{dc}, DC current I_{dc} and IGBT turn-off time t_1.

$$L_{arm} > \frac{V_{dc}}{2} \frac{\left(t_1 - t_0\right)}{\left(\frac{i_{D\_\max}}{2} - \frac{I_{dc}}{3}\right)} \tag{6.19}$$

6.3.3 Challenges

After analyzing the DC cable fault response of the conventional VSCs, it is clear that the fault current cannot be controlled by them. In an MTDC grid, a DC cable fault will therefore result in all converter stations feeding the fault. Each converter will go through different phases of fault response mentioned earlier. Unlike the AC system where the reactance of the transmission lines limit the current, the DC fault current is only limited by the line resistance. The following are the main reasons why protecting a VSC HVDC system is more difficult than protecting AC systems [91]:

1. In VSC HVDC systems, cables are commonly used, which result in lower rise times and higher steady-state short circuit currents.

2. DC converters are very sensitive to overloads and need to be protected against any over-current.

3. Switching DC current is more difficult than switching AC current.
4. Identifying the line in which a fault occurs is not trivial and traditional AC protection methods (e.g., impedance relays) cannot be used.

6.4 Fault-blocking Converters

A few VSC HVDC topologies have been reported in literature which can block or control the DC fault current. This includes the full bridge MMC topology [92, 93] and its variants; some of which are being prototyped by the major manufacturers.

6.4.1 Full-bridge MMC

Figure 6.6 (b) shows the full-bridge MMC submodule. The full-bridge submodule has a full H-bridge configuration (Section 2.3.2). Such full-bridge submodules can be connected in series to form one individual leg of the a full-bridge MMC converter. The full-bridge structure also helps in cell capacitor voltage balancing leading to the reduction in voltage ripple[94].

A converter station with this configuration is shown in Fig. 6.7 where a cable-to-ground-fault is illustrated. A DC-side fault reduces the converter station DC bus voltage, but the submodule capacitors do not discharge completely. Therefore, the

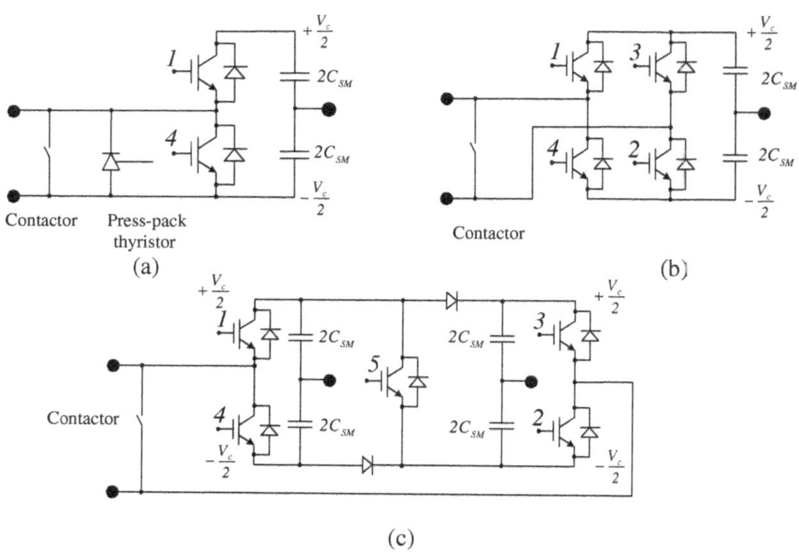

Figure 6.6 (a) The submodule of a half-bridge MMC converter. A press-pack thyristor for the protection of diodes and a contactor is shown. (b) The submodule of a full-bridge MMC converter which does not need any thyristor for protection. (c) The submodule of clamped diode configuration.

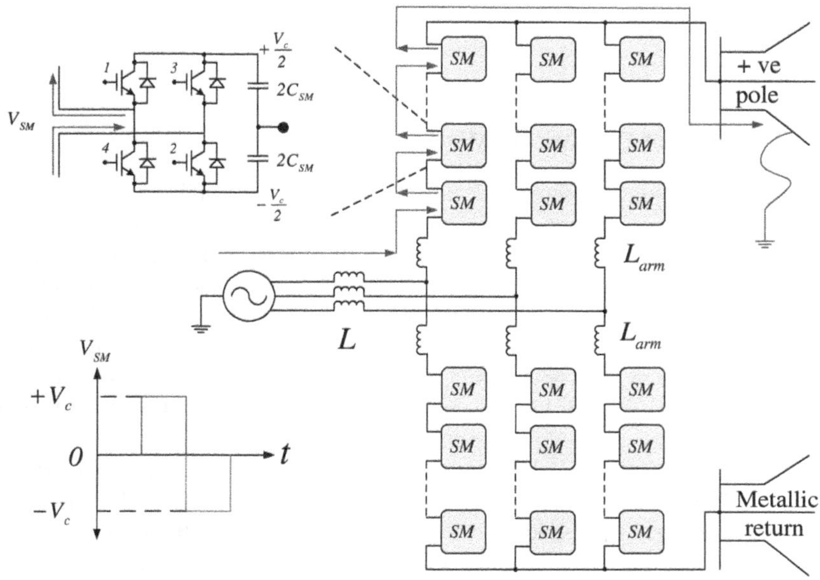

Figure 6.7 A converter station with full-bridge MMC topology. The DC fault current can be blocked or controlled by controlling the full-bridge modules.

full-bridge modules remain operational and can produce positive and negative voltage. In its simplest form the IGBTs can be blocked to stop the flow of the DC current. A negative polarity voltage is particularly advantageous to limit the current surge. The voltage output of the submodules can be modulated to bring down the current through the arm inductors to zero [92, 93]. Since the submodules can control the current, the diodes do not need protection from over current. The full-bridge submodule do not need a need a press-pack thyristor unlike the half-bridge module (Fig. 6.6).

Under such a fault condition, the real (X-axis) and the reactive (Y-axis) component of the AC side current can be controlled independently (see Section 2.4 for a description of converter controls). Thus, the converter station cannot only limit the DC fault current by controlling the X-axis current component, it can also simultaneously provide reactive power support to the AC system by controlling the Y-axis current. In this mode of operation the converter works as a STATCOM.

In spite of the advantages of the full-bridge MMC converter, it suffers from the following disadvantages:

- Cost: It has higher cost compared to the half-bridge MMC since it uses twice the number of diodes and IGBTs.

- Cost of losses: Double the converter losses compared to the half-bridge MMC under normal operating condition.

- Additional switching states: The reverse DC bus polarity is not useful under normal condition.

6.4.2 Variants of Full-bridge MMC

Due to the disadvantages associated with the full-bridge MMC, a few variants are proposed in literature. Figure 6.6(c) shows the clamped double-cell configuration [92, 93]. In this configuration, the converter operates as a series connection of half-bridge cells with an extra switch (#5) in the path of connection. Under normal condition switch #5 is always ON which leads to the half-bridge MMC operation. The total loss is slightly increased due to the conduction loss of switch #5. The clamping diodes do not add to losses as they operate in OFF state. Following the DC fault, switch #5 is blocked resulting in a full-bridge operation. Similar to the full-bridge MMC the clamped modules do not need protective thyristors.

The other variant of the full-bridge topology is called the *alternate-arm multi-level converter* (A2MC) [95]. This topology is shown in Fig. 6.8 where each phase arm of the converter consists of cascaded full-bridge cells which construct the AC voltage from the DC-bus voltage. However, it behaves as a current sourced converter (CSC) as the current is directed to either the upper or the lower arm using a "director switch" consisting of series-connected IGBTs inserted in each arm. Since each arm is used alternatively for each phase, hence the name "alternate-arm." Different features are used to minimize losses in this topology. This includes generating positive (negative)

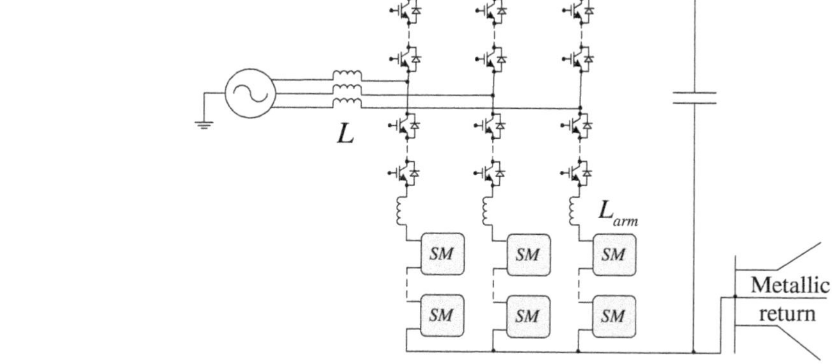

Figure 6.8 A converter station with alternate-arm multi-level converter topology. The DC fault current can be blocked or controlled by controlling the full-bridge modules.

half cycle of the output voltage by the upper (lower) arm, thereby needing half of the total number of full-bridge MMC cells. Moreover, the director switches are used to facilitate zero-current and zero-voltage switching of the outgoing and incoming cells. With adequate number of full-bridge cells, this topology can also control or block the DC current during the DC fault and provide AC network support as a STATCOM.

6.5 DC Circuit Breakers

Lack of commercial availability of high power DC circuit breakers is arguably one of the major barriers toward the practical realization of a VSC-based MTDC grid [96, 97]. Circuit breakers have long been in use in AC transmission systems to interrupt the rated load current or fault current, which could be several times more than the rated current. At high voltage transmission levels, an AC circuit breaker typically needs a minimum of 4–5 cycles (i.e., 80–100 ms) to interrupt the fault current. AC circuit breakers make use of the natural zero crossings of the current which provides an ideal opportunity for interruption with minimal arcing. There is no such natural zero crossing in DC resulting in the fact that conventional AC circuit breaker technology is unable to turn off large DC currents. One way to resolve this is to force a current zero crossing by using a passive [98, 99] or active [99, 100] resonant circuit. Another option is to employ semiconductor switches which can quickly bypass or interrupt the fault current but at the expense of higher on-state power losses.

One challenge is the high rate of rise of DC currents due to the lack of reactance in DC circuits. In this case, an interruption time similar to AC circuit breakers would result in significantly higher maximum current interruption capability and increase in energy to be handled by the arrester leading to prohibitively high cost of DC breakers. Therefore, a DC circuit breaker should be capable of interrupting fault currents much faster (typically within 1–2 ms) compared to their AC counterparts.

DC circuit breakers based on resonant circuit topologies have been prototyped since the mid-1980s to interrupt and commutate current into an alternate path in LCC HVDC systems. Typically, the interruption time for such DC breakers is 30–100 ms. By that time, the DC fault current in a VSC HVDC system could exceed 20 times the rated value [101, 102] making it unsuitable for protection of MTDC grids. Active resonant circuits allow higher (up to 5 kA) current interruption but suffer from similar interruption time as above [99].

DC circuit breakers based on semiconductor switches have been reported and developed over the years. To keep the power losses down under normal condition, a lot of these topologies use a fast mechanical switch in the normal current path while semiconductor switches are used to quickly divert the fault current away from the main path [103]. The semiconductor switches conduct the current until the mechanical switch is able to block the full voltage at which point the former is turned off. Voltage would then increase very fast due to the stored energy in the circuit inductance. Beyond a certain threshold voltage the surge arrestors start conducting and clamps the voltage due to their highly nonlinear *V–I* characteristics. The clamped voltage is higher than the grid voltage which forces the current through the inductor to zero and thereby demagnetizes the line inductance. These circuit breakers can interrupt fault current faster than a resonant breaker but is still slow for high voltage (>100 kV) applications systems primarily due to limited speed of the contact separation of the fast mechanical switch [99]. For medium voltage applications, such mechanical switches with solid-state bypass path could interrupt fault currents within 1–2 ms. However, for higher voltage levels, the only option is to employ semiconductor switches (e.g., IGBTs) in the main current path which can force the current down to zero in less than a millisecond time frame. Although the fastest, such a breaker suffers from high power losses under normal operation. This type of DC circuit breaker is commonly known as *solid-state DC breaker* [103].

The recently proposed proactive hybrid DC breaker is a significant advance toward a fast yet efficient DC breaker. It is hybrid solution combining a fast mechanical switch and less number of semiconductor switches in the main current path together with an alternate path stacked with semiconductors. This offers the best compromise between low losses and fast interruption speed and is currently being developed by ABB [43].

The requirement of an MTDC grid is that the DC fault current should be interrupted within 1–2 ms [97]. Only the solid-state and the proactive hybrid DC breaker are capable of achieving such fast interruption of DC fault current. Use of DC/DC converters have been investigated for their suitability in clearing DC faults within an MTDC grid. The advantage is that the rating of such DC/DC converters is almost independent of the protection processing time unlike the solid state and proactive hybrid breakers. The discussion on DC circuit breakers in this section is limited to only these three types.

6.5.1 Solid-state DC Breaker

The solid-state DC breaker use semiconductor switches with rapid turn-off capability (e.g., IGBTs and GCTs) in the main current flow path. A pair of antiparallel switches is used to allow bidirectional current flow. A surge arrestor is connected in parallel as shown in Fig. 6.9.

During normal operation, the current flows through the semiconductors which lead to high conduction losses. Once a trip signal is received, the semiconductor switches are turned off immediately. As a result, the voltage increases rapidly until the surge arrestor clamps the voltage at a level higher than the DC line voltage in order to demagnetize the DC system inductance. A solid-state breaker does not require any mechanical switch and can therefore interrupt the fault current much faster than other DC circuit breakers. The semiconductor switches are required to be rated to interrupt a smaller fault current and hence lower rated devices may be used. This design also requires fewer components than the others. The major disadvantage, however, is that the power losses under normal operation are significantly higher than other types of DC circuit breakers. Since on-state losses are most important, GCTs instead of IGBTs are often preferred for solid-state DC breakers [103].

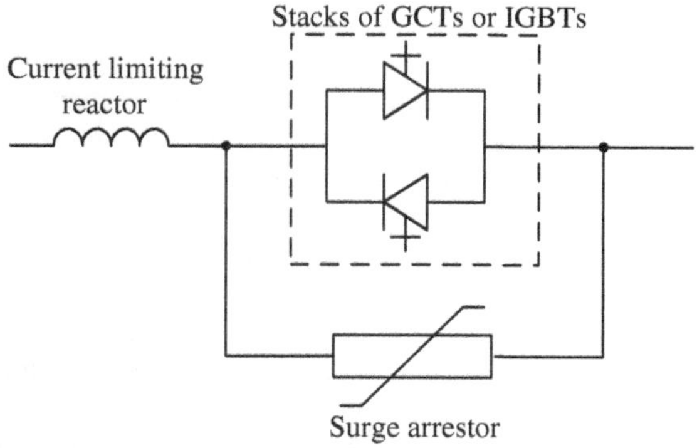

Figure 6.9 Solid-state DC breaker.

6.5.2 Proactive Hybrid DC Breaker

The proactive hybrid DC breaker consists of a main breaker and an auxiliary breaker. The main breaker has several sections each comprising a string of IGBTs connected back-to-back with antiparallel diodes. A surge arrestor in parallel with each section limits the rapid increase of voltage after the main breaker is switched off. The auxiliary breaker has a few IGBTs rated for the forward voltage drop across the sections (less than 1 kV in total) of the main breaker. The path for normal current flow has a fast mechanical disconnector switch in series with an auxiliary DC breaker while the main DC breaker is on the bypass path as shown in Fig. 6.10.

The main breaker is designed for the rated voltage and fault current interruption. During normal operation, the load current flows through the mechanical disconnector and the auxiliary breaker while the current through the main breaker is zero. Once the preset overcurrent threshold is exceeded, the current is immediately commutated from the auxiliary breaker to the main breaker which operates like a solid-state DC breaker. The mechanical disconnector is then opened without an arc. As soon as the disconnector is in a position to block the recovery voltage, the main breaker could be switched off.

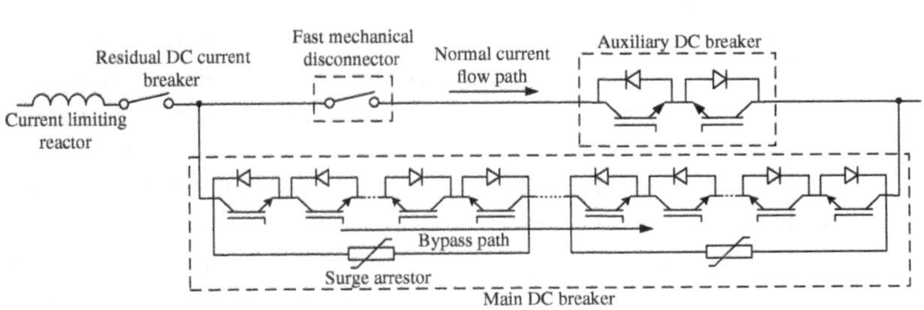

Figure 6.10 Proactive hybrid DC breaker.

The proactive hybrid breaker could be operated in a current limiting mode by controlling the main breaker so that the voltage across the DC inductance is controlled. This allows the protection strategy further time to decide whether or not the breaker needs to be tripped permanently. Switching off the main DC breaker is delayed until a trip command is received or the fault current rises close to the maximum breaking current capability of the main breaker. To reclose the breaker in case the fault is on a different cable, the current is commutated back to the main path by closing the disconnector and turning-on the auxiliary breaker. Otherwise, the main DC breaker is switched-off resulting in rapid increase of voltage until it is clamped by the surge arrestors. The clamped voltage (which is higher than the rated DC voltage) establishes a counter voltage across the DC reactor which forces the fault current to go down to zero. After the fault is cleared, the residual DC breaker isolates the faulty cable form the MTDC to relieve the surge arresters from getting thermally overloaded.

The power losses incurred in a proactive hybrid breaker are much less than a solid-state breaker as there are significantly less number of IGBTs in the main current flow path. Due to the presence of semiconductor switch in the main current path, this particular topology allows much faster transfer of the fault current into the auxiliary path compared to other DC circuit breakers involving semiconductors (with the exception of solid-state DC breakers). The interruption speed is further improved as the disconnector opens without an arc while other types of circuit breakers relies on the arc voltage to transfer the fault current. The current limiting mode which is inherent in such a proactive hybrid breaker allows the protection strategy additional time to select the faulty cable.

If two proactive hybrid breakers are connected in series, the back-up breaker could also be commanded to transfer the fault current to the main breaker, so that it can interrupt the DC fault current very fast (<0.2 ms) even if the primary breaker fails [43].

ABB expects that a proactive hybrid DC breaker would be commercially available for DC voltages up to 320 kV and rated currents of 2.6 kA very soon [43]. It is now well established that the solid-state and the proactive hybrid DC breakers are the only viable options for reliable operation of MTDC grids. The two candidate types are compared next in terms of speed of operation, power losses, cost and footprint and availability and reliability of operation.

Speed of interruption: The solid-state DC circuit breaker is capable of interrupting the DC fault current within approximately 0.2 ms while the proactive hybrid DC circuit breaker requires anywhere between 1–2 ms [43, 97]. Typically, the protection system would need about 1 ms to detect a DC fault and identify the faulty cable. Taking this into consideration, the solid-state breaker could clear the fault within 1.2 ms from the inception of the DC fault. As discussed before, the proactive hybrid breaker has inherent over current protection which allows it to prepare for the trip command by transferring the line current on to the main DC breaker. The overall fault clearing time for a proactive hybrid breaker is therefore expected to be around 2 ms.

Power losses: The power losses in a solid-state circuit breaker could be as high as 30% of the total losses of a VSC HVDC station which is at least 20 times higher than the losses in a proactive hybrid circuit breaker [43]. Power losses in DC circuit breakers are an important consideration as they are likely to be installed at both ends of each of the positive and negative poles. Adding up the lost revenue due to the very high losses of a solid-state breaker, it is highly likely the proactive hybrid breaker would be a more cost-effective solution.

Cost and footprint: The auxiliary DC breaker section of a proactive hybrid breaker can be neglected when comparing the cost and footprint as it is a fraction of the size and cost of the main DC breaker or of a solid-state breaker. For a given nominal current and allowable rate of rise of current, the solid-state breaker is required to interrupt less current than a proactive hybrid breaker as the former interrupts the fault current faster. The components (e.g., semiconductor switches and surge arrestors) for a proactive hybrid breaker are required to have higher ratings than a solid-state breaker. This increases the cost and the footprint. However, all the semiconductor switches in a solid-state breaker carries the load current continuously and hence requires water cooling. For a proactive hybrid breaker, only the auxiliary breaker needs cooling which is a negligible fraction of the overall cost and footprint. This would add cost and size to the solid-state breaker.

Energy availability: DC circuit breakers like any other component could operate abnormally or fail. A failure of the breaker under normal operation ultimately leads to the connected cable being taken out of service. The failure of a proactive hybrid breaker, due to the mechanical disconnector inadvertently opening or the auxiliary DC breaker failing open circuit, would result in the load current commutating to the main DC breaker. Nonetheless, the power transfer is continued with higher on-state losses and the cable is still adequately protected. The failed component in the main current path can be repaired without taking the proactive hybrid breaker out of service. A fault within the solid-state breaker would most likely lead to isolation of the concerned cable. Thus, a proactive hybrid breaker ensures has higher energy availability than a solid-state breaker.

Reliability: In the event of a fault, the solid-state breaker is required to switch off and the surge arrestor must clamp the recovery voltage for successful fault current interruption. The proactive hybrid breaker, however, requires the auxiliary DC breaker to switch-off, the disconnector to open, the main DC breaker to switch off and the surge arrestor to clamp the voltage. The fewer number of states and actions required indicate that the solid-state circuit breaker is more reliable for clearing a DC fault.

6.5.3 DC/DC Converter

One major problem with either solid-state or proactive hybrid DC breaker is the sizing of the semiconductors and the surge arrestor is highly dependent on the processing time of the protection system. System level studies have shown that these DC breakers essentially pushes the fault current to other parts of the MTDC grid and it is therefore required to have a coordinated grid-wide protection within 1–2 ms [101] which is very difficult to achieve especially, with long distance cables. For protection processing times in the order of tens of a few of milliseconds, the semiconductors and surge arrestors become prohibitively expensive and bulky [101].

Use of DC/DC converters have been investigated for their suitability in clearing DC faults within an MTDC grid. Two DC/DC converter topologies—LCL thyristor (shown in Fig. 6.11) and full-bridge DC/DC chopper (shown in Fig. 6.12) converter—have been studied for interruption of DC fault current.

Use of LCL thyrsitor-based converter for DC fault current interruption was proposed in [104]. As shown in Fig. 6.11, there are two back-to-back *LC* resonant circuits connected to a common capacitance C_3. Use of antiparallel thyristors allows current reversal on both sides (I_1,I_2) and hence fast reversal of power flow.

In voltage step-up mode ($V_1 > V_2$), the pair of thyristors $S7, S8$ are fired sequentially with a 50% duty ratio. The combination of inductors L_2 and capacitance C_2 is forced to resonate which causes V_c to rise which allows $S7, S8$ to turn on and off under zero current. Similarly on the other side resonance involving L_1, C_2 enables zero-current switching of the thyristor pair $S3, S4$. The principle of operation remains the same in step-down mode as well but the in this case the thyristor pairs $S5, S6$ and $S1, S2$ are involved instead. The LCL thyristor converter can thus limit the current in case of DC faults on either HV or LV side.

A full-bridge DC/DC chopper as shown in Fig. 6.12 can also be used to interrupt the DC fault current [105]. This is an improved version of the half-bridge DC/DC chopper with a set of additional switches $S2, D2, S4, D4$. Presence of $S2, D2$ provides the additional capability to fully interrupt the current for faults on the HV side. In [101], a simplified topology is proposed which uses only a diode ($D4$) instead of $D4, S4$ combination and uses pulse-width modulation control for only $S1, S3$. Also $S2$ is kept gated continuously unless there is a fault on the HV side. A full-bridge chopper allows full control over DC fault current with modest impact on the healthy

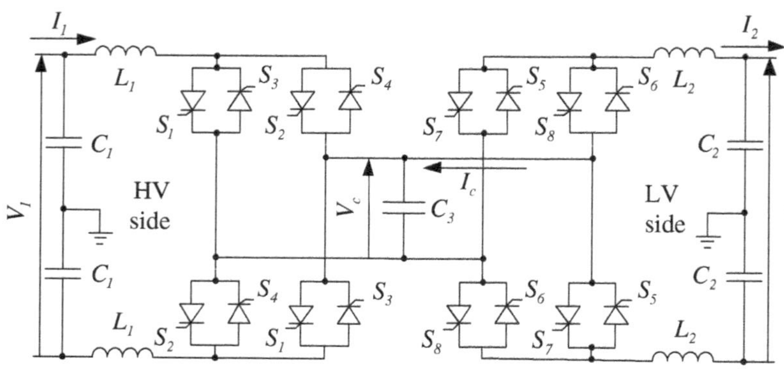

Figure 6.11 LCL thyristor converter.

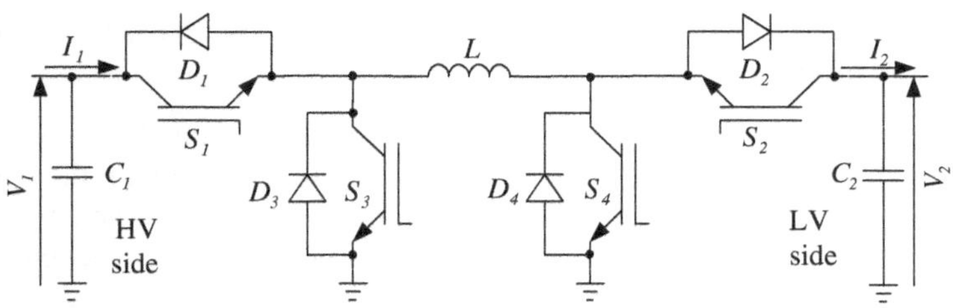

Figure 6.12 Full-bridge DC/DC chopper.

sections of the MTDC grid. Large free-wheeling diodes are required to discharge the energy stored in the inductors.

The LCL thyristor converter on the other hand does not require oversizing of components. The effect of a fault on particular DC cable is not transferred elsewhere and the healthy section of the MTDC grid experiences a load rejection only. Thus, an LCL thyristor converter is expected to perform better and is also likely to be less expensive and more efficient than a DC chopper [101].

Use of LCL thyristor converter or DC/DC chopper for DC fault current interruption is a relatively new area of research. The potential advantage of using DC/DC converters as DC breakers is that they are less dependent on fast protection system and can clear DC faults with local measurements only. A comparison between using a proactive hybrid DC breaker and DC/DC converters (either half-/full-bridge or LCL thyrsitor type) is presented in [101] in terms of size/cost, fault-interruption capability, and losses.

6.6 Protection Strategies

In this section, different protection strategies for MTDC grids proposed in literature are described and analyzed. Any protection system should have the following characteristics [106]:

1. Sensitivity: detect every fault.
2. Selectivity: only operate under fault conditions and only isolate the faulty section.
3. Speed: act before the fault could potentially cause damage to equipment or could no longer be interrupted by the circuit breakers.
4. Reliability: be reliable and have a backup system in case of primary protection system failure.
5. Robustness: be able to act in a degraded mode as well as a normal mode and be able to discriminate between faults and other operations such as set-point changes.

6. Seamless: after the fault clearance the healthy part of the system should continue to operate in a secure state.

The protection system should be able to detect both pole to ground and pole to pole fault. Although the pole to pole fault is rare in cables, it is a possibility if the poles are laid side-by-side in the same trench. The protection scheme should be able to identify the faulted component and isolate only that component. Fault in the protection zone of one breaker should not result in the tripping of a breaker outside that zone. Over and above the primary protection system, there should be a backup protection to ensure reliable operation in case the former fails. Also the healthy components of the grid should adapt to the new condition after the protection system operate in a seamless fashion.

To match these requirements, different protection strategies have been proposed in literature:

1. Strategy I: Conventional two-level or half-bridge MMC converter with AC CBs and DC isolators.
2. Strategy II: Fault-blocking converters with DC isolators.
3. Strategy III: Conventional two-level or half-bridge MMC converter with DC CBs.

6.6.1 Strategy I

For conventional VSC converter topologies like two-level and half-bridge MMC (Sections 2.3.1 and 2.3.3), the DC breaker technology can be used to protect the grid. Since the DC circuit breakers are yet to be commercially available and likely to be expensive and have large footprint (which is critical for offshore applications), an economical way is to use the breakers installed in the AC side of the converter stations for protection. In reference [107] the authors proposed such an approach which they termed as the *handshaking approach*. To illustrate the working principle of this method, let us take an example of the MTDC grid shown in Fig. 6.13. Each pole of the converter station is equipped with an AC CB. Also each DC line is assumed to have two DC isolators at each end of the line. The DC isolators cannot clear faults as they do not have arc quenching capability. In general, for the positive pole, negative pole and the metallic return network a line connecting the ith and the jth converter station has two isolators at each end. The isolator at the DC switchyard of the ith converter station is named as S_{pij} for the positive pole (Fig. 6.13). Subscripts n and m are used for the negative pole and the metallic return circuit respectively (not shown in Fig. 6.13). If a fault occurs in the DC line connecting the positive poles of the converter stations #2 and #4, the following sequence of events take place.

Blocking IGBTs The DC current in the affected pole rises very fast and goes through the different phases mentioned in Section 6.3.1. As soon as the current through the IGBT switches crosses safe limits, they are blocked and the converter stations in the negative pole will loose control over the current. The antiparallel

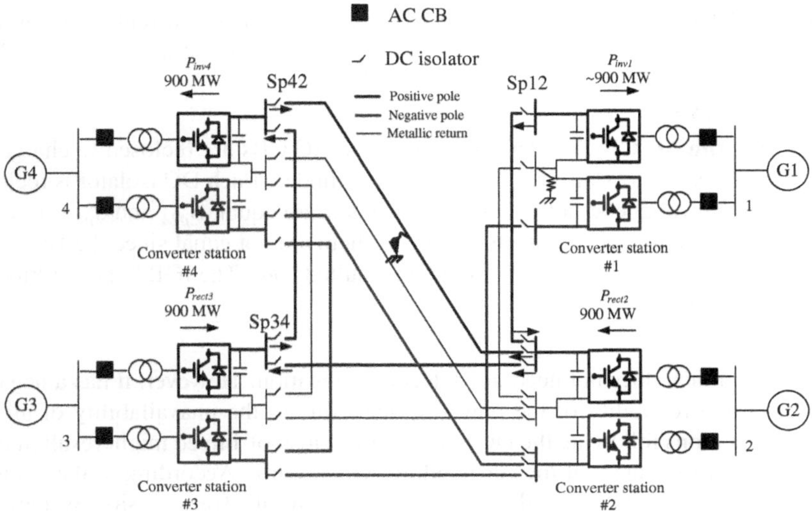

Figure 6.13 Asymmetric bipolar MTDC grid with four terminals. The AC breakers and the DC isolators are shown and the direction of the DC current flowing through the isolators following a DC fault and IGBT blocking is highlighted. Pre-fault power exchange at converter stations are also shown.

diodes conduct the current and the affected converter stations act as diode bridge rectifiers. Phase 3 of fault response mentioned in Section 6.3.1 described this phenomenon.

Selecting the Faulted DC Line The current through the DC isolators are monitored and the direction of the current is a key element in deciding the faulted element. Let us assume a positive sign convention for the current going into the cable through the isolator. At each converter station, the isolator carrying the largest positive current is chosen for opening. As shown in Fig. 6.13 only one DC isolator at each of the stations #1, #3, and #4 carry positive current following the blocking of the converter IGBTs. Therefore, these isolators S_{p12} (station #1), S_{p34} (station #3), and S_{p42} (station #4) are selected for opening. At station #2, the current through the isolators S_{p23} and S_{p24} are both positive. Due to the proximity of the fault location, the current through S_{p24} is higher than the current through S_{p23}. Therefore, S_{p24} is selected for opening.

It is clear that S_{p24} and S_{p42} are both chosen which in effect isolates the faulted line. The other isolators selected for opening are S_{p12} and S_{p34} which will result in opening only one end of the lines connecting stations #1 to #2 and #3 to #4, respectively.

Opening AC Breakers The AC breakers at ***all*** the converter stations of the positive pole are opened. Conventional electromechanical breakers take about 100 ms to open.

Opening Selected Isolators After the AC breakers open, the current in the DC system is allowed to decay to zero before the selected DC isolators are opened.

Re-closing the AC CBs and DC Isolators The AC CBs are reclosed to charge the DC bus capacitors through the antiparallel diodes. Each DC isolator is reclosed when the voltage at both ends of the isolator are equal. S_{p24} and S_{p42} are never reclosed as the voltage on either side of them are not equal since the DC line is isolated and has discharged through the faulted path. The IGBTs are de-blocked to restart normal operation.

This approach does not need any telecommunication. However, it has a negative impact on the reliability of the power systems due to the unavailability of all DC converter stations following the DC faults. A huge loss-of-infeed might result in large excursions in frequency of the surrounding AC systems. According to the Security and Quality of Supply Standard of the National Electricity Transmission System, UK (NETS SQSS), the maximum amount of allowable loss-of-infeed cannot exceed 1.8 GW [108]. Therefore, the MTDC grids with capacity larger than 1.8 GW cannot tolerate complete de-energization. The allowable loss-of-infeed changes based on the capacity of the AC grid.

Another key issue with this approach is the delay involved in opening the AC breakers before the DC grid is de-energized. Commercial AC breakers operate in about five cycles which might result in unacceptably high surge in DC current and damage to the DC cables and other equipments. The fault current is expected to reach about 10 kA in about 2 ms [106]. Therefore, protection using Strategy II and Strategy III appears to be the only realistic option.

6.6.2 Strategy II

In this scheme, the fault-blocking converter topologies (Section 6.4) are used in each converter station. DC isolators are installed at each end of the cables. For the example system shown in Fig. 6.13, when a DC-side fault occurs all converter stations in the affected pole are blocked. This can be done very fast by blocking all the IGBTs of the fault blocking converters. Also the DC-side voltage in each station can be reversed to take the fault energy away from the DC grid to achieve a fast fault clearance. After the DC current goes to zero, the isolators in the DC grid are opened selectively to isolate the faulty component using the same approach as in Strategy I. Reclosing operation of the isolators corresponding to the non-faulty part can follow the "handshaking" method. All converter stations are then de-blocked to restart the system operation.

Although the fault clearance and system restoration can happen much faster than Strategy I, it leads to the de-energization of the entire grid for as low as 300–400 ms. Whether the loss-of-infeed resulting from such an operation is acceptable to the surrounding AC grids has to be determined by the system planners through extensive studies.

6.6.3 Strategy III

In this scheme, the conventional VSC topologies (two-level, half-bridge MMC) and DC breakers are installed at each end of the cables. The most advanced DC circuit breakers are expected to be able to interrupt fault current of approximately 10 kA in about 1 ms [106]. Therefore the fault sensing, identifying the faulty component and sending the command for breaker opening should be done in less than 1 ms. The power rating of the largest VSC HVDC project [19] to be commissioned in 2014 is 1.0 GW at ±320 kV i.e., with an approximate current rating of 1700 A. For a successful interruption the fault current is allowed to rise by about 8.4 kA in 2 ms leading to a maximum allowable rate of rise of 4.2 kA/ms. A 100 mH reactor is needed to meet this requirement—which is several times higher than the reactor normally used for present point-to point VSC HVDC systems. The size of this inductor is not only limited by cost, but also by the grid stability [109]. A root cause analysis based on modal participation factor described in Section 3.11 can reveal such instabilities.

An MTDC protection system following Strategy III should [91]:

1. Detect the fault, insensitive to noise, and normal operational variations.
2. Identify the location of a fault (faulted branch) in a DC meshed system.
3. Send trip commands to appropriate DC breakers at both sides of the affected line in a selective manner within 1–2 ms.
4. Have sufficiently fast DC breakers.
5. Offer a backup in case the former fails.

6.6.3.1 Detection and Identification Let us consider the same example test system as before. Figure 6.14 shows the modified system diagram with DC breakers installed at each end of the cables. Only one pole is shown here for simplicity. In addition to the DC CBs, a mechanical isolator is installed at each end of the cable (not shown in the figure). The mechanical isolators are necessary to open lines under no load condition both following faults and also for maintenance. Following a line-to-ground DC fault on the cable connecting the stations #2 and #4, the first step is to detect it and the identify the faulty component. The objective is to open only DCB_{24} and DCB_{42}. All other breakers should remain closed. Different protection philosophies can be used as described below:

Distance Protection: Traditional ways of protecting the transmission lines in an AC grid include distance protection and directional protection. Distance protection relays monitor the voltages and currents to estimate the fundamental frequency line impedance between the relay location and the fault point, thereby estimating the distance of the fault point. If the distance falls within the zone of protection of the relay, it commands to trip the relevant breaker. The impact of fault resistance can be eliminated by using symmetrical component analysis [110].

In a DC system, fundamental frequency impedance cannot be defined. The impedance is determined primarily by the resistance of the cable which is quite

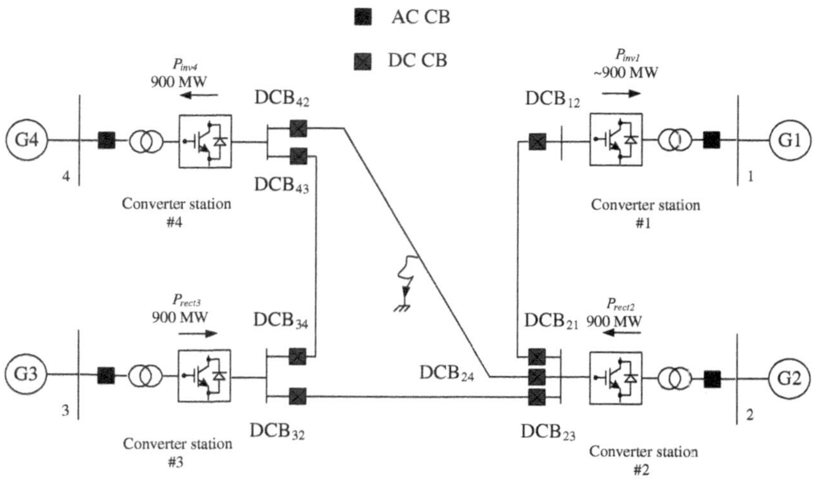

Figure 6.14 Single-line diagram of asymmetric bipolar MTDC grid with four terminals. The AC breakers and the DC breakers are shown. Pre-fault real power flow at converter stations are also shown.

low. Error in measurements makes the problem of estimation of resistance even more challenging. During faults, travelling waves distort the signals to a great extent. Since the distance computation has to be done within a millisecond, it becomes even more challenging to determine the distance of the fault location accurately [91]. Moreover, the fault resistance impacts the calculation a lot.

Differential Protection: In differential protection, if the difference between the currents at both ends of the cable cross a threshold the breakers at both ends are opened. Figure 6.15 shows the differential protection scheme for the cable connecting stations #2 and #4 in for a fault near the converter station #2. Under normal condition, current flows from converter station#2 to #4 (Section 3.10, Chapter 3). The current through breaker DCB_{24} flows in the same direction as that of the pre-fault condition. However, the current through DCB_{42} reverses after the traveling wave of the fault current from the fault end to the receiving end reaches station #4.

The speed of light through the XLPE cable is roughly two-third of the speed of propagation in vacuum [111]. Therefore, it takes about 2.5 ms for the traveling wave to reach converter station #4. The information about the current through DCB_{24} needs to be communicated to the remote-end of the line before making a decision to trip DCB_{42} (i.e., to evaluate the difference between the currents at both ends). Assuming the medium of communication is fiber optic cables, the speed of light through it is almost equal to that through the XLPE cables. Thus, the data travels at the same time to the converter station #4 as that of the travelling wave. Following the reversal of current, the data needs to be communicated

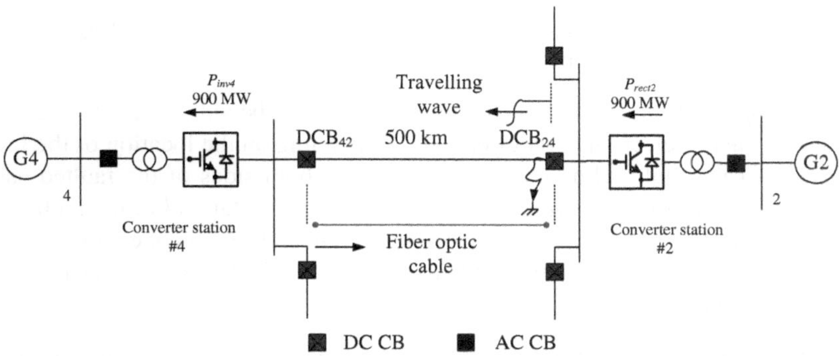

Figure 6.15 Differential protection of DC cables. The pre-fault real power flow at converter stations are also shown.

again to the remote-end of the line for tripping DCB_{24}. Therefore, it will take another 2.5 ms before this information reaches the other end. A delay of 5 ms might not be acceptable for tripping the breakers in the DC grid.

In addition, there could be issues with communication failure. The fiber optic cables for communication are usually embedded in the power transmission cables. If the power cable is faulted or damaged, it is very likely that the fiber optic cable will also be damaged at the same time.

Local Measurement-based Approach: Following the fault, the capacitors at the DC bus of the converter stations using two or three level topologies immediately start discharging through the fault point (see capacitor discharge phase in Section 6.3.1). This results in a very fast rise in the current in the line and fast reduction of DC voltage. For a pole-to-pole fault, the positive pole and the negative pole DC bus will discharge together. The following measurements can be used to detect the DC fault and select the CB to be opened:

- Current magnitude i: If the fault current flowing through a DC CB goes above a pre-determined threshold for more than a pre-defined time duration, the CB can be commanded to open.
- Voltage magnitude v: If the DC bus voltage adjacent to the DC CB falls below a pre-determined threshold for more than a pre-defined time duration, the CB can be commanded to open.
- Rate of change of current $\frac{di}{dt}$: A CB can be commanded to trip, if the rate of rise of fault current flowing through a DC CB goes above a pre-determined threshold for more than a pre-defined time duration.
- Rate of change of voltage $\frac{dv}{dt}$: A CB can be commanded to trip, if the rate of decay of the DC bus voltage adjacent to the DC CB falls below a pre-determined threshold for more than a pre-defined time duration.

- A combination of the above four indices can be used as a "voting scheme" to take the breaker opening decision.

The problem with this approach is that it might not be able to satisfy the "sensitivity" and "selectivity" requirements. Depending on the location of the fault, length of the faulted line etc., the breakers on both sides of the faulted cable might not pass the "sensitivity" criteria i.e., one at the far end of the fault point might not open. Similarly, under certain circumstances, breakers close to the fault location but on a healthy cable might open—which does not meet the "selectivity" requirement.

New Signal Processing Techniques: Different signal processing techniques can be applied on the local measurements described above to improve both selectivity and sensitivity. Both Fourier transform [112] and Wavelet analysis [91] can be used. Fourier transform is not good at analysing aperiodic transient waveforms. This can be overcome by Short-time Fourier transform [113, 114], which in turn suffers from using the same time window for all frequencies. In general, signal processing of fast transients in frequency domain might encounter issues due to low frequency content of the traveling waves.

Wavelet analysis can overcome these issues. It has been widely used to detect sudden changes in waveforms while retaining the time aspect [115]. Reference [91] gives a concise summary of different Wavelet transform methods that could be used for this application and applies the Fast Dyadic Wavelet transform (FDWT) in its proposed protection algorithm. The details of the FDWT can be found in [116].

The proposed protection algorithm is based on three modules:

- Module M1: fault detection by using the voltage wavelet coefficients (criterion 1)
- Module M2: fault detection by using the current wavelet coefficients (criterion 2)
- Module M3: fault detection by using the voltage derivative and magnitude (criterion 3)

These three criteria are needed because the measurements vary with fault location. A triple modular redundancy (TMR) approach could be useful where the final decision is made based on a voting between these three modules. To generate a trip signal, at least a majority of two out of three is necessary.

The success of this approach depends to a large extent on the threshold values of different parameters. Appropriate thresholds should be set through extensive offline time-domain analysis. Worst-case scenarios involving faults at the ends of each cable can be simulated to find out these values. For a complex network, more structured methods might be necessary. As discussed before, the problem of selecting the faulty component becomes more challenging when the fault occurs near a converter station where multiple cables are terminated. Figure 6.14

shows such an example where the fault occurs near the converter station #2. This affects the voltage near all DC breakers DCB_{21}, DCB_{23}, and DCB_{24}. The TMR scheme can be effective in selectively determining the faulty component. It is demonstrated in [91] that the decision on the breakers to be tripped can be determined within 1 ms following the fault.

Wavelet analysis is shown to be quite effective in detecting a fault and selectively determine the faulty component to be isolated. One important aspect that needs attention is the processing time required to compute the wavelet coefficients. A hardware-in loop (HIL) real-time digital simulation with the actual relay should done to validate the proof-of-concept.

Opening DC breakers: After the fault location is detected and the faulty component is identified the respective DC breakers are opened to isolate the fault. This strategy is more advantageous compered to Strategies I and II and does not result in a sudden loss-of-infeed to AC grids.

6.6.4 Backup Protection

Backup protection is absolutely critical in an MTDC grid due to the reliability considerations. As discussed in the previous sections, the DC circuit breaker is commanded to open following the detection of the DC fault and selection of the faulted cable. However, the concern is what if a breaker fails to operate. The concern is even more valid due to the absence of any operational experience with DC breakers. If a DC breaker fails to operate a backup protection system must work to save the grid.

Let us consider the example system shown in Fig. 6.13. If DCB_{24} fails to open, the following options for backup protection can be considered in this context.

Backup with DC and AC breakers After DCB_{24} fails to open the current through this breaker keeps rising. When the current keeps rising beyond a time threshold following the instant of the trip command the adjacent breakers, DCB_{21} and DCB_{23} are commanded to trip. However, the converter station still feeds current to the fault point through DCB_{24}. As illustrated in Fig. 6.16, the AC circuit breaker of the converter station #2 needs to open to stop the current being fed to the fault point. To restart the system the mechanical isolators at each end isolate the faulty cable. This is followed by the closing of the AC CB to charge the DC bus of the converter station. The DC CBs DCB_{21} and DCB_{23} are then closed. The advantages of this scheme is that no extra breaker is required for backup protection. The major disadvantage is due to similar issues mentioned in Section 6.6.2. The system reliability gets affected based on the amount of loss-of-infeed taking place due to the isolation of converter station #2. Also converter station #1 acting as an inverter will not be able to supply the AC grid $G1$ due to the opening of DCB_{21} and DCB_{23}.

Backup with ring bus configuration Backup protection using a ring bus configuration can eliminate some of the drawbacks of the above mentioned approach.

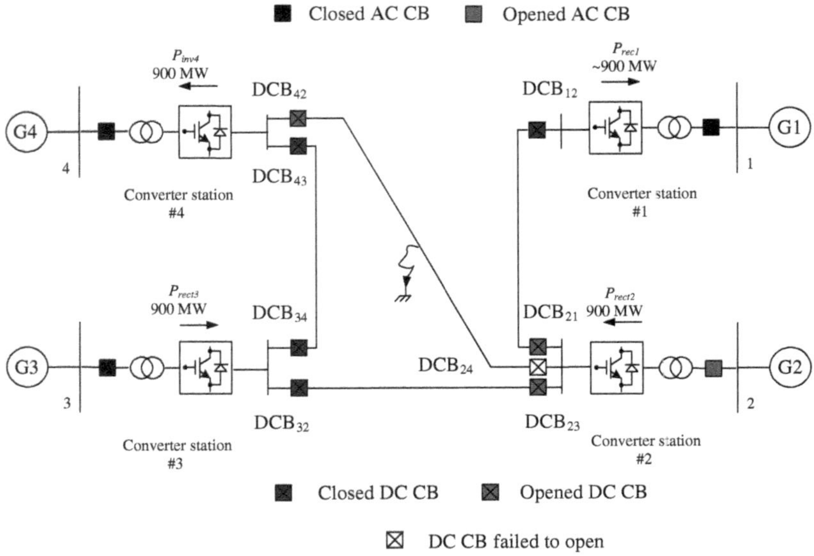

Figure 6.16 Backup protection scheme using DC and AC breakers. DC breaker DCB_{24} fails to operate following a fault in the cable connecting converter stations #2 and #4. Real power exchange at the converter stations for pre-fault condition are highlighted.

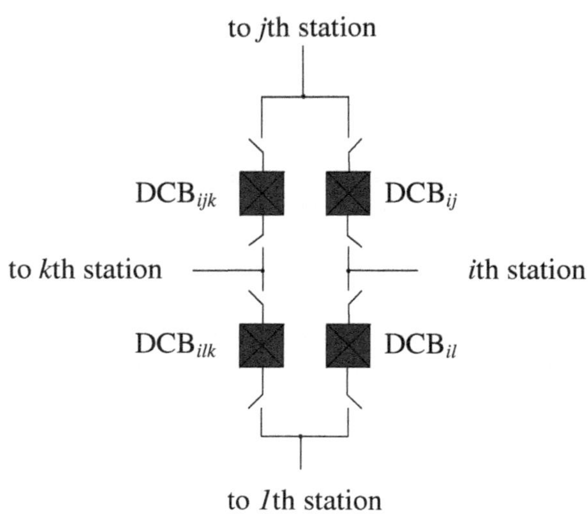

Figure 6.17 DC breakers in ring configuration at the ith converter station. The breakers are connected to three outgoing cables connecting converter stations j, k, and l. The convention for naming the respective breakers is highlighted.

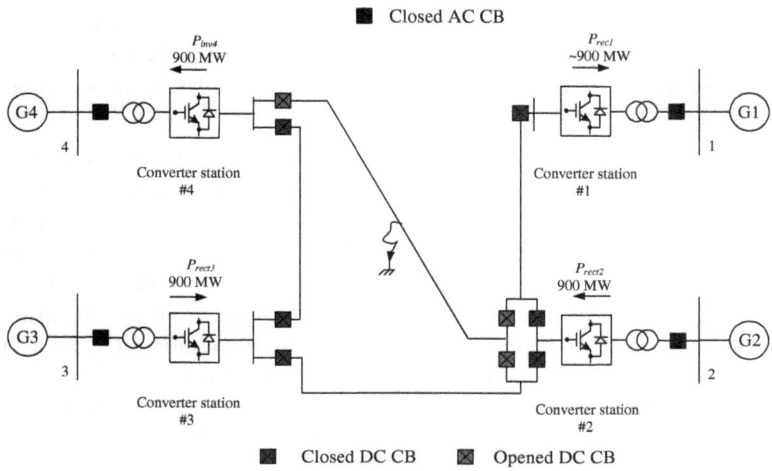

Figure 6.18 Protection scheme with ring configuration of DC breakers at converter station #2. DC breakers DCB_{214} and DCB_{234} are supposed to operate following a fault in the cable connecting converter stations #2 and #4. Pre-fault power exchange at converter stations are also shown.

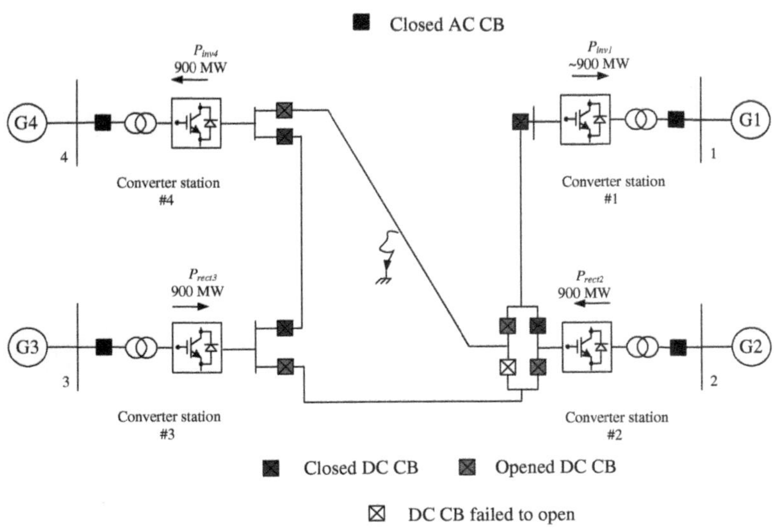

Figure 6.19 Backup protection scheme using ring configuration of DC breakers at station #2. DC breaker DCB_{234} fails to operate following a fault in the cable connecting converter stations #2 and #4. Breakers DCB_{23} and DCB_{32} are opened as a backup protection. Pre-fault power exchange at converter stations are also shown.

The DC CBs in a ring main configuration at the *i*th converter station is shown in Fig. 6.17. The breakers are connected to three outgoing cables connecting converter stations *j*, *k*, and *l*. Mechanical isolators used in the ring bus configuration are shown in this figure. The general notations of DC breakers connecting different converter stations are shown.

Figure 6.18 shows the four-terminal grid with DC CBs in ring main configuration at station #2. To isolate the faulted line, DCB_{214} and DCB_{234} need to open in the ring breaker. Let us consider the scenario when DCB_{234} fails to operate. Under this situation the breakers DCB_{23} and DCB_{32} are opened as a backup protection (Fig. 6.19). Under this condition station #2 serves station #1 through DCB_{21}. Therefore, no converter station is completely isolated from the AC grid. The disadvantage with this approach is that it needs an additional DC breaker and also the primary protection becomes more complicated as two circuit breakers need to be tripped for each cable fault. With more branching and/or a highly meshed topology the complexity will scale up even further.

REFERENCES

1. T. McNichol, *AC/DC : the savage tale of the first standards war.* San Francisco, CA: Jossey-Bass; [Chichester : John Wiley, distributor], 2006.
2. M. N. 21945. (2006) £22bn Pan-European marine energy 'Supergrid' proposed. [Online]. Available: http://www.maritimejournal.com/news101 (accessed Feb 2014)
3. T. Singh. (2010) 10 EU countries pledge to create North Sea renewable energy grid. [Online]. Available: http://inhabitat.com/10-eu-countries-pledge-to-create-north-sea-renewable-energy-grid/ (accessed Feb 2014)
4. Allen and Y. Blog. (2011) The European Wind Energy Association. [Online]. Available: http://allenandyork.wordpress.com/2011/09/ (accessed Feb 2014)
5. W. I. Europe. European Commission. [Online]. Available: http://www.worldwatch-europe.org/es/node/50 (accessed Feb 2014)
6. Zhoushan MTDC project. [Online]. Available: http://www.epri.sgcc.com.cn/ (accessed Feb 2014)
7. E. Kimbark, *Direct current transmission*, ser. Direct current transmission. Wiley-Interscience, 1971, no. v. 1. [Online]. Available: http://books.google.com/books?id=eMMiAAAAMAAJ (accessed Feb 2014)
8. C. Adamson and N. Hingorani, *High voltage direct current power transmission.* Garraway, 1960. [Online]. Available: http://books.google.com/books?id=mQkjAAAAMAAJ (accessed Feb 2014)
9. J. Arrillaga and B. D. Smith, *AC-DC power system analysis.* Stevenage: Institution of Electrical Engineers, 1998.

Multi-terminal Direct-Current Grids: Modeling, Analysis, and Control. First Edition.
Nilanjan Ray Chaudhuri, Balarko Chaudhuri, Rajat Majumder, and Amirnaser Yazdani.

10. J. Arrillaga, Y. H. Liu, and N. R. Watson, *Flexible power transmission: the HVDC options*. John Wiley, 2007.

11. K. R. Padiyar, *HVDC power transmission systems*, 2nd ed. New Delhi: New Age International, 2012.

12. P. Kundur, *Power system stability and control*, ser. The EPRI power system engineering series. New York; London: McGraw-Hill, 1994.

13. V. C. Billon, J. P. Taisne, V. Arcidiacono, and F. Mazzoldi, "The Corsican tapping: from design to commissioning tests of the third terminal of the Sardinia-Corsica-Italy HVDC," *IEEE Transactions on Power Delivery*, vol. 4, no. 1, pp. 794–799, 1989.

14. SACOI: Wikipedia. [Online]. Available: http://en.wikipedia.org/wiki/ (accessed Feb 2014)

15. Hydro Quebec-New England interconnection. [Online]. Available: http://new.abb.com/systems/hvdc/references/quebec-new-england (accessed Feb 2014)

16. Caprivi link. [Online]. Available: http://new.abb.com/systems/hvdc/references/caprivi-link (accessed Feb 2014)

17. C. M. Franck, "HVDC circuit breakers: a review identifying future research needs," *IEEE Transactions on Power Delivery*, vol. 26, no. 2, pp. 998–1007, 2011.

18. East-west interconnector project. [Online]. Available: http://www.eirgridprojects.com/projects/east-westinterconnector/ (accessed Feb 2014)

19. J. Glasdam, J. Hjerrild, L. H. Kocewiak, and C. L. Bak, "Review on multi-level voltage source converter based HVDC technologies for grid connection of large offshore wind farms," in *IEEE International Conference on Power System Technology (POWERCON), 2012*, 2012, pp. 1–6.

20. Planned HVDC projects in China. [Online]. Available: www.siemens.com/energy/hvdc

21. Z. Shu, L. Jun, J. B. Ekanayake, and N. Jenkins, "Control of multi-terminal VSC-HVDC transmission system for offshore wind power generation," in *Proceedings of the 44th International Universities Power Engineering Conference (UPEC), 2009*, 2009, pp. 1–5.

22. Cigre study committee B4 website. [Online]. Available: http://b4.cigre.org/ (accessed Feb 2014)

23. A. Yazdani and R. Iravani, *Voltage-sourced converters in power systems: modeling, control, and applications*. Oxford: Wiley, 2010.

24. W. Leonhard, *Control of electrical drives*. Berlin, Heidelberg: Springer, 2001.

25. MATLAB®, *Version 7.10.0 (R2010a)*. Natick, MA: The MathWorks Inc., 2010.

26. PSCAD/EMTDC, *Version 4.2.1*. Winnipeg, MB, Canada: Manitoba HVDC Research Centre.

27. J. Blau, "Europe plans a North Sea grid," *IEEE Spectrum*, vol. 47, no. 3, pp. 12–13, 2010.

28. S. Lefebvre, W. K. Wong, J. Reeve, M. Baker, and D. Chapman, "Considerations for modeling MTDC systems in transient stability programs," *IEEE Transactions on Power Delivery*, vol. 6, no. 1, pp. 397–404, 1991.

29. S. Lefebvre, W. K. Wong, J. Reeve, J. M. Gagnon, and B. K. Johnson, "Experience with modeling MTDC systems in transient stability programs," *IEEE Transactions on Power Delivery*, vol. 6, no. 1, pp. 405–413, 1991.

30. S. Cole, J. Beerten, and R. Belmans, "Generalized dynamic VSC MTDC model for power system stability studies," *IEEE Transactions on Power Systems*, vol. 25, no. 3, pp. 1655–1662, 2010.

31. C. Zheng, X. Zhou, and L. Ruomei, "Dynamic modeling and transient simulation for VSC based HVDC in multi-machine system," in *International Conference on Power System Technology, 2006. PowerCon 2006*, 2006, pp. 1–7.

32. C. Hairong, W. Chao, Z. Fan, and P. Wulue, "Control strategy research of VSC based multiterminal HVDC system," in *IEEE PES Power Systems Conference and Exposition, 2006*, 2006, pp. 1986–1990.

33. X. Lie, Y. Liangzhong, and M. Bazargan, "DC grid management of a multi-terminal HVDC transmission system for large offshore wind farms," in *International Conference on Sustainable Power Generation and Supply, 2009. SUPERGEN '09*, 2009, pp. 1–7.

34. L. Fei, M. Zhiwen, Y. Xiaojie, and Z. Trillion, "The grid connected converter control of multi-terminal DC system for wind farms," in *Proceedings of the Eighth International Conference on Electrical Machines and Systems, 2005. ICEMS 2005*, vol. 2, 2005, pp. 1021–1023.

35. T. Lianxiang and O. Boon-Teck, "Locating and isolating DC faults in multi-terminal DC systems," *IEEE Transactions on Power Delivery*, vol. 22, no. 3, pp. 1877–1884, 2007.

36. N. Chaudhuri, R. Majumder, B. Chaudhuri, and J. Pan, "Stability analysis of VSC MTDC grids connected to multimachine AC systems," *IEEE Transactions on Power Delivery*, vol. 26, no. 4, pp. 2774–2784, 2011.

37. C. Schauder and H. Mehta, "Vector analysis and control of advanced static var compensators," *IEE Proceedings on Generation, Transmission and Distribution*, vol. 140, no. 4, pp. 299–306, 1993.

38. P. W. Sauer and M. A. Pai, *Power system dynamics and stability.* Upper Saddle River, NJ: Prentice Hall, 1998.

39. K. R. Padiyar, *Power system dynamics: stability and control*, 2nd ed. Tunbridge Wells: Anshan, 2004.

40. "IEEE recommended practice for excitation system models for power system stability studies," *IEEE Std 421.5-2005 (Revision of IEEE Std 421.5-1992)*, pp. 1–85, 2006.

41. J. J. Grainger and W. D. Stevenson, *Elements of power system analysis.* New York; London: McGraw-Hill, 1994.

42. S. A. Soman, S. A. Khaparde, and S. Pandit, *Computational methods for large sparse power systems analysis: an object oriented approach.* Boston, MA; London: Kluwer Academic, 2002.

43. J. Hafner and B. Jacobson, "Proactive hybrid HVDC breakers - a key innovation for reliable HVDC grids," *Cigre Bologna Symposium*, pp. 1–9, 2011.

44. K. Rudion, A. Orths, P. B. Eriksen, and Z. A. Styczynski, "Toward a benchmark test system for the offshore grid in the North Sea," in *IEEE Power and Energy Society General Meeting, 2010*, 2010, pp. 1–8.

45. N. Chaudhuri, R. Majumder, B. Chaudhuri, J. Pan, and R. Nuqui, "Modeling and stability analysis of MTDC grids for offshore wind farms: a case study on the North Sea

benchmark system," in *IEEE Power and Energy Society General Meeting, 2011*, 2011, pp. 1–7.

46. X. Lie, B. W. Williams, and Y. Liangzhong, "Multi-terminal DC transmission systems for connecting large offshore wind farms," in *IEEE Power and Energy Society General Meeting - Conversion and Delivery of Electrical Energy in the 21st Century, 2008*, 2008, pp. 1–7.
47. B. Berggren, R. Majumder, C. Sao, and K. Linden, "Method and control device for controlling power flow within a DC transmission network," World IPO Patent WO 2012/000 549 A1, international filing 30 June, 2010; international publication 5 January, 2012.
48. T. M. Haileselassie and K. Uhlen, "Primary frequency control of remote grids connected by multi-terminal HVDC," in *IEEE Power and Energy Society General Meeting, 2010*, 2010.
49. G. P. Adam, O. Anaya-Lara, and G. Burt, "Multi-terminal DC transmission system based on modular multilevel converter," in *44th International Universities Power Engineering Conference, 2009*, 2009, pp. 961–965.
50. C. Dierckxsens, K. Srivastava, M. Reza, S. Cole, J. Beerten, and R. Belmans, "A distributed DC voltage control method for VSC MTDC systems," *Electric Power Systems Research*, vol. 82, no. 1, pp. 54–58, 2012.
51. O. Gomis-Bellmunt, A. Egea-Alvarez, A. Junyent-Ferre, J. Liang, J. Ekanayake, and N. Jenkins, "Multiterminal HVDC-VSC for offshore wind power integration," in *IEEE Power and Energy Society General Meeting, 2011*, 2011.
52. T. M. Haileselassie and K. Uhlen, "Precise control of power flow in multiterminal VSC-HVDCs using DC voltage droop control," in *IEEE Power and Energy Society General Meeting, 2012*, 2012.
53. R. T. Pinto, S. F. Rodrigues, P. Bauer, and J. Pierik, "Description and comparison of DC voltage control strategies for offshore MTDC networks: steady-state and fault analysis," *EPE Journal*, vol. 22, no. 4, pp. 31–39, 2012.
54. T. K. Vrana, J. Beerten, R. Belmans, and O. B. Fosso, "A classification of DC node voltage control methods for HVDC grids," *Electric Power Systems Research*, vol. 103, pp. 137–144, 2013.
55. L. Xu, L. Yao, and M. Bazargan, "DC grid management of a multi-terminal HVDC transmission system for large offshore wind farms," in *International Conference on Sustainable Power Generation and Supply, 2009*, vol. vols 1–4, 2009, pp. 1660–1666.
56. J. Cao, W. Du, H. F. Wang, and S. Q. Bu, "Minimization of transmission loss in meshed AC/DC grids with VSC-MTDC networks," *IEEE Transactions on Power Systems*, vol. 28, no. 3, pp. 3047–3055, 2013.
57. J. Beerten, D. V. Hertem, and R. Belmans, "VSC MTDC systems with a distributed DC voltage control - a power flow approach," in *IEEE Trondheim PowerTech 2011*, 2011, pp. 1–7.
58. J. Beerten, S. Cole, and R. Belmans, "Modeling of multi-terminal VSC HVDC systems with distributed DC voltage control," *IEEE Transactions on Power Systems*, vol. 29, no. 1, pp. 34–42, 2014.

59. J. Beerten and R. Belmans, "Analysis of power sharing and voltage deviations in droop-controlled DC grids," *IEEE Transactions on Power Systems*, vol. 28, no. 4, pp. 4588–4596, 2013.

60. O. Despouys, "Offshore DC grids: impact of topology on power flow control," in *10th IET International Conference on AC and DC Power Transmission ACDC 2012*, 2012, pp. 1–7.

61. N. Pogaku, M. Prodanovic, and T. Green, "Modeling, analysis and testing of autonomous operation of an inverter-based microgrid," *IEEE Transactions on Power Electronics*, vol. 22, no. 2, pp. 613–625, 2007.

62. R. Majumder, B. Chaudhuri, A. Ghosh, R. Majumder, G. Ledwich, and F. Zare, "Improvement of stability and load sharing in an autonomous microgrid using supplementary droop control loop," *IEEE Transactions on Power Systems*, vol. 25, no. 2, pp. 796–808, 2010.

63. A. J. Wood and B. F. Wollenberg, *Power generation, operation and control*, 2nd ed. New York; Chichester: Wiley, 1996.

64. N. Hatziargyriou, H. Asano, R. Iravani, and C. Marnay, "Microgrids," *IEEE Power and Energy Magazine*, vol. 5, no. 4, pp. 78–94, 2007.

65. F. Katiraei and M. Iravani, "Power management strategies for a microgrid with multiple distributed generation units," *IEEE Transactions on Power Systems*, vol. 21, no. 4, pp. 1821–1831, 2006.

66. M. Delghavi and A. Yazdani, "An adaptive feedforward compensation for stability enhancement in droop-controlled inverter-based microgrids," *IEEE Transactions on Power Delivery*, vol. 26, no. 3, pp. 1764–1773, 2011.

67. "IEEE standard for interconnecting distributed resources with electric power systems," *IEEE Std 1547-2003*, pp. 1–16, 2003.

68. T. Nakajima and S. Irokawa, "A control system for HVDC transmission by voltage sourced converters," in *IEEE Power Engineering Society Summer Meeting, 1999*, vol. 2, 1999, pp. 1113–1119.

69. T. Haileselassie, "Control of multi-terminal VSC-HVDC systems," *MS Thesis*, June 2008.

70. N. Chaudhuri and B. Chaudhuri, "Adaptive droop control for effective power sharing in multi-terminal DC (MTDC) grids," *IEEE Transactions on Power Systems*, vol. 28, no. 1, pp. 21–29, 2013.

71. L. Xu, B. Williams, L. Yao, and M. Bazargan, "Control and operation of multi-terminal DC systems for integrating large offshore wind farms," in *Proceedings of the 7th International Workshop on Large-Scale Integration of Wind Power into Power Systems, 2008*, Madrid, 2008.

72. L. Xu, L. Yao, and M. Bazargan, "Dc grid management of a multi-terminal HVDC transmission system for large offshore wind farms," in *International Conference on Sustainable Power Generation and Supply, 2009*, 2009.

73. H. T. Ma and B. H. Chowdhury, "Working towards frequency regulation with wind plants: combined control approaches," *IET Renewable Power Generation*, vol. 4, no. 4, pp. 308–316, 2010.

74. Y. G. Rebours, D. S. Kirschen, M. Trotignon, and S. Rossignol, "A survey of frequency and voltage control ancillary services part I: technical features," *IEEE Transactions on Power Systems*, vol. 22, no. 1, pp. 350–357, 2007.

75. (2007) Wind farm transmission grid code provisions: a direction by the commission for energy regulation. [Online]. Available: http://www.cer.ie /GetAttachment.aspx?id=994a0a8b-7c96-44f4-ac50-dced85a45434 (accessed Feb 2014)

76. (2004) Wind turbines connected to grids with voltages above 100kV - technical regulation for the properties and the regulation for wind turbines. [Online]. Available: http://www.energinet.dk/SiteCollectionDocuments/Engelske (accessed Feb 2014)

77. (2008) Technical requirements for wind power and photovoltaic installations and any generating facilities whose technology does not consist of a synchronous generator directly connected to the grid. [Online]. Available: http://www.ree.es/en/publications (accessed Feb 2014)

78. (2005) Requirements for the connection of generation facilities to the hydro-quebec transmission system âŁ" supplementary requirements for wind generation. [Online]. Available: http://www.hydroquebec.com/transenergie/fr/commerce/pdf/eolienne-transport-en.pdf (accessed Feb 2014)

79. J. Aho, A. Buckspan, J. Laks, P. Fleming, J. Yunho, F. Dunne, M. Churchfield, L. Pao, and K. Johnson, "A tutorial of wind turbine control for supporting grid frequency through active power control," in *American Control Conference (ACC), 2012*, 2012, pp. 3120–3131.

80. I. M. de Alegra, J. Andreu, J. L. Martn, P. Ibaez, J. L. Villate, and H. Camblong, "Connection requirements for wind farms: a survey on technical requirements and regulation," *Renewable and Sustainable Energy Reviews*, vol. 11, no. 8, pp. 1858–1872, 2007.

81. I. Erlich and M. Wilch, "Primary frequency control by wind turbines," in *IEEE Power and Energy Society General Meeting, 2010*, 2010, pp. 1–8.

82. X. Juankorena, I. Esandi, J. Lopez, and L. Marroyo, "Method to enable variable speed wind turbine primary regulation," in *International Conference on Power Engineering, Energy and Electrical Drives, 2009*, 2009, pp. 495–500.

83. R. G. de Almeida and J. A. Peas Lopes, "Participation of doubly fed induction wind generators in system frequency regulation," *IEEE Transactions on Power Systems*, vol. 22, no. 3, pp. 944–950, 2007.

84. N. Chaudhuri, R. Majumder, and B. Chaudhuri, "System frequency support through multi-terminal DC (MTDC) grids," *IEEE Transactions on Power Systems*, vol. 28, no. 1, pp. 347–356, 2013.

85. F. Rahmatian and J. Blake, "Applications of high-voltage fiber optic current sensors," in *IEEE Power Engineering Society General Meeting, 2006*, 2006.

86. P. Anderson, *Power system protection*, ser. IEEE Press Series on Power Engineering. Wiley, 1998.

87. J. Yang, J. Fletcher, and J. O'Reilly, "Multiterminal DC wind farm collection grid internal fault analysis and protection design," *IEEE Transactions on Power Delivery*, vol. 25, no. 4, pp. 2308–2318, 2010.

88. ——, "Short-circuit and ground fault analyses and location in VSC-based DC network cables," *IEEE Transactions on Industrial Electronics*, vol. 59, no. 10, pp. 3827–3837, 2012.

89. B. Gemmell, J. Dorn, D. Retzmann, and D. Soerangr, "Prospects of multilevel VSC technologies for power transmission," in *IEEE/PES Transmission and Distribution Conference and Exposition, 2008*, 2008.

90. Y. Li and F. Wang, "Analysis on fault current stress for power switches in modular multilevel converter based DC system," in *CURENT 2013 Site Visit and Industry Conference, UTK, Knoxville*, 2013.

91. K. De Kerf, K. Srivastava, M. Reza, D. Bekaert, S. Cole, D. Van Hertem, and R. Belmans, "Wavelet-based protection strategy for DC faults in multi-terminal VSC HVDC systems," *IET Generation, Transmission Distribution*, vol. 5, no. 4, pp. 496–503, 2011.

92. R. Marquardt, "Modular multilevel converter: an universal concept for HVDC-networks and extended DC-bus-applications," in *International Power Electronics Conference (IPEC), 2010*, 2010, pp. 502–507.

93. ——, "Modular multilevel converter topologies with DC-short circuit current limitation," in *IEEE 8th International Conference on Power Electronics and ECCE Asia (ICPE ECCE), 2011*, 2011, pp. 1425–1431.

94. F. D. Alireza Nami, Liwei Wang and A. Shukla, "Five level cross connected cell for cascaded converters," in *EPE, Lille, France, September 3, 2013*, 2013.

95. M. Merlin, T. Green, P. Mitcheson, D. Trainer, D. Critchley, and R. Crookes, "A new hybrid multi-level voltage-source converter with DC fault blocking capability," in *9th IET International Conference on AC and DC Power Transmission, ACDC, 2010*, 2010.

96. K. Bullis, "Supergrids a high-power circuit breaker could finally make DC power grids practical," *Technology Review*, vol. 116, no. 3, pp. 56–57, 2013, times Cited: 0.

97. D. Jovcic, D. Van Hertem, K. Linden, J. P. Taisne, and W. Grieshaber, "Feasibility of DC transmission networks," in *2nd IEEE PES International Conference and Exhibition on Innovative Smart Grid Technologies (ISGT Europe), 2011*, 2011, pp. 1–8.

98. B. Bachman, G. Mauthe, H. P. Lipps, E. Ruoss, J. Porter, and J. Vithayathil, "Development of a 500kV airblast HVDC circuit breaker," *IEEE Power Engineering Review*, vol. PER-5, no. 9, pp. 43–43, 1985.

99. D. Andersson and A. Henriksson, "Passive and active DC breakers in the three Gorges-Changzhou HVDC project," in *International Conference on Power Systems, 2001*, 2001.

100. H. A. Darwish, M. A. Izzularab, and N. I. Elkalashy, "Enhanced commutation circuit design of HVDC circuit breaker using EMTP," in *IEEE PES Transmission and Distribution Conference and Exhibition*, 2005/2006, pp. 978–985.

101. M. Hajian, D. Jovcic, and B. Wu, "Evaluation of semiconductor based methods for fault isolation on high voltage DC grids," *IEEE Transactions on Smart Grid*, vol. 4, no. 2, pp. 1171–1179, 2013.

102. D. Jovcic and B. Wu, "Fast fault current interruption on high-power DC networks," in *IEEE Power and Energy Society General Meeting, 2010*, 2010, pp. 1–6.

103. C. Meyer, M. Kowal, and R. W. De Doncker, "Circuit breaker concepts for future high-power DC applications," in *Industry Applications Conference, 2005*, vol. 2, 2005, pp. 860–866.

104. D. Jovcic, "Bidirectional, high-power DC transformer," *IEEE Transactions on Power Delivery*, vol. 24, no. 4, pp. 2276–2283, 2009.

105. C. Zhan, C. Smith, A. Crane, A. Bullock, and D. Grieve, "DC transmission and distribution system for a large offshore wind farm," in *9th IET International Conference on AC and DC Power Transmission, ACDC 2010*, 2010, pp. 1–5.

106. D. Jovcic, D. Van Hertem, K. Linden, J.-P. Taisne, and W. Grieshaber, "Feasibility of DC transmission networks," in *2nd IEEE PES International Conference and Exhibition on Innovative Smart Grid Technologies (ISGT Europe), 2011*, 2011, pp. 1–8.

107. L. Tang and B.-T. Ooi, "Locating and isolating DC faults in multi-terminal DC systems," *IEEE Transactions on Power Delivery*, vol. 22, no. 3, pp. 1877–1884, 2007.

108. N. Grid, "2011 offshore development information statement," *Report*, 2011.

109. J. Hafner and B. Jacobson, "Proactive hybrid HVDC breakers - a key innovation for reliable HVDC grids," in *Cigre Bologna 2011*, 2011.

110. X. Yang, M.-S. Choi, S.-J. Lee, C.-W. Ten, and S.-I. Lim, "Fault location for underground power cable using distributed parameter approach," *IEEE Transactions on Power Systems*, vol. 23, no. 4, pp. 1809–1816, 2008.

111. A. Rogers, *Understanding optical fibre communications*. Artech House Publishers, 2001.

112. D. V. Hertem, M. Ghandhari, J. B. Curis, O. Despouys, and A. Marzin, "Protection requirements for a multi-terminal meshed DC grid," *Cigre Bologna 2011*, 2011.

113. E. Jacobsen and R. Lyons, "The sliding DFT," *IEEE Signal Processing Magazine*, vol. 20, no. 2, pp. 74–80, 2003.

114. J. B. Allen, "Short term spectral analysis, synthesis, and modification by discrete Fourier transform," *IEEE Transactions on Acoustics, Speech and Signal Processing*, vol. 25, no. 3, pp. 235–238, 1977.

115. S. G. Mallat and G. Peyre, *A wavelet tour of signal processing: the sparse way*, 3rd ed. Amsterdam; London: Academic Press, 2009.

116. S. Mallat, *A wavelet tour of signal processing*. Academic Press, 1999.

INDEX

Multi-terminal Direct-Current Grids: Modeling, Analysis, and Control. First Edition.
Nilanjan Ray Chaudhuri, Balarko Chaudhuri, Rajat Majumder, and Amirnaser Yazdani.